Application and Management of Smart Energy for Smart Cities

Application and Management of Smart Energy for Smart Cities

Guest Editors

Daniele Menniti
Giovanni Brusco
Pasquale Vizza
Anna Pinnarelli
Nicola Sorrentino

MDPI

Basel • Beijing • Wuhan • Barcelona • Belgrade • Novi Sad • Cluj • Manchester

Guest Editors
Daniele Menniti
Department of Mechanical,
Energy and Management
Engineering
University of Calabria
Rende
Italy

Giovanni Brusco
Creta Energie Speciali
Cosenza
Italy

Pasquale Vizza
Department of Mechanical,
Energy and Management
Engineering
University of Calabria
Rende
Italy

Anna Pinnarelli
Department of Mechanical,
Energy and Management
Engineering
University of Calabria
Rende
Italy

Nicola Sorrentino
Department of Mechanical,
Energy and Management
Engineering
University of Calabria
Rende
Italy

Editorial Office
MDPI AG
Grosspeteranlage 5
4052 Basel, Switzerland

This is a reprint of the Special Issue, published open access by the journal *Energies* (ISSN 1996-1073), freely accessible at: https://www.mdpi.com/journal/energies/special_issues/YVA72RM0N4.

For citation purposes, cite each article independently as indicated on the article page online and as indicated below:

Lastname, A.A.; Lastname, B.B. Article Title. *Journal Name* **Year**, *Volume Number*, Page Range.

ISBN 978-3-7258-5891-0 (Hbk)
ISBN 978-3-7258-5892-7 (PDF)
https://doi.org/10.3390/books978-3-7258-5892-7

Contents

About the Editors

Daniele Menniti

Daniele Menniti is Professor in the Department of Mechanical, Energy and Management Engineering (DIMEG) of the University of Calabria, Italy. His research activity focuses on power generation, operation, stability and control, with particular attention to power electronics and FACTS technology. He has extensive expertise in renewable energy integration, distributed generation, and advanced grid architectures such as smart grids, microgrids and nanogrids. His interests also include demand response modeling, energy market analysis and aggregator frameworks for energy districts and renewable energy communities.

Giovanni Brusco

Giovanni Brusco is a researcher at Creta Energie Speciali in Cosenza, Italy. His scientific interests span power systems, microgrids, energy storage, and distributed generation. He has contributed to the development of smart grid solutions, with a focus on smart metering, smart charging, and power quality. His work emphasizes the role of renewable energy integration and renewable energy communities as enablers of future sustainable energy systems.

Pasquale Vizza

Pasquale Vizza is affiliated with the Department of Mechanical, Energy and Management Engineering (DIMEG) at the University of Calabria, Italy. His research is mainly devoted to power converters, renewable energies, and distributed generation. He is particularly engaged in the study of energy communities, storage systems, and power forecasting techniques, as well as innovative approaches such as power cloud solutions for energy management.

Anna Pinnarelli

Anna Pinnarelli is a researcher in the Department of Mechanical, Energy and Management Engineering (DIMEG), University of Calabria, Italy. Her expertise covers FACTS technology, harmonic analysis, and electrical system automation. She works on decentralized control and management of power systems, with particular focus on the impacts of electricity market scenarios. Her interests also include smart grids, microgrids, nanogrids, and demand response modeling, as well as the development of market models and aggregator frameworks for energy districts and renewable energy communities.

Nicola Sorrentino

Nicola Sorrentino is a researcher in the Department of Mechanical, Energy and Management Engineering (DIMEG), University of Calabria, Italy. His research focuses on renewable energy technologies, power systems analysis and simulation, and distributed generation. He is actively engaged in energy management strategies and the study of electricity markets, with the aim of fostering sustainable and efficient energy systems.

energies

MDPI

Article

Optimizing Integration of Fuel Cell Technology in Renewable Energy-Based Microgrids for Sustainable and Cost-Effective Energy

Manish Kumar Singla [1], Jyoti Gupta [2], Mohammed H. Alsharif [3,*] and Abu Jahid [4,*]

1 Department of Interdisciplinary Courses in Engineering, Chitkara University Institute of Engineering & Technology, Chitkara University, Punjab 140401, India
2 Department of Computer Science, Shree Guru Gobind Singh Tricentenary University, Gurugram 122505, India
3 Department of Electrical Engineering, College of Electronics and Information Engineering, Sejong University, 209 Neungdong-ro, Gwangjin-gu, Seoul 05006, Republic of Korea
4 School of Electrical Engineering and Computer Science, University of Ottawa, 25 Templeton St., Ottawa, ON K1N 6N5, Canada
* Correspondence: malsharif@sejong.ac.kr (M.H.A.); ajahi011@uottawa.ca (A.J.)

Abstract: This article presents a cost-effective and reliable solution for meeting the energy demands of remote areas through the integration of multiple renewable energy sources. The proposed system aims to reduce dependence on fossil fuels and promote sustainable development by utilizing accessible energy resources in a self-contained microgrid. Using the Hybrid Optimization Model for Electric Renewable (HOMER) software, the study examined the optimal combination of energy sources and storage technologies for an integrated hybrid renewable energy system (IHRES) in the Patiala location of Punjab. The total life cycle cost (TLCC) is the main objective of this manuscript. The HOMER result is taken as a reference, and the results are compared with the optimization hybrid algorithm (PSORSA). From this, it is clear that the proposed algorithm has less TLCC as compared to others. Two combinations of energy sources and storage technologies were considered, namely solar photovoltaic (PV)/battery and solar PV/fuel cell (FC). The results showed that the solar PV/FC combination is more cost-effective, reliable, and efficient than the solar PV/battery combination. Additionally, the IHRES strategy was found to be more economically viable than the single energy source system, with lower total life cycle costs and greater reliability and efficiency. Overall, the proposed IHRES model offers a promising solution for meeting energy demands in remote areas while reducing dependence on fossil fuels and promoting sustainable development.

Keywords: solar photovoltaic; PV; FC; fuel cell; sustainable energy; green energy; energy harvesting; HOMER; battery; smart grids; TLCC

1. Introduction

One of the essential components for a nation's economic development is the availability of power. In many countries, 17% of people, particularly those in remote areas, still have no access to power [1]. The high costs of network transmission in remote regions often necessitate the use of diesel generators to meet load requirements. The high cost of diesel fuel, environmental pollution, and scarcity of fossil fuels make diesel power generation not always an affordable and effective option. The use of renewable energy has been investigated as a potential solution to this problem by many scholars in the past few years. The use of renewable energy systems, particularly solar PV systems, can be a viable alternative in isolated areas [2]. Solar PV systems can provide renewable and sustainable energy with minimum environmental impact. Furthermore, they are relatively easy to install, maintain, and operate. As a result, they are increasingly being adopted in rural and remote areas. Solar PV systems are also cost-effective and can reduce energy costs in the long run. Moreover, they are an important part of a clean energy future, helping to reduce carbon emissions and combat climate change. However, it is difficult to provide a steady load in remote regions since solar radiation is unpredictable, and solar systems depend on the climate. Therefore, for standalone locations, a PV system with an energy storage unit is a feasible approach. The battery is one of the most widely used forms of storage. In order to meet the demand for the load at the lowest possible cost and with the highest level of reliability, it is important to design the ideal configuration of the power scheme components for remote areas. In order to tackle these issues, effective modelling and an optimization technique are crucial.

Numerous studies have investigated the mathematical modeling, optimal size, and techno-economic analysis of hybrid energy designs based on solar energy. These studies have helped to identify cost-effective hybrid energy designs that can provide reliable energy supply. Additionally, they have led to an increased understanding of the potential of solar energy and its integration into existing energy systems. For a remote island, Javed et al. [2] optimized a hybrid solar/wind system with storage using a genetic algorithm. In comparison, the results were compared to those obtained using the software HOMER (hybrid optimization of multiple energy sources). This optimization showed significant improvement in energy efficiency, cost, and reliability. The study concluded that hybrid systems are more reliable and cost-effective than traditional energy systems. It also highlighted the potential of renewable energy sources to replace fossil fuels. In an Indian radio transmitter station, Das et al. [3] created a techno-economically ideal stand-alone hybrid solar/biogas/energy storage scheme using metaheuristic optimization.

The optimal sizing of hybrid solar energy systems has been investigated using a number of optimization techniques, such as HOMER software [4], genetic algorithms, tabu searches, simulated annealing, particle swarm searches, gray wolf searches, harmony searches, and global dynamic harmony searches. It has been studied that various theories and methods can be applied to power systems. A harmony search-based approach was suggested by Yu et al. [5] for determining the optimal capacity and location for off-grid PV/battery systems. An optimization algorithm based on harmony search performs better than one based on simulation annealing. The majority of renewable energy sources are not available 24 h a day in nature, so they cannot provide continuous power supply. Batteries are therefore essential to all renewable energy power generation and conversion systems [6]. The batteries available on the market today are lithium-based, which is heavy, toxic, and expensive to recycle. It is possible to solve all such problems by generating hydrogen from renewable energy sources (e.g., solar PVs), storing it, and using it as a fuel cell to create electricity when renewable sources are not available. However, current hydrogen technology is not efficient enough to be commercially viable. Research and development of a more efficient and cost-effective hydrogen generation and storage technology is needed to make hydrogen fuel cells a viable alternative to lithium-based batteries. Governments should invest in R&D to make hydrogen fuel cells and storage more affordable and efficient. This technology could then be used to power electronic devices and transportation with minimal environmental impact [7,8].

Prior studies have mostly concentrated on finding the most cost-effective hybrid solar energy system configurations. The effect of the reliability index (RI) on the hybrid energy system has also been studied in some research. In the past, studies typically performed techno-economic assessments using the HOMER software tool based on the input data of hybrid systems. Although changes in the modelling of the hybrid system's components are limited, the HOMER software tools enable a fast assessment of hybrid energy systems. The HOMER tool has some limitations regarding the capacity to alter mathematical models and input data for various renewable energy systems. It is rare in earlier studies to examine the effects of important economic parameters and the reliability index when optimizing hybrid systems with a robust metaheuristic algorithm. This type of optimization can be used to improve the reliability and performance of hybrid systems. Furthermore, it can be used to reduce the cost of maintenance and operation of such systems. The main contribution of the manuscript is given below:

- A hybrid system configuration incorporating solar PV/battery energy storage and solar PV/fuel cell based methods is presented in this manuscript for an optimal hybrid system configuration.
- The optimization model presented in this article is used to accomplish a case study in Patiala.
- An altitude of 257 m places Patiala in southeastern Punjab, northwestern India at a position of 30.3398° N, 76.3869° E.
- The purpose of the paper is to minimize the total life cycle costs (TLCC).
- The TLCC is calculated by considering the cost of generation, installation, and operation of the system.
- The results show that the optimal hybrid system configuration can reduce the total cost as compared to the standalone solar PV system.

Section 2 describes the system model and the methodology. Section 3 presents the results and discussion. The conclusion is in Section 4.

2. System Description

To analyze and quantify the cost of both systems, two models of HRES (Solar PV/Battery and Solar PV/FC) have been evaluated with HOMER (Version 3.14.0). The hourly electricity demand of the micro-grid was simulated for a period of twenty years, and the results were compared in terms of total life cycle cost (TLCC). Cost optimization was performed to identify the optimal configuration and evaluate the economic feasibility of both systems. For HOMER simulation to assess the optimization outcomes of these two models, additional input data is required, which is provided in the following section.

- Load Profile Data: As can be seen from Table 1, the baseline and scaled load on a daily basis has an average value of 175.47 kWh, the average (kW) load level with a scaled load level of 7.3 kW, the peak capacity load level with a scaled load level of 25.47 kW, and the load factor for the baseline, and scaled load level at 0.5 is assumed. Based on the annual load data presented in Tables 2 and 3, the weekday and weekend loads are calculated.

Table 1. Baseline and scaled load data.

Parameters	Baseline Load	Scaled Load
Average (kWh/day)	175.47	175.47
Average (kW)	7.3	7.3
Peak (kW)	25.47	25.47
Load Factor	0.5	0.5

- Battery Bank: The conventional approach to storing electrical energy is to use a battery bank. Assembling one or more batteries in order to store energy electrochemically is also considered electrochemical storage [9]. Using HOMER software, a battery is modelled as an energy storage system that can store a specified amount of direct current (DC) with a fixed energy efficiency round-trip. It determines whether a battery needs to be replaced based on the number of charge and discharge cycles [10]. The battery system can be used in various applications such as energy storage in renewable energy systems, power-grid support, uninterruptible power supply, and others. It is important to note that batteries have a finite lifespan and, thus, need to be replaced at regular intervals. To ensure batteries are replaced on time, a battery system can be used to monitor their condition. This system can also be used to alert when the battery needs to be replaced. Additionally, the system can be used to optimize battery performance. Table 4 represents the battery description which is used in the proposed model.

Table 2. Weekdays load data profile.

Hour	Jan	Feb	Mar	Apr	May	Jun	Jul	Aug	Sep	Oct	Nov	Dec
0	0.109	0.109	0.109	0.109	0.109	0.109	0.109	0.109	0.109	0.109	0.109	0.109
1	0.095	0.095	0.095	0.095	0.095	0.095	0.095	0.095	0.095	0.095	0.095	0.095
2	0.095	0.095	0.095	0.095	0.095	0.095	0.095	0.095	0.095	0.095	0.095	0.095
3	0.095	0.095	0.095	0.095	0.095	0.095	0.095	0.095	0.095	0.095	0.095	0.095
4	0.327	0.327	0.327	0.327	0.327	0.327	0.327	0.327	0.327	0.327	0.327	0.327
5	0.500	0.500	0.500	0.500	0.500	0.500	0.500	0.500	0.500	0.500	0.500	0.500
6	0.550	0.550	0.550	0.550	0.550	0.550	0.550	0.550	0.550	0.550	0.550	0.550
7	0.500	0.500	0.500	0.500	0.500	0.500	0.500	0.500	0.500	0.500	0.500	0.500
8	0.420	0.420	0.420	0.420	0.420	0.420	0.420	0.420	0.420	0.420	0.420	0.420
9	0.430	0.430	0.430	0.430	0.430	0.430	0.430	0.430	0.430	0.430	0.430	0.430
10	0.495	0.495	0.495	0.495	0.495	0.495	0.495	0.495	0.495	0.495	0.495	0.495
11	0.533	0.533	0.533	0.533	0.533	0.533	0.533	0.533	0.533	0.533	0.533	0.533
12	0.691	0.691	0.691	0.691	0.691	0.691	0.691	0.691	0.691	0.691	0.691	0.691
13	0.519	0.519	0.519	0.519	0.519	0.519	0.519	0.519	0.519	0.519	0.519	0.519
14	0.418	0.418	0.418	0.418	0.418	0.418	0.418	0.418	0.418	0.418	0.418	0.418
15	0.397	0.397	0.397	0.397	0.397	0.397	0.397	0.397	0.397	0.397	0.397	0.397
16	0.409	0.409	0.409	0.409	0.409	0.409	0.409	0.409	0.409	0.409	0.409	0.409
17	0.658	0.658	0.658	0.658	0.658	0.658	0.658	0.658	0.658	0.658	0.658	0.658
18	1.231	1.231	1.231	1.231	1.231	1.231	1.231	1.231	1.231	1.231	1.231	1.231
19	1.003	1.003	1.003	1.003	1.003	1.003	1.003	1.003	1.003	1.003	1.003	1.003
20	0.676	0.676	0.676	0.676	0.676	0.676	0.676	0.676	0.676	0.676	0.676	0.676
21	0.480	0.480	0.480	0.480	0.480	0.480	0.480	0.480	0.480	0.480	0.480	0.480
22	0.300	0.300	0.300	0.300	0.300	0.300	0.300	0.300	0.300	0.300	0.300	0.300
23	0.204	0.204	0.204	0.204	0.204	0.204	0.204	0.204	0.204	0.204	0.204	0.204

Table 3. Weekend Load Data Profile.

Hour	Jan	Feb	Mar	Apr	May	Jun	Jul	Aug	Sep	Oct	Nov	Dec
0	0.109	0.109	0.109	0.109	0.109	0.109	0.109	0.109	0.109	0.109	0.109	0.109
1	0.095	0.095	0.095	0.095	0.095	0.095	0.095	0.095	0.095	0.095	0.095	0.095
2	0.095	0.095	0.095	0.095	0.095	0.095	0.095	0.095	0.095	0.095	0.095	0.095
3	0.095	0.095	0.095	0.095	0.095	0.095	0.095	0.095	0.095	0.095	0.095	0.095
4	0.327	0.327	0.327	0.327	0.327	0.327	0.327	0.327	0.327	0.327	0.327	0.327
5	0.500	0.500	0.500	0.500	0.500	0.500	0.500	0.500	0.500	0.500	0.500	0.500
6	0.550	0.550	0.550	0.550	0.550	0.550	0.550	0.550	0.550	0.550	0.550	0.550
7	0.500	0.500	0.500	0.500	0.500	0.500	0.500	0.500	0.500	0.500	0.500	0.500
8	0.462	0.462	0.462	0.462	0.462	0.462	0.462	0.462	0.462	0.462	0.462	0.462
9	0.473	0.473	0.473	0.473	0.473	0.473	0.473	0.473	0.473	0.473	0.473	0.473
10	0.545	0.545	0.545	0.545	0.545	0.545	0.545	0.545	0.545	0.545	0.545	0.545
11	0.586	0.586	0.586	0.586	0.586	0.586	0.586	0.586	0.586	0.586	0.586	0.586
12	0.760	0.760	0.760	0.760	0.760	0.760	0.760	0.760	0.760	0.760	0.760	0.760
13	0.571	0.571	0.571	0.571	0.571	0.571	0.571	0.571	0.571	0.571	0.571	0.571
14	0.460	0.460	0.460	0.460	0.460	0.460	0.460	0.460	0.460	0.460	0.460	0.460
15	0.437	0.437	0.437	0.437	0.437	0.437	0.437	0.437	0.437	0.437	0.437	0.437
16	0.450	0.450	0.450	0.450	0.450	0.450	0.450	0.450	0.450	0.450	0.450	0.450
17	0.658	0.658	0.658	0.658	0.658	0.658	0.658	0.658	0.658	0.658	0.658	0.658
18	1.231	1.231	1.231	1.231	1.231	1.231	1.231	1.231	1.231	1.231	1.231	1.231
19	1.003	1.003	1.003	1.003	1.003	1.003	1.003	1.003	1.003	1.003	1.003	1.003
20	0.676	0.676	0.676	0.676	0.676	0.676	0.676	0.676	0.676	0.676	0.676	0.676
21	0.480	0.480	0.480	0.480	0.480	0.480	0.480	0.480	0.480	0.480	0.480	0.480
22	0.300	0.300	0.300	0.300	0.300	0.300	0.300	0.300	0.300	0.300	0.300	0.300
23	0.204	0.204	0.204	0.204	0.204	0.204	0.204	0.204	0.204	0.204	0.204	0.204

Table 4. Battery Description.

Nominal Voltage (V)	55
Nominal Capacity (kWh)	3.26
Nominal Capacity (Ah)	63
Roundtrip efficiency (%)	94.9
Maximum Charge Current (A)	63
Maximum Discharge Current (A)	63

- Solar PV Panel: Electricity is generated by solar panels by absorbing sunlight. However, there are certain variables that affect the PV's output, such as temperature and sunlight. Solar panel output is also affected by shading, dust, and dirt. To maximize the output, it is important to install the panels in a location that receives direct sunlight and is not affected by shade. Regularly cleaning the panels can help maintain their efficiency. Panels should also be angled correctly to maximize the amount of sunlight received. Additionally, panels should be installed in areas with low temperatures to ensure optimal performance.
- Fuel Cell: An electro-chemical device that converts chemical energy into electrical energy is called a fuel cell (FC). Fuel cells are available in a variety of configurations based on the types of electrodes, operating characteristics, and power ranges they operate at. Fuel cells can be used to generate power for a wide range of applications from home heating and cooling systems to electric vehicles. Additionally, fuel cells are clean and efficient sources of energy, producing no emissions and creating no waste. Among the variety of FCs available on the market, Proton Exchange Membrane Fuel Cells (PEMFCs) are the most commonly used due to their excellent start-up time and low operating temperature. PEMFCs are also known for their high energy efficiency and low emissions, making them an ideal choice for clean energy applications. Additionally, they are relatively compact and lightweight, making them suitable for powering vehicles, portable electronics, and other applications. PEMFCs are also relatively inexpensive compared to other types of fuel cells, making them a cost-effective option for many applications. Furthermore, they are relatively low-maintenance, making them an attractive option for many users.
- Electrolyzer: Carbon-free hydrogen can be produced from renewable energy sources by electrolysis. The electrolysis process involves splitting water into hydrogen and oxygen using electricity. During this reaction, an electrolyzer is used. An electrolyzer can range in size from a small appliance that can be used to manufacture hydrogen at a small scale to a large scale, central production facility that can be connected directly to renewable or non-greenhouse gas-emitting electricity sources. The hydrogen produced by electrolysis is considered to be a clean energy source, as there are no carbon emissions generated during the production process. This makes it an attractive alternative to traditional fossil fuels. Additionally, hydrogen can be stored and used as a fuel for transportation. Furthermore, hydrogen can be produced from renewable sources such as solar, wind, and water, making it an even greener option. Hydrogen also has a high energy density and can be used in stationary power plants.
- Hydrogen Storage Tank: Hydrogen is stored in hydrogen tanks and supplied to fuel cells for energy generation after it is produced by the electrolyzer [11]. During a day, hydrogen is produced by the electrolyzer and stored in the hydrogen tank, which is used when there is no PV generation. The stored hydrogen is then used as a source of renewable energy, which is converted into electrical energy by the fuel cell. This electrical energy can then be used to power the electric vehicle or stored in the battery for later use.

2.1. Methodology

The proposed system is optimized, simulated, and modeled using HOMER software. The results from this process allowed us to analyze the performance of the system. We then used this information to make adjustments and improve the system's efficiency. Finally, we validated the system to ensure its accuracy and reliability. As discussed below, two cases are examined in this manuscript:

Case 1: Solar PV/Battery: Figure 1 shows the IHRES consisting of solar photovoltaics and a battery bank. In this case, solar PV is the primary source of supply, while batteries serve as a storage medium.

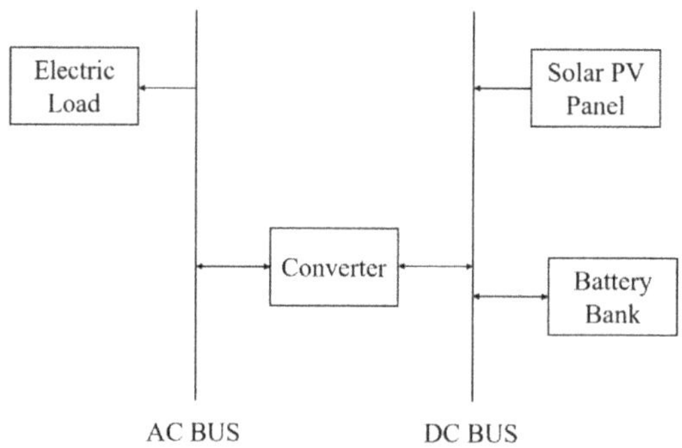

Figure 1. Proposed diagram for Case 1.

Case 2: Solar PV/Fuel Cell: As shown in Figure 2, IHRES consists of solar photovoltaic and a fuel cell. In this scenario, solar PV and fuel cells make up the majority of the supply. Fuel cells are also being used for energy storage.

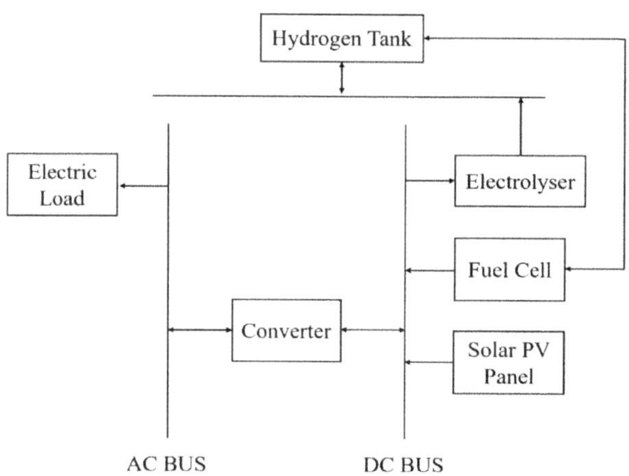

Figure 2. Proposed diagram for Case 2.

2.2. *System Modelling*

- Solar PV Model: Based on Equation (1), a solar PV panel's power output (depends on solar radiation) is proportional to the amount of sunlight it receives.

$$P_{solar}(t) = P_s^t f_{loss} \frac{H_i(t)}{H_t} \tag{1}$$

Solar PV panel rating is represented by P_s^t; loss factor or derating is represented by f_{loss}, due to dirt, shadows, temperature, etc.; the hourly solar radiation incident at the surface of solar PV panels (W/m^2) is represented by $H_i(t)$; and standard incident radiation (1000 W/m^2) by H_t.

Converter: When a system contains AC and DC components, DC to AC and AC to DC power converters are required. Fuel cells, solar PV panels, and batteries produce DC output while AC load is considered. Upon the peak load demand $P_L^m(t)$, converter size is taken. As shown in Equation (2), the inverter rating $P_{in\ v}$ is calculated as follows:

$$P_{in\ v}(t) = P_L^m(t) / \eta_{in\ v} \quad (2)$$

where the efficiency of the inverter is denoted by $\eta_{in\ v}$.

- Battery Bank: During times when the power from renewable systems is insufficient or absent, a battery is used to store excess energy. An accurate estimation of the state of charge (SOC) can lead to the measurement of energy. Equation (3) can be used to calculate the SOC of a battery:

$$\frac{SOC(t)}{SOC(t-1)} = \int_{r-1}^{r} \frac{P_b(t)\eta_{battery}}{V_{bus}} dt \quad (3)$$

A battery's input/output voltage is denoted by $P_b(t)$, the battery's efficiency by $\eta_{battery}$, and its voltage by V_{bus}. A positive value indicates that the battery is charging, while a negative value indicates it is discharging. Equation (4) gives you the round-trip efficiency of a battery:

$$\eta_{battery} = \sqrt{\eta_{battery}^{c} \eta_{battery}^{d}} \quad (4)$$

The charging efficiency $\eta_{battery}^{c}$ and discharging efficiency $\eta_{battery}^{d}$ of the battery are depicted in [12]. It is estimated that the battery has a round-trip efficiency of 92%. Additionally, discharging and charging efficiencies are assumed to be different and calculated as 100% and 85%, respectively. As shown in Equation (5), SOC_{max} is the maximum value of SOC and equals the aggregate capacity of the battery bank C_n:

$$C_n = \frac{N_{battery}}{N_{battery}^{s}} C_b \quad (5)$$

Single battery capacity should be denoted as C_b, total battery number should be denoted as $N_{battery}$, and number of batteries connected in series should be denoted as $N_{battery}^{s}$. SOC_{min} is the minimum discharge limit that can be reached by the battery bank. System constraints can be applied according to the usage of the battery bank. Batteries are connected in series in order to achieve desired bus voltage. As shown in Equation (6), a series of batteries is calculated as follows:

$$N_{battery}^{s} = \frac{V_{bus}}{V_{battery}} \quad (6)$$

An individual battery's voltage is represented by $V_{battery}$. In battery modeling, the maximum charge or discharge power at any given time is also a major factor. Equation (7) calculates maximum charge current based on maximum charge current:

$$P_b^{max} = \frac{N_{battery} V_{battery} I_{max}}{1000} \quad (7)$$

where the maximum charge current in amperes is denoted as I_{max}.

- Fuel Cell: In hydrogen fuel cells, chemical energy is converted directly into electric power, while at the same time, heat and water are produced as by-products as long as fuel is available. Hydrogen fuel cells are considered to be a clean and renewable source of energy, as they do not emit any greenhouse gases [13]. They are also highly efficient, with some fuel cells achieving up to 80% efficiency. The anode and cathode chemical reactions are shown in equations 8 to 10. By combining reactions on both electrodes, the entire reaction may be obtained.

The anode's chemical reaction is as follows:

$$H_2 \rightarrow 2H^+ + 2e^- \tag{8}$$

The cathode's chemical reaction is as follows:

$$\frac{1}{2}O_2 + 2H^+ + 2e^- \rightarrow H_2O \tag{9}$$

The overall chemical reaction is as follows:

$$H_2 + \frac{1}{2}O_2 \rightarrow H_2O \tag{10}$$

Across the membrane, hydrogen gas dissociates into protons and electrons at an anode on the left, forming the concentration gradient between the electrodes. The gradient causes protons to diffuse across the membrane while leaving electrons behind. In relation to the anode, the cathode becomes positively charged when protons drift towards it. Since the membrane prevents electrons from passing through, the remaining electrons are drawn to the positively charged cathode, causing current to flow the other way. According to Equation (11), fuel cells produce the following voltage:

$$V_{FC} = E_{Nernst} - V_{act} - V_{ohm} - V_{con} \tag{11}$$

Equation (12) shows the maximum voltage that can be generated by a cell given a reversible open circuit voltage, an ohmic voltage drop, an activation voltage drop, and a concentration voltage drop.

$$E_{Nernst} = -\frac{\Delta G^{\circ}}{nF} + \frac{\Delta S}{nF}\left(T - T_{ref}\right) + \frac{R.T}{n.F} ln\left(\frac{P_{O_2}^{0.5}.P_{H_2}^{1}}{P_{H_2O}^{1}}\right) \tag{12}$$

where ΔG^o is an electric work which is called Gibbs free energy; the number of moles is denoted by n; F is a Faraday's constant; specific entropy is denoted by ΔS; absolute temperature is denoted by T; T_{ref} is taken as 25 °C; $P_{O_2}^{0.5}, P_{H_2}^{1}, P_{H_2O}^{1}$ are pressure of pure oxygen, hydrogen, and H_2O as fuel; and R is gas constant, respectively. Equation (13) shows the total power output of a stack of fuel cells P_{FC}:

$$P_{FC} = p_{FC} V_{FC} I_{FC} \tag{13}$$

In a stack of fuel cells, p_{FC} and I_{FC} represent rating and current, respectively.

A major component of fuel cell modeling for power generation is hydrogen mass flow rate $\left(Q_{H_2}\right)_1^D$. According to Equation (14), 1 kW fuel cells consume hydrogen in kg/h at the following mass flow rate:

$$\left(Q_{H_2}\right)_1^D = \alpha_1 P_{FC}^s + \alpha_2 P_{FC} \tag{14}$$

The intercept coefficient t α_1, α_2. and P_{FC}^s of a fuel cell is measured in kilograms per hour per kW, and its rated capacity is measured in kilowatts. In the case of a fuel cell rating of 1 kW, taking into account α_1 and α_2 to be 0.0003 and 0.58 kg/h/kW, $\left(Q_{H_2}\right)_1^D$ calculated is 0.059 kg/h. For a fuel cell to generate 1 kW of power, 0.059 kg of hydrogen is required per hour [14].

- Electrolyzer: The electrolyzer converts the surplus electricity from the PV panels into hydrogen, which is stored in a hydrogen tank and used to power the fuel cell. Hydrogen production by the electrolyzer is the primary aspect to consider. Equation (15) shows the mass flow rate $(m_{H_2})_1^Q$ of hydrogen produced by a 1 kW electrolyzer in kilograms per hour:

$$(m_{H_2})_1^Q = \frac{3600 P_{el}^s n_{el}}{HV_{H_2}} \tag{15}$$

The heating value of hydrogen in MJ/kg, and the efficiency of the electrolyzer P_{el}^s, HV_{H_2} and n_{el} are 1 kW, MJ/kg, and kWh/kg, respectively. Electrolyzers are estimated to be 90% efficient [15]. The mass flow rate of hydrogen in an electrolyzer with 142 MJ/kg heating value will be 0.02268 kg/h/kW. Based on this, 1 kW electrolyzers produce 0.02268 kg/h hydrogen.

- Hydrogen Tank: The electrolyzer produces hydrogen that is stored in hydrogen tanks for use in fuel cells. As the electrolyzer produces hydrogen during the day, the hydrogen tank stores it, as well as supplying hydrogen to the fuel cell when no PV power is generated. A decision variable for the proposed system was the size of the hydrogen tank, T_{H_2} in kilograms. t is $m_{H_2}(t)$, the time at which hydrogen is available to flow. As shown in Equation (16), hydrogen is available at a given time after being produced by the electrolyzer.

$$m_{H_2}(t) = P_{el}(m_{H_2})_1^Q \tag{16}$$

In Equation (15), P_{el} is the power consumed by the electrolyzer, and $(m_{H_2})_1^Q$ is the mass flow rate of hydrogen produced in kg/h by a 1 kW electrolyzer. Equation (17) calculates this available hydrogen stored in a hydrogen tank:

$$T_{H_2}(t) = m_{H_2} + T_{H_2}(t-1) \tag{17}$$

where $T_{H_2}(t-1)$ is the hydrogen in the tank at the previous hour. Hydrogen is used to generate power in a fuel cell. Equation (18) represents the mass flow rate of hydrogen taken from hydrogen tank by a fuel cell at a particular time:

$$Q_{H_2}(t) = P_{FC}(Q_{H_2})_1^D \tag{18}$$

Based on Equation (14), the mass flow rate of hydrogen consumed by 1 kW fuel cell is calculated in kg/h. Equation (19) calculates the remaining hydrogen in the hydrogen tank after taking hydrogen:

$$T_{H_2}(t) = T_{H_2}(t-1) - Q_{H_2} \tag{19}$$

The amount of hydrogen $T_{H_2}(t-1)$ in the tank at the time is $t-1$.

2.3. Proposed Algorithm

2.3.1. Rat Search Algorithm (RSA)

Various sizes and weights of rats belonging to two species are studied in this study, which provides insights into their behavior as intelligent and social rodents. Among the activities they engage in within their territorial communities are grooming, tumbling, hopping, boxing, and chasing. It is not uncommon for rats to exhibit violent behavior during competitions for prey, resulting in fatalities. Mathematically modeling the aggressive behavior of rats during chases and battles is the focus of this research.

Mathematical Modeling:

Chasing:

Animals such as rats engage in social agonistic behaviors to pursue prey in groups. For math modeling, we assume the most effective search agent knows where the prey is, and other search agents adjust their positions based on the best agent. In order to explain this mechanism, Equation (20) is introduced.

$$\vec{Q} = B.\vec{Q_j}(z) + D.\left(\vec{Q_s}(z) - \vec{Q_j}(z)\right) \quad (20)$$

The rat's position is denoted by $\vec{Q_j}(z)$, while the best optimal solution is denoted by $\vec{Q_s}(z)$. The values of B and D are specified by Equations (21) and (22).

$$B = S - z \times \left(\frac{S}{Max_{iter}}\right) \quad (21)$$

where, z = 0, 1, 2, ..., Max_{iter}

$$D = 2.\ \text{rand}\ () \quad (22)$$

A parameter S represents a random number, while a parameter D represents a random number, with values [0, 2] and [1, 5], respectively. In the iterative process, B and D are responsible for optimizing the performance of these parameters during exploration and exploitation.

Fighting:

Rats engage in battles with their prey using Equation (23) as a mathematical representation.

$$\vec{Q_j}(z+1) = \left|\vec{Q_s}(z) - \vec{Q}\right| \quad (23)$$

The rat search algorithm updates the next position of the rat by utilizing equation $\vec{Q_j}(z+1)$. As a result, the optimal solution is maintained, and the locations of other search agents are adjusted in relation to it. In order to facilitate exploration and exploitation, the parameters B and D have been adjusted. RSA proposes a solution that can be obtained with a minimum amount of operators. The pseudo-code of the rat search algorithm is presented in Algorithm 1, and the flow chart of that algorithm is shown in Figure 3.

Algorithm 1 Rat search algorithm pseudo-code

```
Input: Population of rat
Output: Optimal Search agent
Initialize the parameter B, D, and S.
Calculate the fitness value of all search agents.
Qs is the best search agent.
   While (x < Maxiter) do.
         For each search agent do.
         Update the position of current search agent.
         End for
         Update parameter B, D, and S.
Check if there is any search agent which goes beyond the boundary condition.
Calculate the fitness of each search agent.
Update Qs, if the solution is better than the previous one.
End while
return Qs,
End
```

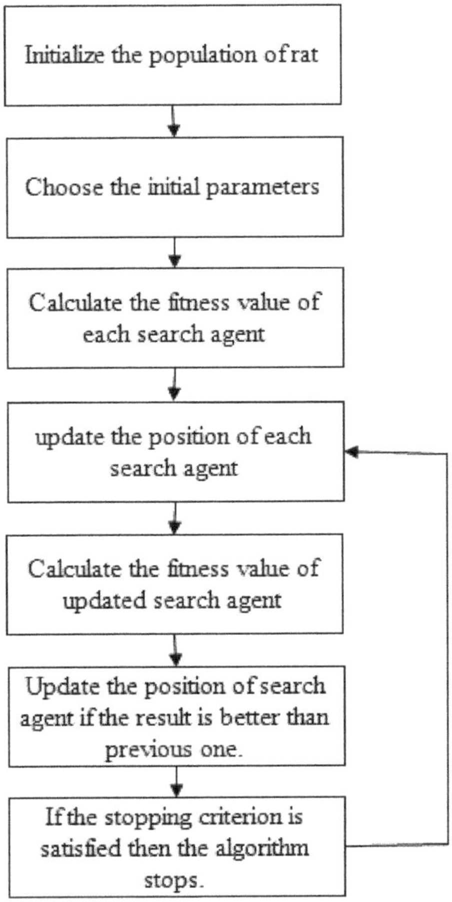

Figure 3. Flow chart of RSA.

2.3.2. Hybrid Particle Swarm Optimization and Rat Search Algorithm (PSORSA)

We propose a hybrid approach to combine two algorithms for increased efficiency. Our approach enhances the precision of results achieved by RSA by leveraging PSO, which is a precursor algorithm to the class of optimization algorithms known as swarm algorithms. As a result of our hybrid method, the system is not trapped in local minimums and is also enhanced in accuracy. This enhances the speed of the system and ensures the system finds the global optimum within reasonable time. The combination of RSA and PSO thus provides a powerful hybrid approach to optimize the system. The hybrid method is also extremely cost effective, since it does not require a large amount of computational power to achieve the desired result. The combination of RSA and PSO thus provides a powerful and cost-effective solution to optimize the system. The flow chart of that algorithm is shown in Figure 4.

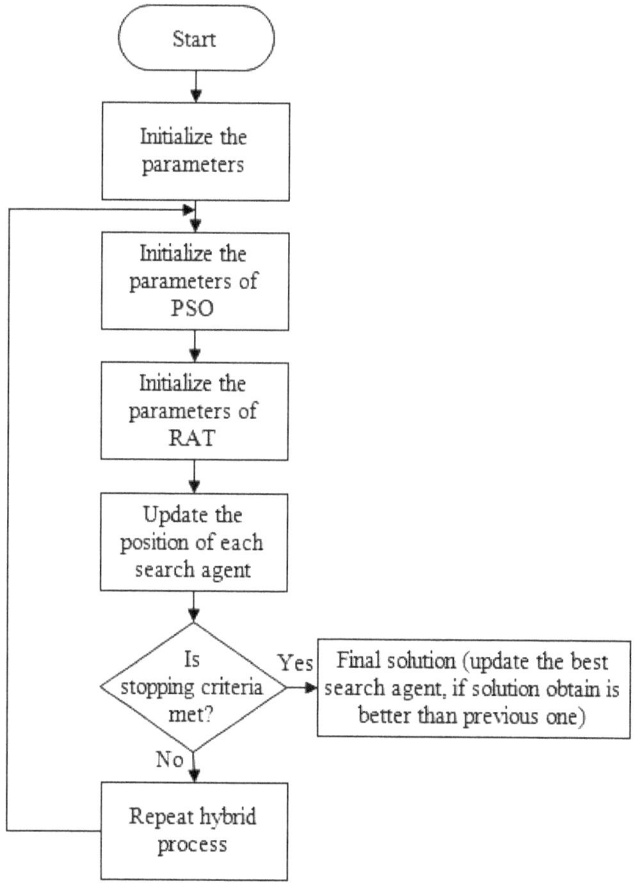

Figure 4. Flow chart of PSORSA.

2.4. Objective Function and Constraints

Objective Function: A value that minimizes the total life cycle cost (TLCC) is the objective function. Costs associated with ongoing operation and maintenance (C&R) and capital and replacement (O&M) will be included in the total life cycle cost of the project. These costs should be weighed against the expected benefits of the project to determine the optimal value of the objective function. A variety of techniques, such as cost-benefit analysis, can be used to ensure a thorough evaluation of the project, which is given in Equation (24):

$$\text{TLCC} = \text{C\&R} + \text{O\&M} \quad (24)$$

In a mathematical model, the capital recovery cost (CRF) is determined based on the interest rate (j) and system life span (m), which occur at the beginning and during the duration of the project. The CRF is used to calculate the value of the project by taking into account the amount of money needed to pay off the initial investment. It is a key factor in determining the profitability of a project and is given in Equation (25):

$$CRF\ (j,m) = \frac{j(1+j)^m}{(1+j)^m - 1} \quad (25)$$

Constraints: The number of solar PV panel, fuel cell, electrolyzer, hydrogen tank, and battery constraints and storage units are given in Equation (26a)–(26e):

$$N_{solar}^{min} \leq N_{solar} \leq N_{solar}^{max} \tag{26a}$$

$$N_{fuel\ cell}^{min} \leq N_{fuel\ cell} \leq N_{fuel\ cell}^{max} \tag{26b}$$

$$N_{electrolyzer}^{min} \leq N_{electrolyzer} \leq N_{electrolyzer}^{max} \tag{26c}$$

$$N_{hydrogen\ tank}^{min} \leq N_{hydrogen\ tank} \leq N_{hydrogen\ tank}^{max} \tag{26d}$$

$$N_{battery}^{min} \leq N_{battery} \leq N_{battery}^{max} \tag{26e}$$

where maximum number of solar PV panel is denoted by N_{solar}^{max}, maximum rating of fuel cell is denoted by $N_{fuel\ cell}^{max}$; maximum rating of electrolyzer is denoted by $N_{electrolyzer}^{max}$; maximum rating of size of hydrogen tank is denoted by $N_{hydrogen\ tank}^{max}$; and maximum rating of size of battery is denoted by $N_{battery}^{max}$.

3. Results and Discussion HOMER Based

A case study is presented in Patiala using the optimization model. In the south-eastern part of Punjab, northwestern India, Patiala has an altitude of 257 m and is situated at 30.3398° N, 76.3869° E. The work focuses on two case studies, namely Solar PV/Battery-based system and Solar PV/Fuel Cell-based system [16,17]. Table 5 lists the specifications of the components of hybrid systems. To develop and implement a renewable-based hybrid energy system, the TLCC reliability index is of the utmost importance. In this manuscript, some parts of the suggested IHRES are evaluated for their ability to fulfill a community's AC (Alternating Current) and DC (Direct Current) primary load requirements. The size of the bi-directional plant is determined by its peak load. Bi-directional converters are assumed to be 90% efficient. The HOMER software is used to calculate the cost of all the system components. In this study, we used five main parameters, such as the rating of the solar PV panel, the fuel cell, the battery, the electrolyzer, and the size of the hydrogen tank. Inverter size was not considered as an optimization parameter; it was selected based on peak load demand. It was assumed that the project would last 20 years at a 5% interest rate. The two cases are considered as discussed below:

Table 5. System component specifications and cost.

Component	Parameter	Value	Unit
PV Panel	Maximum power (P_{max})	100	W
	Maximum power voltage (V_{mp})	18	V
	Maximum power current (I_{mp})	5.56	A
	Open circuit voltage (V_{OC})	22.3	V
	Short circuit voltage (I_{SC})	6.1	A
	Number of cells	36	-
	Nominal operating cell temperature	45	°C
	Capital cost and replacement cost	1084	USD/kW
	O&M cost	5/kW	USD/yr
	Life time	20	yr

Table 5. *Cont.*

Component	Parameter	Value	Unit
Inverter	Rated power	1	kW
	Inverter efficiency (η_{inv})	90	%
	Capital and replacement cost	127	USD/kW
	O&M cost	1	USD/yr
	Life time	20	yr
Fuel cell	Rated power	1	kW
	Capital and replacement cost	600	USD/kW
	O&M cost	0.01	USD/h/kW
	Fuel cell coefficient	0.0003	USD/h/kW
	Hydrogen to electricity conversion by fuel cell $(QH_2)_c^1$	0.059	USD/h/kW
	Life time of FC	20	yr
Electrolyzer	Capital and replacement cost	150	USD/kW
	O&M cost	8	USD/yr/kW
	Efficiency	90	%
	Life time	20	yr
Hydrogen tank	Capital cost	1.3	USD/kg
	Replacement cost	0.5	USD/kg
	O&M cost	0.6	USD/yr/kg
	Life time	20	yr
Battery	Nominal capacity	360	Ah
	Nominal voltage	6	V
	Maximum charging current	18	A
	Minimum state of charge	30	%
	Maximum state of charge	100	%
	Round trip efficiency	92	%
	Capital cost	300	USD
	Replacement cost	200	USD
	O&M cost	3.67	USD/yr
	Life-time	20	yr
Others	Project life (N)	20	yr
	Interest rate (i)	5	%

Case 1: Solar PV/Battery Based System: According to Figure 1, IHRES consists of a solar photovoltaic system and a battery bank. A battery serves as a storage device in this case, with Solar PV as the main source of supply. Table 6 and Figure 5 show the overall system component cost for this case. The total capital, replacement, and O&M cost are 252,251, 68,375, and 31,153. From Table 6, it is observed that the TLCC of battery, converter, and solar PV is 136,213, 67,682, and 147,884, respectively, but the overall TLCC of the system is 351,779. The Figure 6 shows the TLCC of all the system for case 1.

Table 6. Overall system component cost for Case 1.

Component	Capital (USD)	Replacement (USD)	O&M (USD)	TLCC (USD)
Battery	102,150	30,750	3313	136,213
Converter	28,958	37,625	1099	67,682
Solar PV	121,143	0	26,741	147,884
System	252,251	68,375	31,153	351,779

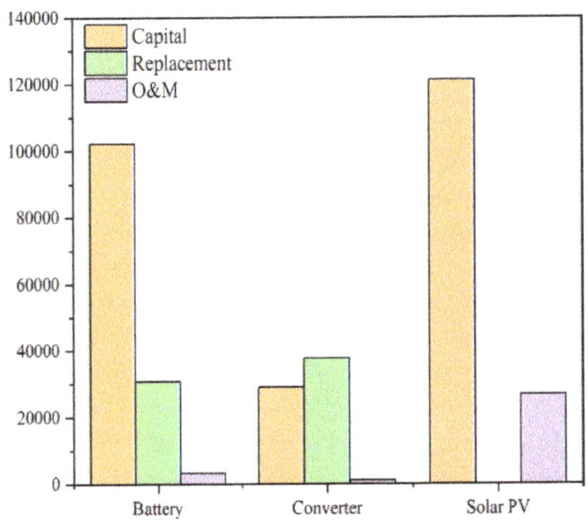

Figure 5. Overall system component cost for Case 1.

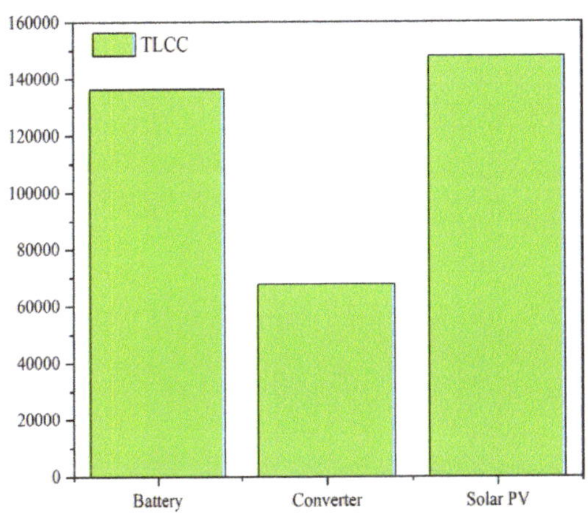

Figure 6. Total life cycle cost for Case 1.

Using HOMER software, the costs of the entire component were calculated of the proposed system. These parameters were then compared with the proposed algorithm, as well as their parent algorithms, as previously mentioned. The feasible and optimal solution was determined by ranking the parameters based on TLCC, as shown in Table 7. Table 7 displays the results in TLCC format for HOMER, RAT, PSO, GWO, and HHO case studies. While the proposed hybrid system took hours to simulate with HOMER, the simulation time was significantly reduced with the help of developed algorithms. From the results, it was concluded that the proposed algorithm outperformed both the parent algorithms, compared algorithm, and the HOMER simulation.

Table 7. Optimization results.

Algorithm	TLCC (USD)
PSO	350,229
RAT	330,125
GWO	345,115
HOMER	351,779
HHO	343,451
PSORSA	320,224

Case 2: Solar PV/Fuel Cell Based System: According to Figure 2, IHRES consists of solar photovoltaics and fuel cells. As a storage device, fuel cell is used in this case as a main source of supply in addition to Solar PV. Table 8 and Figure 7 show the overall system component cost for this case. The total capital, replacement, and O&M cost are 244,910, 62,569, and 23,893. From Table 8, it is observed that the TLCC of fuel cell, electrolyzer, hydrogen tank, converter, and solar PV is 112,825, 29,448, 890, 62,940, and 125,269, respectively, but the overall TLCC of the system is 331,372. Figure 8 shows the TLCC of all the system for Case 2.

Table 8. Overall system component cost for Case 2.

Component	Capital (USD)	Replacement (USD)	O&M (USD)	TLCC (USD)
Fuel Cell	88,524	21,147	3154	112,825
Electrolyzer	14,000	6321	9127	29,448
Hydrogen Tank	140	0	750	890
Converter	26,989	35,101	850	62,940
Solar PV	115,257	0	10,012	125,269
System	244,910	62,569	23,893	331,372

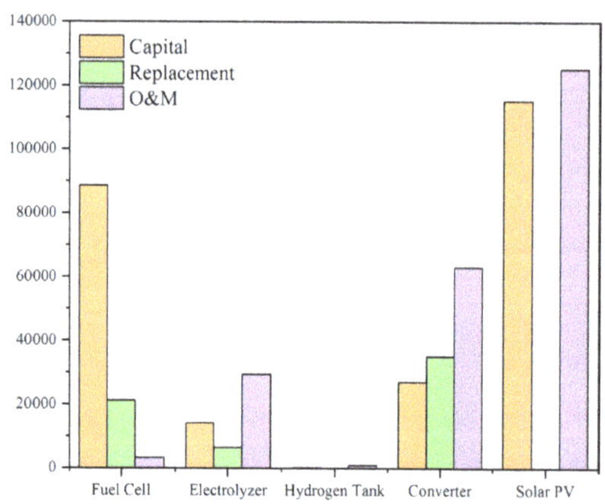

Figure 7. Overall system component cost for Case 2.

Using HOMER software, the costs of the entire component were calculated for the proposed system. These parameters were then compared with the proposed algorithm, as well as their parent algorithms, as previously mentioned. The feasible and optimal solution was determined by ranking the parameters based on TLCC, as shown in Table 9. Table 9 displays the results in TLCC format for HOMER, RAT, PSO, GWO, and HHO case studies. While the proposed hybrid system took hours to simulate with HOMER, the simulation time was significantly reduced with the help of developed algorithms. From the results, it was concluded that the proposed algorithm outperformed both the parent algorithms, compared algorithm, and the HOMER simulation.

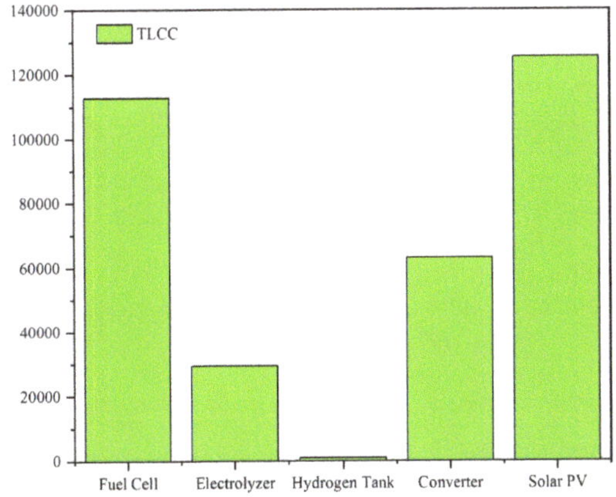

Figure 8. Total life cycle cost for Case 2.

Table 9. Optimization results.

Algorithm	TLCC (USD)
PSO	325,152
RAT	306,124
GWO	323,522
HOMER	331,372
HHO	318,189
PSORSA	302,198

Based on the results obtained, the performance of the system presented in Case 2 is significantly better than that in Case 1. Based on the above comparison, the Case 2 is the more economical system. Case 2 has a lower cost of all the components, a lower TLCC overall, and a fuel cell instead of a battery, which is more economical.

Fuel cell systems with hydrogen as a fuel are gradually replacing lithium-ion batteries. It is possible for fuel cells to replace lithium-based batteries, which cause water and air pollution. It costs approximately five times more to produce a lithium-based battery that has been recycled than one that has been extracted from the ground. All battery manufacturing units are unconsciously extracting more and more lithium from the Earth, causing the Earth to become a 'lithium dump'. There is real potential for fuel cells to qualify as technology that generates electricity with harmless byproducts. Hydrogen fuel cells have the advantages of scalability, low greenhouse gas emissions, and silent operation. Unlike a battery, the fuel cell does not self-discharge. It does not release any hazardous gases either. This makes it an ideal choice for powering electric vehicles, homes, and businesses. It can also be used in stationary applications such as backup power and grid-scale energy storage.

4. Conclusions and Outcomes

The aim of this manuscript is to provide an optimal configuration for stand-alone hybrid solar PV/battery systems and solar PV/fuel cell systems. According to the reliability index, the objective function is to minimize the total life cycle cost (TLCC). With the HOMER software, it has been possible to determine the optimal size of different components such as solar PV, fuel cells, converters, batteries, hydrogen tanks, and electrolyzers. It is obvious from the TLCC that Case 2 has a higher economic benefit than Case 1. Through proper sizing of components and efficient energy flow management between system components, a cost-effective AC/DC hybrid energy system can be designed. The HOMER software can help to identify the most cost-effective combination of components to meet the user's energy needs. By optimizing the size and configuration of the system, the user can maximize the economic benefits and ensure a reliable energy supply. The HOMER software is an invaluable tool for designing AC/DC hybrid energy systems, as it can quickly analyze the various options and determine the most cost-effective solution. The HOMER software is used as a reference in the manuscript. The main disadvantage of HOMER software is that it takes lot of time to simulate the things to overcome that disadvantage optimization results are conducted. Here, it is seen that the proposed algorithm TLCC is lower than the rest of the compared algorithms as well as HOMER software. Furthermore, it can accurately predict the system's performance and energy output.

Author Contributions: Conceptualization, M.K.S. and J.G.; methodology, M.K.S.; software, M.K.S. and J.G.; validation, M.H.A., A.J. and M.K.S.; formal analysis, M.H.A. and A.J.; investigation, A.J.; resources, M.K.S. and J.G.; writing—original draft preparation, M.K.S.; writing—review and editing, J.G., M.H.A. and A.J. All authors have read and agreed to the published version of the manuscript.

Funding: This research received no external funding.

Data Availability Statement: Not applicable.

Conflicts of Interest: The authors declare no conflict of interest.

References

1. Ghasemi, A.; Enayatzare, M. Optimal energy management of a renewable-based isolated microgrid with pumped-storage unit and demand response. *Renew. Energy* **2018**, *123*, 460–474. [CrossRef]
2. Javed, M.S.; Song, A.; Ma, T. Techno-economic assessment of a stand-alone hybrid solar-wind-battery system for a remote island using genetic algorithm. *Energy* **2019**, *176*, 704–717. [CrossRef]
3. Das, M.; Singh, M.A.K.; Biswas, A. Techno-economic optimization of an off-grid hybrid renewable energy system using metaheuristic optimization approaches–case of a radio transmitter station in India. *Energy Convers. Manag.* **2019**, *185*, 339–352. [CrossRef]
4. Mokhtara, C.; Negrou, B.; Settou, N.; Bouferrouk, A.; Yao, Y. Design optimization of grid-connected PV-Hydrogen for energy prosumers considering sector-coupling paradigm: Case study of a university building in Algeria. *Int. J. Hydrogen Energy* **2021**, *46*, 37564–37582. [CrossRef]
5. Yu, X.; Li, W.; Maleki, A.; Rosen, M.A.; Birjandi, A.K.; Tang, L. Selection of optimal location and design of a stand-alone photovoltaic scheme using a modified hybrid methodology. *Sustain. Energy Technol. Assess.* **2021**, *45*, 101071. [CrossRef]
6. Sawle, Y.; Gupta, S.C.; Kumar Bohre, A. PV-wind hybrid system: A review with case study. *Cogent Eng.* **2016**, *3*, 1189305. [CrossRef]
7. Mahato, D.P.; Sandhu, J.K.; Singh, N.P.; Kaushal, V. On scheduling transaction in grid computing using cuckoo search-ant colony optimization considering load. *Clust. Comput.* **2020**, *23*, 1483–1504. [CrossRef]
8. Rani, S.; Babbar, H.; Kaur, P.; Alshehri, M.D.; Shah, S.H.A. An optimized approach of dynamic target nodes in wireless sensor network using bio inspired algorithms for maritime rescue. *IEEE Trans. Intell. Transp. Syst.* **2022**, *24*, 2548–2555. [CrossRef]
9. Hina, F.; Prabaharan, N.; Palanisamy, K.; Akhtar, K.; Saad, M.; Jackson, J.J. *Hybrid-Renewable Energy Systems in Microgrids*; Woodhead Publishing: Cambridge, UK, 2018.
10. Lambert, T.; Gilman, P.; Lilienthal, P. Micropower system modeling with HOMER. *Integr. Altern. Sources Energy* **2006**, *1*, 379–385.
11. Singh, S.; Chauhan, P.; Aftab, M.A.; Ali, I.; Hussain, S.S.; Ustun, T.S. Cost optimization of a stand-alone hybrid energy system with fuel cell and PV. *Energies* **2020**, *13*, 1295. [CrossRef]
12. Singh, S.; Singh, M.; Kaushik, S.C. Feasibility study of an islanded microgrid in rural area consisting of PV, wind, biomass and battery energy storage system. *Energy Convers. Manag.* **2016**, *128*, 178–190. [CrossRef]
13. Hemi, H.; Ghouili, J.; Cheriti, A. A real time fuzzy logic power management strategy for a fuel cell vehicle. *Energy Convers. Manag.* **2014**, *80*, 63–70. [CrossRef]
14. Wu, K.; Zhou, H.; An, S.; Huang, T. Optimal coordinate operation control for wind–photovoltaic–battery storage power-generation units. *Energy Convers. Manag.* **2015**, *90*, 466–475. [CrossRef]
15. Nehrir, M.H.; Wang, C.; Strunz, K.; Aki, H.; Ramakumar, R.; Bing, J.; Miao, Z.; Salameh, Z. A review of hybrid renewable/alternative energy systems for electric power generation: Configurations, control, and applications. *IEEE Trans. Sustain. Energy* **2011**, *2*, 392–403. [CrossRef]
16. Nijhawan, P.; Singla, M.K.; Gupta, J. A proposed hybrid model for electric power generation: A case study of Rajasthan, India. *IETE J. Res.* **2021**, *69*, 1952–1962. [CrossRef]
17. Nijhawan, P.; Singla, M.K.; Gupta, J. Site selection of solar PV power plant at Bathinda. *Int. J. Recent Technol. Eng.* **2020**, *8*, 5032–5038. [CrossRef]

energies

MDPI

Review

The Risks and Challenges of Electric Vehicle Integration into Smart Cities

Oluwagbenga Apata *, Pitshou N. Bokoro and Gulshan Sharma

Department of Electrical Engineering Technology, University of Johannesburg, Johannesburg 2028, South Africa; pitshoub@uj.ac.za (P.N.B.)
* Correspondence: g.apata@ieee.org

Abstract: The integration of electric vehicles (EVs) into smart cities presents a promising opportunity for reducing greenhouse gas emissions and enhancing urban sustainability. However, there are significant risks and challenges associated with the integration of EVs into smart cities, which must be carefully considered. Though there are various reviews available on the challenges of integrating EVs into smart cities, the majority of these are focused on technical challenges, thereby ignoring other important challenges that may arise from such integration. This paper therefore provides a comprehensive overview of the risks and challenges associated with the integration of EVs into smart cities in one research paper. The different challenges associated with the integration of EVs into smart cities have been identified and categorized into four groups, namely: technical, economic, social, and environmental, while also discussing the associated risks of EV integration into smart cities. The paper concludes by highlighting the need for a holistic approach to EV integration into smart cities that considers these challenges and risks. It also identifies possible future trends and outlooks to address these challenges and promote the successful integration of EVs into smart cities. Overall, this paper provides valuable insights for policymakers, city planners, and researchers working towards sustainable urban transportation systems.

Keywords: electric vehicle; gasoline vehicle; grid integration; smart city; greenhouse gas emissions; information and communication technologies

1. Introduction

The concept of smart cities has been around for several decades, but it has gained significant attention in recent years due to advances in technology and growing urbanization. The idea of using technology to improve urban living can be traced back to the 1970s, when researchers began to explore the potential of using computer networks to manage urban infrastructure. In the 1990s, the concept of the "digital city" emerged, which focused on using digital technologies to improve urban services and public participation. The term "smart city" was first coined in the early 2000s, and it has since become a popular buzzword in urban planning and development. The smart city concept gained momentum in 2008 when IBM launched its Smarter Planet initiative [1,2], which focused on using technology and data to address global challenges such as climate change, energy management, and urbanization. Since then, smart city initiatives have been implemented in cities around the world, and the concept has continued to evolve with advances in technology. These cities use sensors, devices, and networks to collect and analyse data, which is used to inform decision-making processes and optimize resource usage. One of the key challenges faced by smart cities is reducing their carbon footprint and promoting sustainable transportation. Electric vehicles (EVs) have emerged as a promising solution to this challenge, as they offer a more efficient and cleaner alternative to conventional gasoline-powered vehicles. This can be seen in the exponential growth of EV sales globally when compared to conventional gasoline-powered vehicles [3]. As seen from Figure 1, statistics available from [4] show that Germany had the highest EV share at 26% in 2021 compared to other countries such as Canada, the United Kingdom, France, and Korea.

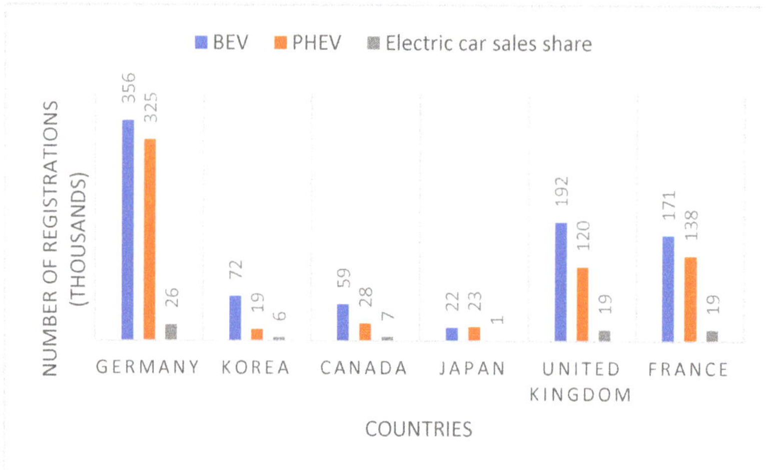

Figure 1. EV registrations for 2021.

From Figure 2, it is observed that the Tesla Y model was the highest selling EV for the year 2022 [5]. By integrating EVs into smart cities, urban planners can create a more sustainable and efficient transportation system, while also reducing traffic congestion and improving air quality, and increasing energy efficiency.

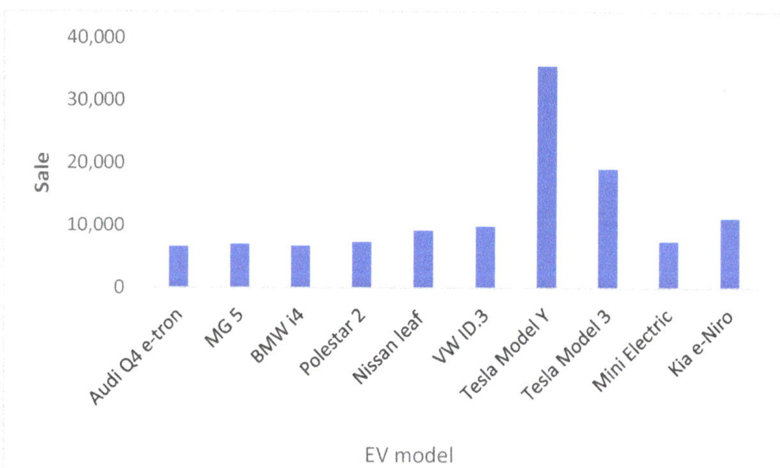

Figure 2. Models of EV sold in 2022 [5].

The integration of EVs into smart cities offers numerous benefits. First and foremost, EVs reduce the carbon footprint of cities. According to the International Energy Agency (IEA), the transportation sector accounts for 24% of global CO_2 emissions [6], with road transport being the largest contributor. By replacing conventional vehicles with EVs, cities can significantly reduce their carbon emissions and improve air quality since EVs are mostly powered by electricity generated from renewable sources such as wind and solar power. This makes them a sustainable alternative to conventional vehicles that rely on fossil fuels such as gasoline and diesel. The different benefits of EVs have been well documented in various literature amongst which are [7–10]. The authors in [11] carried out a study using a life cycle assessment (LCA) to compare the environmental impacts of conventional gasoline-powered vehicles and electric vehicles in Hong Kong. The results showed that electric vehicles have lower environmental impacts in terms of greenhouse gas emissions, air pollution, and resource depletion compared to conventional fossil-fuelled vehicles. A comparative analysis of different results obtained by various authors in terms of LCA of EVs and conventional fossil-fuel-powered vehicles was carried out in [12]. The findings revealed that with the adoption of EVs, there are reduced greenhouse gas emissions (GHG) compared to conventional vehicles even though there is an increase in the human toxicity levels. Despite their numerous benefits, the integration of EVs into smart cities also poses several challenges. These challenges include the need for adequate charging infrastructure [13], efficient energy management [14,15], effective communication and data management [16,17], and addressing social and cultural factors. Addressing these challenges requires a coordinated and collaborative approach involving government agencies, private companies, and other stakeholders. With careful planning and implementation, the integration of EVs into smart cities can help create a more sustainable, efficient, and liveable urban environment. These challenges have been discussed in various literature. The majority of research papers discussing the challenges associated with the integration of EVs into smart cities, smart grids, or conventional grids are focused on the technical challenges. In [18], the authors presented some challenges such as the cost of EV batteries and charger issues. The authors in [19] discussed the status of charging infrastructure development for EVs in the UK. The issues of increased load demand due to charging are discussed in [20], where the authors reviewed various smart charging strategies to tackle this challenge. A review of the critical impacts of grid-tied EVs is presented in [21], where the authors focused on the interaction of EVs in a smart grid environment. In [22], the issues of charging infrastructure are also presented and the authors presented a critical review on applications of machine learning for solving the issues around infrastructure planning.

A systematic and comprehensive review of the charging infrastructure planning for smart cities is presented in [23], while ref. [24] discusses the possibility of increased stress on the electricity grid during peak load as a result of high penetration of EVs in residential low-voltage networks. A centralized control algorithm was proposed to manage the EV charging points to mitigate congestion while the challenge of a reduction in battery life as a result of an extreme case of full daily battery discharge is reported in [25]. A systematic literature review on the integration of EVs into the smart grid is also presented in [26], where the authors presented various results from different research works on the subject of EV integration into smart cities and smart grids. However, the bulk of this systematic review done in [26] also focused on the various technological challenges associated with integrating EVs into smart grids and smart cities, particularly the problems of charging infrastructure. As seen in [13–24], most reviews and research papers concerning the risks and challenges of integrating electric vehicles into smart cities are focused on technical challenges. However, there are other associated risks and challenges associated with EV integration into smart cities, such as the security risks of electric vehicle infrastructure. Charging stations could be hacked, leading to significant damage to the energy management systems of smart cities. Another risk factor associated with electric vehicle adoption is the potential for battery disposal. Electric vehicle batteries have a limited lifespan, and if not appropriately disposed, can cause significant environmental damage. Therefore, implementing disposal strategies that minimize the environmental impact is essential.

Unlike so many other review papers focused on individual risk factors or challenges associated with EV integration into smart cities, this paper has grouped all of the various challenges and associated risks of EV integration into smart cities into four groups, namely: technical, economic, social, and environmental. By categorizing the challenges into technical, economic, social, and environmental aspects, the paper provides a structured framework for understanding the multiple dimensions involved in the integration process.

The technical challenges include issues related to EV infrastructure, such as charging stations and battery management systems, as well as interoperability and standardization. The economic challenges involve costs associated with EV deployment and infrastructure, as well as the potential impact on traditional transportation systems and industries. The social challenges involve the need to address issues related to equity and accessibility, as well as concerns about user acceptance and behavioural change. The environmental challenges involve the potential impact of EVs on the electricity grid and the need to address the life cycle emissions associated with EV production and disposal. The paper concludes by highlighting the need for a holistic approach to EV integration into smart cities that considers these challenges and risks. It also identifies potential research directions to address these challenges and promote the successful integration of EVs into smart cities. Overall, this paper provides valuable insights for policymakers, city planners, and researchers working towards sustainable urban transportation systems.

2. Methodology

In order to carry out a comprehensive overview of the different risks and challenges associated with the integration of EVs into smart cities, this research has identified existing review papers, research articles, reports, and relevant publications on the topic to gain insights into the current state of knowledge. Based on the identified existing literature, a technical, economic, social, and environmental analysis has been carried out and the different challenges associated with the integration of EVs into smart cities have been broadly categorized into four groups. This makes this review outstanding from the different reviews already conducted in existing literature since it incorporates all of these risks and challenges into one research paper. From a technical standpoint, challenges include the availability and scalability of charging infrastructure, interoperability of charging stations, battery range limitations, and the capacity and stability of the power grid. Economic risks encompass the high costs of EVs, charging infrastructure deployment, and maintenance, along with potential impacts on industries such as automotive manufacturing, energy, and transportation services. Social challenges involve addressing public acceptance and perception of EVs, addressing range anxiety, ensuring equitable access to charging infrastructure, and accommodating diverse user needs. Environmental risks include the need to carefully evaluate the lifecycle environmental impacts of EVs and their batteries, while also considering the potential benefits of reduced greenhouse gas emissions and improved air quality. Addressing these risks is crucial for successful integration of EVs into smart cities, requiring collaborative efforts among stakeholders to develop sustainable solutions that foster technological advancements, economic viability, social equity, and environmental sustainability.

3. Risks of EV Integration into Smart Cities

The integration of EVs into smart cities is critical in promoting sustainable transportation solutions and reducing the impact of climate change. However, this integration process presents several risks. The major risks associated with integrating EVs into smart cities are the risks of cybersecurity and data privacy. Policymakers, energy managers, and other stakeholders must address these risks and develop effective strategies to mitigate these risks.

3.1. Cybersecurity Risks

As EVs gain traction in smart cities, the risks and challenges associated with their integration are becoming increasingly evident. Cybersecurity risks are one of the most significant challenges that need to be addressed during the integration of EVs into smart cities. With the integration of information and communication technologies (ICT) in the energy sector, the power grid components and utility servers of a smart city are susceptible to malware attacks from EVs [27]. Previous research on cybersecurity of the smart grid has largely focused on local attacks that impact specific components of the grid. This is because local attacks are often more straightforward to execute and require less skill and fewer resources than global attacks that impact the entire grid. As a result, local attacks remain the most common cyber threats to the smart grid, with malicious actors employing various techniques such as jamming, denial of service (DoS), controller malfunction, and load alteration attacks. Jamming attacks involve degrading the quality of communication signals between grid components, resulting in information loss and system malfunction [28]. Similarly, DoS attacks aim to disrupt normal communication channels by flooding them with traffic, and degrading the components of the grid. The issue of cybersecurity in smart cities has been well documented in different pieces of literature. In [29], the authors categorized the risk of cybersecurity in smart cities into passive and active attacks. A passive cyberattack in a smart city refers to a type of cyber threat where an attacker attempts to gain unauthorized access to sensitive information or systems without actively disrupting or altering their normal functioning. These attacks often target the vulnerabilities in the smart city's infrastructure, such as communication networks, sensors, data repositories, and control systems. The attacker aims to exploit weaknesses in these systems to gain access to sensitive data, monitor activities, or gather information for further malicious activities. The active cyberattack typically involves actions taken by an attacker to gain unauthorized access, disrupt operations, manipulate data, or cause damage to critical infrastructure components. The attacker may employ various techniques, such as hacking, malware, social engineering, or denial-of-service attacks, to compromise the security and integrity of the smart city's systems. Ref. [30] carried out a comprehensive literature review detailing the various privacy and security concerns in a smart city and proposed various solutions to addressing them.

A model was designed in [31] to detect attacks on the industrial control system (ICS) of smart cities, and by using interactive visualization, false alarms can be filtered out. The risk of cyberattacks originating from online social networks was studied in [32], where focal structure analysis (FSA) and deviant cyber flash mob detection (DCFM) techniques were used to develop a model to prevent damage from such attacks. Ref. [33] presented the Application, Communication, Infrastructure, Data, and Stakeholders (ACIDS) security framework to develop security measures and identify possible threats in the various layers of a smart city system. The overall security of a smart city system was improved by this proposed layered approach. Different cybersecurity issues and their associated risks in the operations of a smart city as well as different strategies for reducing such cyber-attacks in a smart city have been well documented in [34,35], respectively. The charging infrastructure of EVs are typically connected to a network, making them potential targets for cyber-attacks. The authors in [36] conducted a semi-quantitative analysis of the vulnerabilities between the various end-to-end subsystems of an EV. Malicious actors can exploit the connection between EVs and the grid to disrupt power supply, potentially causing widespread blackouts or other disruptions. In [37], the authors presented the various challenges and issues that can be exploited to seriously harm the charging stations of EVs, the power grid, or both. The communication protocols between EVs, chargers, and back-end systems are vulnerable to tampering. The protocols used to communicate the state of charge, the authentication of the charging process, and the billing systems can be manipulated by hackers to steal personal data and query the charging status of the vehicle. The report presented by the Kaspersky Lab in [38] indicated that the threats of cyber-attacks were linked to the charge points of home electric vehicle supply equipment (EVSE) units. This report validated the previous research done by the authors in [39], which outlined the risks associated with the deployment of EVSE devices. The various authors in [40–43] identified potential areas of vulnerabilities associated with EVSE. These could lead to security breaches in a smart city such as spoofing, data loss, and DoS. The phasor measurement unit (PMU) networks and electric vehicle infrastructure (EVI) in a smart city are vulnerable to cyberattacks, with potential consequences ranging from grid instability and power disruptions due to compromised PMU data, to unauthorized access and control of EV charging infrastructure leading to service disruptions and potential safety risks. This is well discussed and highlighted in [44] where the authors researched on the vulnerability of EVI and PMU networks to cyberattacks. Overall, cyberattacks on phasor measurement units (PMUs) in smart cities can lead to inaccurate power system measurements, faulty control strategies, and pose a significant threat to the overall reliability and safety of the power grid.

The use of third-party libraries in implementing various features such as navigation and communication in an EV makes it vulnerable to cyberattacks, thereby posing a significant risk in smart cities. Attackers may target vulnerabilities in the EV's software or network connectivity to gain unauthorized remote access to the vehicle's systems [45]. Once inside, they can potentially manipulate or disrupt critical functions, such as braking, acceleration, or steering. Electric vehicles often use keyless entry systems or wireless key fobs for authentication and remote access. Attackers may use various techniques to intercept or clone these key fobs, allowing them to gain unauthorized access to the vehicle without physical contact. The vulnerability of EVs with regards to third party libraries is well documented in [46], where a vulnerable application in a Tesla car was well exploited by a hacker to remotely access the features of the car and even bypass its keyless entry to start the vehicle. In [47], the infotainment system of an EV was attacked using third-party libraries to remotely perform software updates on the vehicle. While the integration of third-party libraries in EVs integrated into a smart city offers numerous advantages, it is crucial to carefully consider and address the associated risks. By prioritizing security measures, conducting thorough evaluations, and fostering collaboration, we can harness the benefits of these libraries while safeguarding the integrity, privacy, and safety of the smart city ecosystem. Only through diligent risk management can we ensure the successful integration of third-party libraries in EVs within smart cities, paving the way for a sustainable and secure future of transportation.

3.2. Data Privacy Risks

With the increasing adoption of EVs in smart cities, there has been a significant increase in the amount of data generated by these vehicles and their supporting infrastructure. The data generated include information about the location of the vehicle, battery life, driving patterns, and more. All of these data provide valuable information that can be used in many ways, including optimizing traffic management, predicting maintenance needs, and developing new business models. However, there are significant privacy risks associated with collecting these data. In particular, the collection of personal data and driving patterns is a major concern. This information can be used to identify an individual's behavior, movements, and preferences, creating a high risk of identity theft, stalking, and other privacy violations. Moreover, as seen in [48], these data can also be used for targeted marketing or insurance pricing, exposing drivers to financial risks. The data collected from EVs, when combined with other datasets, can lead to the identification and tracking of individuals. For example, by correlating location data from EVs with other publicly available information, it may be possible to determine an individual's home address, workplace, or frequently visited locations. This personal identifiability increases the risk of privacy breaches and surveillance. In some cases, EV data may be shared with third parties, such as service providers, government entities, or researchers, for various purposes, such as traffic management or infrastructure planning. However, the sharing of data raises concerns about who has access to the data, how they will be used, and whether individuals' privacy rights are adequately protected. According to [49], it is estimated that by the year 2030, global profits from vehicle generated information could reach a staggering USD 75 billion, making EVs an ideal target for hackers to steal data from.

As EVs become increasingly connected and integrated into smart city ecosystems, they generate vast amounts of data that can include personally identifiable information (PII), vehicle telemetry, location data, and more. Understanding the sensitivity of these data is essential for implementing appropriate security measures and ensuring privacy protection in the smart city. It is important to note that different data from the EV require different amounts of protection based on their sensitivity. It is therefore important to determine the sensitivity of EV data. Data classification involves categorizing data based on their levels of sensitivity, enabling organizations to allocate appropriate security controls and determine access privileges. By applying data classification techniques to EV data, it becomes possible to identify and differentiate between different types of information based on their potential impact if compromised. The authors in [50] determined the sensitivity of EV data using data classification techniques. These data have been grouped into confidential, sensitive, unclassified, and secret by the authors in [45,51] and depicted in Figure 3.

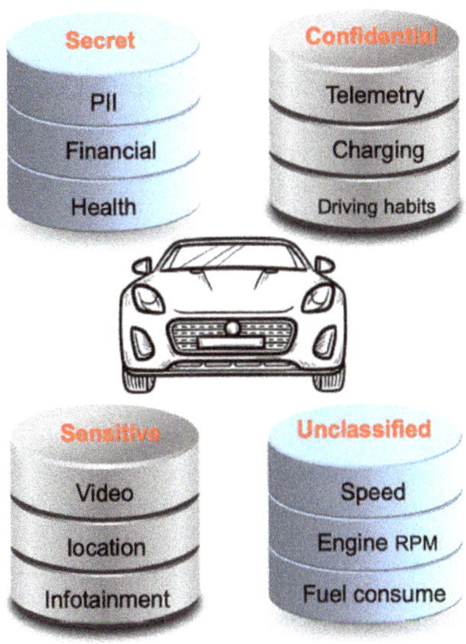

Figure 3. Data classification in EV [52].

One approach to data classification is the use of metadata tags [53], which can be attached to individual data elements to indicate their sensitivity level. For example, PII, such as driver's license numbers, social security numbers, or credit card information, would be classified as highly sensitive, while anonymized vehicle telemetry data may be considered less sensitive. Another technique is automated machine learning (AutoML) [54,55], which can analyze large volumes of EV data and classify them based on patterns, features, or predefined rules. By training machine learning models on labeled datasets, it becomes possible to automate the classification process and identify sensitive data accurately. By leveraging data classification techniques, EV manufacturers, smart city operators, and other stakeholders can gain insights into the sensitivity of EV data. This knowledge enables them to implement appropriate security measures, such as encryption, access controls, and data anonymization, to protect sensitive information effectively. Additionally, data classification aids in compliance with data protection regulations, such as the General Data Protection Regulation (GDPR) or other regional privacy laws.

4. Challenges of EV Integration into Smart Cities

The integration of electric vehicles (EVs) into smart cities represents a transformative shift in urban transportation and sustainability. With the potential to reduce greenhouse gas emissions, decrease reliance on fossil fuels, and enhance overall energy efficiency, EVs offer numerous benefits for smart cities. However, this integration also presents a range of complex challenges that must be addressed to ensure the successful adoption and integration of EVs within the urban fabric. This section will explore and analyze the key challenges associated with EV integration into smart cities, covering aspects such as infrastructure, charging networks, grid capacity, policy and regulations, consumer adoption, and environmental considerations. These challenges have been broadly categorized into four groups in this research paper: technical, economic, social, and environmental. Each category presents unique hurdles and considerations that need to be addressed for successful integration of EVs into the smart city.

4.1. Technical Challenges

One of the primary technical challenges of integrating EVs into a smart city is establishing an extensive and robust charging infrastructure to meet the charging needs of EVs. This includes deploying a network of charging stations that are conveniently located, easily accessible, and equipped with the necessary charging equipment. There has been a lot of focus in this research area lately. The charging infrastructure for electric vehicles is a critical component of their integration into smart cities. The charging infrastructure of an EV comprises a communication and power system [56]. It encompasses the network of charging stations, equipment, and supporting systems necessary to provide convenient and reliable charging options for EVs. EVs are an integral part of sustainable transportation and reducing carbon emissions. In a smart city, where the focus is on environmentally friendly and efficient mobility solutions, the lack of charging infrastructure can hinder the adoption of EVs. Insufficient charging stations may discourage people from choosing electric vehicles as their primary mode of transportation, limiting the city's progress towards achieving sustainable mobility goals. Ensuring an adequate number of charging stations to meet the increasing demand as the number of EVs grows requires strategic planning, coordination with stakeholders, and proactive expansion of charging infrastructure to prevent bottlenecks. One of the challenges for widespread electric vehicle adoption, especially in the context of a smart city, is the availability and accessibility of charging infrastructure. This has been well researched and highlighted by the various authors in [57–60]. Insufficient public charging stations can also create range anxiety among potential electric vehicle owners, as they may be concerned about the availability of charging points and the possibility of running out of power during their daily commutes or longer journeys. The authors in [61] identified the charging infrastructure location for EVs as an optimization problem and approached it by proposing a genetic algorithm to solve this problem.

In as much as it is important to ensure an adequate number of charging stations, there is also a concern about the type of charge available to the EV in a smart city as well as charging time and waiting time of such charging stations. Offering charging stations with various charging speeds (e.g., level 1, level 2, and DC fast charging) to cater to different EV models and user preferences, ensuring compatibility and interoperability between charging connectors and vehicle charging systems, is crucial for user convenience. The different types of charging available to EVs have been well documented in [62]. Level 1 charging refers to using a standard household outlet (120 volts AC) to charge an EV. It is the slowest charging option, typically providing around 2 to 5 miles of range per hour of charging. Level 1 charging is suitable for overnight charging at home or in locations where the vehicle remains stationary for an extended period. However, there is no support for communication control and this can negatively impact the grid such as grid power congestion. Level 2 charging operates at higher voltages (240 volts AC) and provides faster charging compared to Level 1. It typically offers 10 to 30 miles of range per hour of charging, depending on the EV and charging equipment. Level 2 charging is commonly found in residential settings, workplaces, public parking areas, and commercial establishments. As seen in [17], though level 2 charging has a lot of advantages, there are drawbacks with it, such as a surge in power consumption up to 25%. DC fast charging (also known as Level 3 charging) provides the fastest charging speeds and is typically found along highways, at rest areas, and in commercial areas. DC fast chargers can charge an EV up to 80% in 30 min or less, significantly reducing charging time. These chargers operate at higher voltages and convert AC power directly into DC power, bypassing the vehicle's onboard charger. A significant challenge is the slow charging speeds of certain charging stations. While EVs can be charged using a regular household power outlet, it can take several hours to fully charge the vehicle. However, the deployment of fast-charging stations can mitigate this issue. Though DC fast charging has numerous advantages such as a high-power output, the chargers are quite bulky [62]. The authors in [63,64] have documented the different drawbacks of such charging infrastructure. The type of charging infrastructure for an EV impacts a smart city by influencing charging speed, availability, load management, grid stability, demand response capabilities, integration with mobility solutions, and data-driven decision-making. Despite the attractiveness of fast charging for EVs, this presents a significant complex problem that requires traffic flow, utility considerations, and integration of energy supply in the smart city.

The integration of EVs into smart cities holds great promise for sustainable transportation and improved energy efficiency. However, this integration also introduces a significant challenge into the distribution network of the smart city. As the number of EVs on the road increases and their charging demands grow, the electrical grid may face unprecedented challenges in meeting the energy requirements of these vehicles. The increased energy demand from charging EVs requires careful management to ensure grid stability, prevent overloads, and accommodate the growing energy needs of transportation. Without careful planning and infrastructure upgrades, the influx of EVs can strain the existing power grid. The studies carried out in [65–67] emphasize the need for long-term smart grid investment planning, showing the effects of EVs on the distribution network planning. Table 1 presents a summary of the significant challenges that the integration of EVs can pose for electric utilities in smart cities.

Table 1. Impacts of grid integration of EVs in a smart city.

Impacts	**Description**
Power loss [66,68]	The significant penetration of EVs into the smart grid can lead to a large consumption of real power, leading to power loss in the distribution system. This can reach a high of 40% during off-peak hours.
Voltage and phase unbalance [69]	Since the chargers used in EVs are majorly single phased, charging large numbers of EVs simultaneously and using same phase can lead to phase unbalance and current unbalances, which create voltage unbalances.
Increase in load demand [70]	The uncontrolled charging of EVs during peak times can lead to an increase in load levels.
Stability [71–73]	Since EVs are regarded as nonlinear loads and can draw large amounts of power within a short time, they can cause the power system to become unstable. Additionally, a significant penetration of EVs into the grid makes the power system more susceptible to disturbances.
Injection of harmonics [69,72,74–76]	A higher penetration of EVs in the grid can lead to the injection of harmonics in the grid, causing harmonic pollution if not well managed. Although some research has concluded that the total harmonic distortion (THD) level caused by EV charging is well below 1%, this can increase with the number of chargers connected per time in the smart city.
Network component overloading [77–79]	Without a corresponding upgrade of the network infrastructure in the power system, the high energy demand of EVs can lead to a reduction in the lifespan of network equipment such as transformers and cables due to overload.

Establishing standardized communication protocols, charging connectors, and interoperability among various charging networks and EV models is essential to enable seamless charging experiences and compatibility across different systems within a smart city. The challenge of interoperability and standardization is a significant hurdle when integrating electric vehicles (EVs) into a smart city environment. EV charging infrastructure comes in various types, such as different plug designs, power levels, and communication protocols. The lack of interoperability and standardization means that EV owners may encounter compatibility issues when trying to charge their vehicles at different charging stations. This can lead to inconvenience, decreased user experience, and even situations where certain EV models are incompatible with specific charging stations, limiting the options for EV owners. Different regions and countries may have varying regulations, standards, and protocols for EV charging, making it challenging for manufacturers, service providers, and policymakers to align their efforts. The lack of harmonization can slow down the deployment of charging infrastructure and hinder the adoption of EVs on a broader scale. The United States, Europe, Japan, and China all have different standards for EV charging as reported in [80]; however, the two most widely used standards that deal with EV charging are the International Electro-technical Commission (IEC) and Society of Automotive Engineers (SAE) standards used in Europe and the United States, respectively. Table 2 describes the different globally recognized regulatory bodies and their established standards, which oversee different aspects of EVs. Since different countries follow different charging standards, different EV manufacturers are trying to avoid conflicts in charging standards by coming up with a common charging connector [81], as shown in Figure 4a,b [80]. From Figure 4, it can be concluded that there is a need to urgently harmonize the various charging standards and have a universal solution of EV charging devices. These standards play a very crucial role in the grid integration of EVs into a smart city.

Table 2. Standards associated with EV.

Organization	Standard	Detail
Institute of Electrical and Electronics Engineers (IEEE) [82]	P1547	Grid connection of EVs
	P2030	Standard for the interoperability of smart grids
	P2030.1	Electrified transportation infrastructure draft
	P2100.1	Charging system standardization and wireless power transfer
National Electric Code (NEC) [83]	625	Standard for offboarding charging system
	626	Requirements for parking lots for electrified trucks
Deutsches Institut fuer Normung (DIN, Germany) [84]	43538	Specifications for battery systems
	EN50620	Specifications on charging cable
	VDEO510-11	Specifications on testing procedures of Li-ion batteries
Society of Automotive Engineers (SAE) [85]	J2293	Requirements for EV and offboard EV supply equipment for charging from the utility grid.
	J2847	Communication standard between EVs and utility grid.
	J2931	Standard for digital communication between EV and utility grid.
	J2894	Power quality requirements and testing procedure for EVs
	J1772	Conductive charging standards
	J1773	Contactless charging standards
Japan Electric Vehicle Association (JEVA) [86]	C601	Standard for charging plugs and receptacles.
	D001-002	Standardizes the battery characteristics for EV.
	G106-109	Contactless charging standards
	G101-105	Quick charging standards
Standardization Administration of China [87]	GB/T 20234	Standards for plugs, sockets, and connectors for EV conductive charging

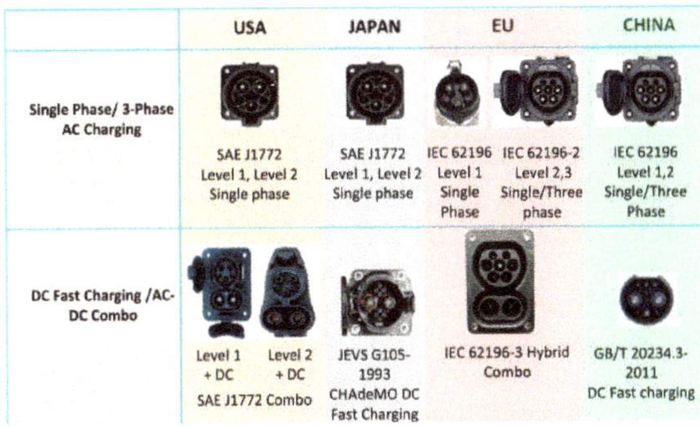

(a)

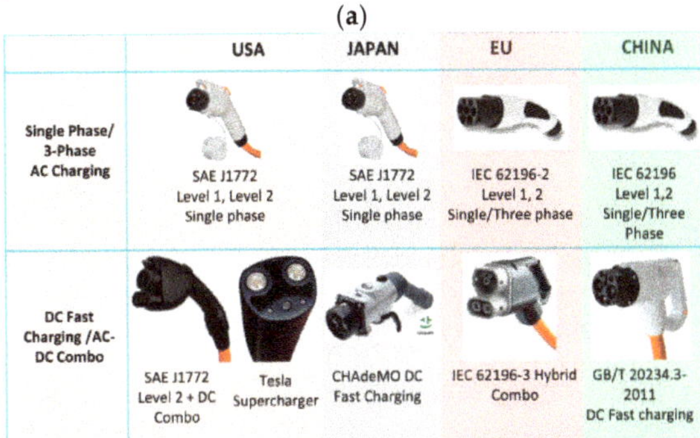

(b)

Figure 4. Schematic of (**a**) charging ports, (**b**) charging connectors [80].

4.2. Economic Challenges

Historically, one of the main barriers to widespread EV adoption has been the higher upfront cost compared to conventional internal combustion engine vehicles (ICEVs). Affordability and cost competitiveness will play a crucial role in encouraging consumer adoption. The initial cost of an EV is one factor; another factor is the lifetime cost of such vehicles, often expressed as the total cost of ownership (TCO) [88]. It is worth noting that while the purchase cost of EVs has been declining, there may still be a price premium compared to similar-sized ICEVs. The TCO of an EV includes factors such as the purchase price, maintenance costs, fuel/charging costs, and resale value. However, the lower operating costs, potential savings in maintenance, and the long-term benefits of reduced fuel consumption can help offset the initial investment and make EV ownership economically viable for many consumers. The specific pricing and cost dynamics of electric vehicles can vary across manufacturers, models, and regions, so it is advisable to research and compare different options to find the best fit for individual budgets and requirements. A lower TCO makes EVs more affordable and financially attractive to consumers. In a smart city, where sustainable transportation is a priority, a lower TCO encourages more individuals and businesses to consider purchasing EVs. It increases the likelihood of EV adoption and integration into the city's transportation ecosystem. The authors in [89] carried out a detailed literature review on the total cost of owning an EV while the research in [90] predicts that by 2030, the prices of EVs will be significantly similar to the prices of conventional vehicles. The prices of EVs will to a large extent determine their integration into smart cities.

The significant investment required to build a comprehensive charging infrastructure network can pose financial challenges towards integrating EVs into a smart city. Balancing the costs associated with installation, operation, and maintenance of charging stations is a key consideration. Underdeveloped public charging infrastructure, particularly rapid charging, remains a challenge in the deployment of EVs in smart cities. One of the reasons for this is the cost of building charging stations or upgrading existing infrastructure. The authors in [91] discussed the possibility of smart or managed charging. Managed or smart charging entails scheduling EV charging during periods when the cost of generating and delivering power is less expensive, while still meeting the needs of the vehicle owner. Electricity pricing, sophisticated technology, and the best location of charging infrastructure can all be used to control EV charging. Incorporating more renewable energy sources into EV charging can maximize the use of the current network infrastructure and reduce the need for additional investment. Managed charging can help EVs operate more affordably and give them a competitive edge over gas-powered vehicles. The authors in [92] identified available charging infrastructure as a significant factor in the deployment of EVs. The costs associated with installing a charging infrastructure on private premises to enable stationary charging adds further costs to the purchase and installation of EVs. An analysis of the costs of stationary charging infrastructure is carried out in [93] where the authors identified and carried out an analysis of various factors influencing the economic suitability of different charging infrastructure and the impact on EV deployment. The cost of charging infrastructure can influence the rate of EV adoption within a smart city. If the cost is relatively low, more businesses, homeowners, and municipalities may be willing to install charging stations. This expanded charging network can help increase the confidence of individuals considering purchasing EVs, knowing that they will have convenient access to charging facilities. Conversely, high infrastructure costs may slow down the adoption rate, as potential EV owners may hesitate to invest in a vehicle without an adequate charging infrastructure.

The cost of batteries is a crucial component of the overall price of an electric vehicle. Batteries serve as the primary energy storage system, providing power to propel the vehicle and support various onboard systems. The price of batteries has been a significant concern for consumers and a barrier to the widespread adoption of EVs, especially in smart cities. However, there have been notable advancements and cost reductions in battery technology in recent years. Historically, lithium-ion batteries, which are the most commonly used batteries in EVs, are expensive to produce due to several factors. Firstly, the materials used in lithium-ion batteries, such as lithium, cobalt, nickel, and manganese, have limited availability and are subject to price fluctuations in the global market. Secondly, the manufacturing processes for battery cells are complex and require specialized equipment and expertise, further driving up costs. The authors in [94] emphasized the importance of reducing the cost of lithium-ion batteries. The authors noted that despite the reduction in costs of batteries for EVs, these costs still remain high when compared to the set targets, and therefore remain a hinderance to the wide scale adoption of EVs. It is therefore expected that as the cost of lithium-ion batteries continues to decline, the number of EVs in smart cities will increase.

Grid upgrades and integration costs play a crucial role in the successful integration of EVs into a smart city. EVs require charging infrastructure, which puts an additional load on the electric grid. The existing grid infrastructure may not be capable of handling the increased demand for electricity that comes with widespread EV adoption. Grid upgrades are necessary to accommodate the charging needs of EVs. These upgrades can involve increasing the capacity of power distribution networks, installing new transformers, upgrading substations, and implementing smart grid technologies. Insufficient grid upgrades can lead to challenges such as voltage drops, power outages, and increased stress on the grid during peak charging times. These issues can hamper the widespread adoption of EVs and create a negative user experience. Adequate grid upgrades ensure a reliable and robust infrastructure capable of supporting the charging needs of EVs, facilitating their integration into a smart city. To promote the successful integration of EVs into a smart city, it is essential to address both grid upgrades and integration costs. Governments, utility companies, and stakeholders need to collaborate to ensure that the grid infrastructure is upgraded to meet the increased demand from EV charging. Simultaneously, efforts should be made to reduce integration costs through economies of scale, incentives, and public–private partnerships. By effectively managing these factors, the integration of EVs into a smart city can be accelerated, leading to a cleaner, more sustainable transportation system.

4.3. Environmental Challenges

The integration of EVs into smart cities holds tremendous potential for reducing greenhouse gas emissions, improving air quality, and mitigating the environmental impacts associated with traditional transportation systems. However, this transition is not without its environmental challenges. One of the primary environmental challenges is the upstream impact of EVs. While EVs themselves produce zero tailpipe emissions, the electricity used to charge them is generated from various sources, which may include fossil fuels such as coal or natural gas. The carbon intensity of the electricity grid plays a crucial role in determining the overall environmental benefit of EVs. Therefore, a key challenge is to ensure that the electricity powering EVs is generated from renewable and low-carbon sources [95].

The production of EVs requires the extraction of raw materials such as lithium, cobalt, and rare earth elements. Irresponsible extraction practices can have severe environmental and social consequences, including deforestation, water pollution, and human rights violations. Additionally, the end-of-life management of EV batteries raises concerns about recycling and proper disposal. As EVs reach the end of their lifespan, proper disposal and recycling becomes crucial. EV batteries contain valuable materials that can be recovered and reused. Without a proper recycling infrastructure and processes, there is the risk of hazardous waste and environmental contamination as reported in [96]. Recycling of lithium-ion batteries is a research area that has continued to gain attention. The authors in [97] have presented a comprehensive overview of various technologies relevant for recycling lithium-ion batteries. It is also important to note that improper recycling or reuse of lithium-ion batteries can generate a lot of toxic waste, as recorded in [98]. However, in [99], the author noted that the price of permanently disposable materials used in an EV battery is far less than the price of the fully used battery. EV batteries can be reused, especially for less demanding applications as shown in [100], prompting the concept of second demonstration use of EV.

Typically, when an EV reaches the end of its useful life for personal transportation, it still retains a significant portion of its battery capacity. Instead of retiring the vehicle entirely, the second demonstration use concept aims to repurpose these EVs to serve other functions, maximizing their lifespan and optimizing resource utilization. The batteries in EVs can store surplus electricity generated from renewable sources, such as solar or wind power, and discharge it back to the grid during times of high demand. By integrating EVs into energy storage systems, they can help balance the grid, smooth out fluctuations in renewable energy generation, and improve overall grid stability. This approach promotes renewable energy integration and enhances the efficiency and reliability of the electricity system. The research in [101] showed that batteries with potential reuse had more ratings than the ratings of batteries increasingly suitable for recycling [102]. However, several challenges remain, including technical considerations, regulatory frameworks, and standardization of interfaces and protocols. Ensuring compatibility and interoperability among different EV models and systems is crucial for the widespread adoption and success of the second demonstration use concept.

Choosing between reuse and recycling depends on factors such as battery condition, feasibility of secondary applications, recycling infrastructure availability, and market demand for recycled materials. A comprehensive approach may involve a combination of both strategies, maximizing the lifespan of batteries through reuse and recovering valuable resources through recycling. This approach will help to optimize resource utilization, minimize waste, and contribute to a more sustainable and circular economy for EV batteries within a smart city.

4.4. Social Challenges

The integration of EVs in smart cities can be limited by societal obstacles just like the technical and economic challenges. A consumer's choice of an EV can be affected by a mix of various emotions and practicality. It is therefore important to investigate the perceptions of consumers to support large-scale acceptance of EVs within the smart city. The social challenges associated with integrating EVs into smart cities encompass various aspects of society, including accessibility, equity, and behavioral changes. These challenges arise from the need to ensure that the benefits of EV adoption are accessible to all members of the community and that the transition is inclusive and fair.

One of the primary social challenges of adopting EVs in a smart city is vehicle safety. Consumers prioritize safety when considering any vehicle purchase, and EVs are no exception. Perceptions of EV safety can be influenced by factors such as battery performance, crash test ratings, and overall vehicle design. Providing accurate information about the safety features and crashworthiness of EVs, along with conducting rigorous safety testing and certification processes, helps build trust and confidence among consumers. Battery safety is a significant concern for EV consumers. While lithium-ion batteries used in EVs have a strong safety record, isolated incidents of battery fires or thermal runaway events can create negative perceptions. Enhancing battery safety through robust design, engineering, and advanced thermal management systems is crucial. Transparent reporting and effective communication about battery safety measures, emergency response protocols, and real-world safety data can help address consumer concerns.

Resiliency and range anxiety is another social challenge that has impacted negatively on consumers embracing EVs [103]. Range anxiety, the fear of running out of battery charge with limited charging infrastructure, can impact consumer perceptions of EV resiliency. Consumers may worry about the availability and accessibility of charging stations, particularly during long-distance travel or in areas with limited charging infrastructure. The time required to charge an EV can also contribute to range anxiety. While charging technology is continuously improving, longer charging times compared to refueling a conventional vehicle can create concerns about travel delays or inconvenience, especially during long trips. The estimated range provided by EVs is based on various factors such as driving conditions, weather, and battery condition. If drivers perceive that the estimated range is not accurate or may not account for real-world driving scenarios, it can exacerbate range anxiety. Uncertainty about how far the vehicle can travel on a single charge can lead to concerns about being stranded.

Equity is another important social consideration in the integration of EVs into smart cities. The benefits of EV adoption, such as reduced emissions and improved air quality, should be shared equitably across communities. It is essential to avoid exacerbating existing socio-economic disparities and ensure that the transition to electric transportation does not disproportionately affect vulnerable populations or contribute to environmental injustice. This requires proactive policies and programs that address the specific needs and concerns of disadvantaged communities, promoting access to EVs, charging infrastructure, and associated benefits for all.

Furthermore, integrating EVs into smart cities requires significant behavioral changes from both individuals and organizations. The widespread adoption of EVs necessitates a shift in attitudes, habits, and infrastructure planning [104]. Encouraging public acceptance and awareness of EV benefits, supporting behavior change initiatives, and providing education on charging infrastructure and maintenance are crucial to facilitate a smooth transition. Additionally, fostering collaborations between public and private sectors, including automakers, utilities, and urban planners, is essential for effective integration and to address the complex social dynamics involved.

5. Case Studies of Successful EV Integration into Smart Cities

In recent years, the integration of EVs into smart cities has emerged as a key strategy for promoting sustainable urban transportation. Cities around the world are actively exploring innovative solutions to reduce carbon emissions, enhance energy efficiency, and improve the overall quality of life for their residents. Though the integration of EVs into smart cities is not without its challenges, there have been notable success stories from various regions across the globe. These success stories showcase the potential and benefits of EV integration despite the complexities involved. Through case studies of successful EV integration into smart cities, we can gain insights into the transformative impact of this technology on urban mobility.

Amsterdam, the capital city of the Netherlands, has emerged as a leading example of successful EV integration into a smart city. The city has embraced sustainable urban planning and has taken bold steps to promote electric mobility as a key component of its transportation system. One of the key factors behind Amsterdam's success is the establishment of an extensive charging infrastructure. The city boasts over 3000 public charging points, ensuring that EV owners have convenient access to charging facilities throughout the city. This widespread availability of charging infrastructure has helped alleviate range anxiety and encourage the adoption of EVs. In addition to the charging network, Amsterdam has implemented various smart mobility solutions to optimize EV usage. Real-time data analytics are utilized to monitor and manage the charging infrastructure effectively. This data-driven approach enables the city to identify high-demand areas, optimize charging schedules, and efficiently allocate resources. Furthermore, Amsterdam has embraced dynamic charging infrastructure, allowing EVs to be charged while driving. This innovative approach eliminates the need for extended charging stops and provides continuous power to EVs, further enhancing their convenience and usability.

San Diego, located on the southern coast of California, is another city that has made significant strides in integrating EVs into its smart city infrastructure. The city has been proactive in promoting sustainable transportation and has implemented various initiatives to support EV integration. One of the key achievements of San Diego is the establishment of a robust charging infrastructure. The city has partnered with public and private stakeholders to install thousands of EV charging stations across the region. These charging stations are strategically located in public spaces, workplaces, and residential areas, ensuring convenient access for EV owners. To maximize the efficiency of EV charging, San Diego has implemented a smart grid infrastructure. This intelligent system optimizes energy usage and load balancing, allowing for intelligent charging management. EVs can be charged during off-peak hours when electricity demand is lower, reducing strain on the grid and minimizing the overall energy consumption. San Diego has also emphasized the integration of renewable energy sources into its charging infrastructure. The city has a strong commitment to clean energy, and a significant portion of its electricity comes from renewable sources such as solar and wind. By integrating renewables into the charging infrastructure, San Diego ensures that EVs contribute to a lower carbon footprint and align with the city's sustainability goals. Furthermore, San Diego has implemented various incentive programs to encourage EV adoption. These include financial incentives, such as rebates and grants for EV purchases, as well as perks like free or discounted parking for EVs. By providing these incentives, the city aims to make EV ownership more accessible and attractive to residents.

Shenzhen, a bustling metropolis in southern China, has emerged as a global leader in the successful integration of EVs into its smart city infrastructure. The city has made remarkable strides in electrifying its transportation system and has become a prominent example of sustainable urban mobility. One of the standout achievements of Shenzhen is its electrification of the entire bus fleet. With over 16,000 electric buses in operation, Shenzhen boasts the world's largest electric bus fleet. This transformation has not only significantly reduced air pollution and greenhouse gas emissions, but has also enhanced the overall quality of life for residents. Shenzhen's success in EV integration can be attributed to its comprehensive charging infrastructure. The city has established an extensive network of charging stations to support the growing number of electric buses and private EVs. These charging facilities are strategically located throughout the city, including bus terminals, public parking lots, and residential areas, providing convenient access to charging for EV owners.

To facilitate the efficient management of EV charging, Shenzhen has implemented advanced technologies. This includes the use of smart charging stations that can monitor and control the charging process, optimize charging schedules, and ensure a stable power supply. Additionally, the city has developed intelligent dispatch systems for its electric bus fleet, enabling efficient routing and reducing downtime. Shenzhen's success in EV integration can be attributed to strong government support and incentives. The city provides financial subsidies for EV purchases, making them more affordable for residents. It has also implemented favorable policies such as preferential treatment for electric vehicles, including access to dedicated bus lanes and exemption from certain traffic restrictions. The integration of EVs into Shenzhen's smart city ecosystem has had significant environmental and social benefits. The electric buses have improved air quality and reduced noise pollution, contributing to a cleaner and healthier urban environment. The increased adoption of EVs has also spurred the development of a vibrant ecosystem of charging infrastructure and related services, generating employment opportunities and fostering technological advancements.

Yokohama, a vibrant city in Japan, has achieved significant success in EVs into its smart city initiatives. The city has been at the forefront of sustainable urban mobility and has implemented a range of innovative measures to promote the adoption of EVs and build a greener transportation system. One of the notable achievements of Yokohama is the successful integration of electric buses into its public transportation network. The city has introduced a considerable number of electric buses, reducing emissions and noise pollution. This transition has not only improved the overall air quality but has also enhanced the commuting experience for residents and visitors. To support the growing number of EVs, Yokohama has developed a comprehensive charging infrastructure. The city has deployed numerous charging stations across key locations, including public parking lots, residential areas, and commercial centers. This extensive network ensures that EV owners have convenient access to charging facilities, promoting the use of electric vehicles and alleviating range anxiety. Yokohama has also implemented an intelligent transportation system that optimizes the management of EV charging and overall traffic flow. Through the integration of smart technologies, the city can monitor the status of charging stations, provide real-time information to drivers regarding the availability of charging points, and optimize the routing of electric buses for maximum efficiency. Furthermore, Yokohama has been proactive in utilizing renewable energy sources to power its EV charging infrastructure. The city has integrated solar panels and other clean energy generation systems into selected charging stations, reducing the reliance on conventional energy sources and contributing to a lower carbon footprint. The successful integration of EVs into Yokohama's smart city framework has had numerous benefits. It has significantly reduced greenhouse gas emissions, enhanced energy efficiency, and improved the overall sustainability of the transportation system. Additionally, the transition to electric mobility has increased the use of renewable energy and helped Yokohama move closer to its ambitious environmental targets.

Oslo, the capital city of Norway, has been a pioneer in the successful integration of EVs into its smart city infrastructure. The city has implemented a comprehensive set of measures and incentives to promote EV adoption, making it one of the leading EV-friendly cities in the world. One of the key factors behind Oslo's success is its strong policy support for EVs. The city has implemented a range of financial incentives to encourage EV purchases, including tax exemptions, reduced toll fees, and access to bus lanes. These incentives have made EV ownership more attractive and affordable for residents, resulting in a significant increase in EV sales. Oslo has also focused on developing a robust charging infrastructure. The city has established an extensive network of public charging stations, making it convenient for EV owners to charge their vehicles throughout the city. Moreover, the charging stations are strategically located in public spaces, parking areas, and residential buildings, ensuring easy access for EV users. To further support EV integration, Oslo has prioritized the use of renewable energy for charging infrastructure. The city has increased the share of renewable energy sources in its electricity grid, ensuring that EVs are charged using clean energy. This approach contributes to a significant reduction in carbon emissions from the transportation sector, aligning with Oslo's commitment to sustainability and climate change mitigation. In addition to these measures, Oslo has actively promoted electric public transportation. The city has electrified a significant portion of its bus fleet, making electric buses a common sight on the streets of Oslo. This transition has not only reduced emissions, but has also improved air quality and provided a quieter and more pleasant commuting experience for residents.

6. Discussion

One of the key aspects of successfully integrating EVs into smart cities is comprehensive planning and collaboration among various stakeholders, including city authorities, utility providers, transportation agencies, and private businesses. This approach ensures that all parties work together towards a shared vision of sustainable mobility. Comprehensive planning involves assessing the current infrastructure and identifying the gaps and requirements for EV integration. This includes evaluating the existing charging infrastructure, grid capacity, and transportation networks. It also involves developing long-term strategies and roadmaps for scaling up the charging infrastructure and aligning it with the city's renewable energy goals. Collaboration is essential to leverage the expertise and resources of different stakeholders. Public–private partnerships can play a crucial role in funding and implementing charging infrastructure projects. Additionally, collaboration with utility providers helps in managing grid capacity and integrating EV charging with smart grid technologies. Engaging with transportation agencies can facilitate the integration of EVs into public transportation systems, promoting multimodal options, and seamless connectivity.

Public–private partnerships (PPPs) have proven to be effective in accelerating the integration of EVs into smart cities. These partnerships bring together the resources, expertise, and innovation of both the public and private sectors, leading to faster and more efficient deployment of charging infrastructure and supportive services. PPPs can provide funding and investment for charging infrastructure projects, which often require significant capital investment. Private companies can contribute their knowledge and experience in building and operating charging stations, while the government provides regulatory support and incentives. This collaboration can result in a more extensive and diversified charging network, catering to the needs of different user groups, such as residential, commercial, and public transport.

Integrating EV charging with the smart grid is essential for managing the increased demand for electricity and optimizing energy usage. Smart grid technologies enable efficient charging management, load balancing, and demand response capabilities. Smart grid integration involves utilizing advanced technologies such as communication systems, sensors, and control mechanisms to monitor and manage EV charging. This integration allows for dynamic load management, optimizing charging schedules to avoid peak demand periods and balance the grid's load. Additionally, demand response programs incentivize EV owners to adjust their charging patterns based on grid conditions, promoting grid stability and reducing strain during peak hours. A case study is Tokyo, Japan, which has made significant progress in smart grid integration for EV charging. The city has implemented demand response programs that encourage EV owners to charge during off-peak hours by offering time-of-use pricing incentives. This approach not only helps manage grid demand but also supports renewable energy integration by aligning EV charging with periods of high renewable energy generation.

Future-proofing EV charging infrastructure is crucial to ensure its long-term viability and scalability. As the EV market continues to grow, it is essential to design and deploy infrastructure that can accommodate the increasing demand for charging services and adapt to technological advancements. Future-proof infrastructure entails considering factors such as charging station capacity, power supply requirements, and flexibility for upgrades. It involves deploying charging infrastructure that can support various charging standards, including fast charging and high-power charging, to cater to the evolving needs of different EV models. Scalability is another critical aspect of future-proofing infrastructure. As EV adoption increases, the charging network must expand to meet the growing demand. This may involve deploying charging stations in strategic locations, such as highways, public parking lots, and commercial areas, to ensure convenient access for EV owners. Furthermore, future-proofing infrastructure involves integrating smart features and connectivity. This includes implementing technologies such as Internet of Things (IoT) connectivity, real-time data monitoring, and remote management capabilities. These features enable efficient operation, maintenance, and monitoring of the charging infrastructure and support advanced functionalities such as dynamic pricing, user authentication, and grid integration.

The integration of EVs into smart cities involves various aspects, and one crucial element for seamless operation is interoperability [105]. Interoperability refers to the ability of different systems, technologies, or devices to communicate, exchange data, and work together effectively [106]. In the context of EV integration into smart cities, interoperability plays a pivotal role in enabling efficient charging infrastructure, grid management, and overall sustainable transportation systems. Interoperability in charging infrastructure ensures that EVs can be charged at different charging stations, regardless of the vehicle manufacturer or charging infrastructure provider. Standardization of charging protocols, such as CHAdeMO, CCS (Combined Charging System), and Tesla's Supercharger network, facilitates interoperability by allowing EVs to charge at any compatible charging station. These protocols ensure that EVs can access charging services irrespective of the charging station's operator, ensuring convenience and eliminating the need for multiple memberships or access cards. Interoperability extends beyond physical charging connections. It also involves the seamless exchange of data and communication between EVs, charging stations, the power grid, and other components of the smart city ecosystem. Standardized communication protocols, such as Open Charge Point Protocol (OCPP) and OpenADR (Automated Demand Response), enable interoperability between different devices and systems. This allows for dynamic pricing, load management, demand response programs, and efficient grid integration of EVs. EVs have the potential to act as distributed energy resources, providing storage capacity and grid stabilization services. Interoperability enables bidirectional communication between EVs and the power grid, allowing for vehicle-to-grid (V2G) capabilities. V2G systems enable EVs to feed surplus energy back into the grid during peak demand periods or provide ancillary services, promoting grid stability and renewable energy integration. Interoperable systems enable seamless coordination between EV charging, energy management systems, and the power grid, optimizing energy utilization and reducing strain on the grid. Interoperability must also address the critical aspects of data security and privacy. As various systems exchange data and communicate with each other, it is essential to establish robust cybersecurity measures to protect sensitive information. Standardization of security protocols and encryption methods ensures secure data transmission and prevents unauthorized access or manipulation of EV-related data. To achieve interoperability in EV integration into smart cities, standardization plays a vital role. Governments, industry stakeholders, and standards organizations collaborate to develop common technical standards, protocols, and interfaces. These standards enable interoperability between different vendors, manufacturers, and service providers. Policy frameworks and regulations also play a crucial role in driving interoperability by mandating compliance with specific standards and encouraging open access to charging infrastructure and data. The need for stakeholder engagement and collaboration cannot be overemphasized: collaboration among various stakeholders, including EV manufacturers, charging infrastructure providers, utility companies, government agencies, and technology developers. These stakeholders must work together to define and implement interoperable solutions, ensuring compatibility across different systems and technologies. Collaborative efforts foster innovation, enhance user experience, and drive the adoption of EVs in smart cities.

7. Future Trends and Outlook

One of the key future trends in EV integration into smart cities is the deployment of high-power charging infrastructure. As the demand for EVs continues to grow and battery technology improves, high-power charging solutions will be necessary to address the issue of charging time and provide a seamless experience for EV owners. High-power chargers allow for faster charging times and increased convenience for EV owners. As EV battery technology improves, high-power chargers will become essential to meet the growing demand for fast and efficient charging. These chargers will be capable of delivering significantly higher power levels, reducing charging times to a matter of minutes rather than hours. This advancement will encourage greater EV adoption by addressing the issue of charging convenience and range anxiety. High-power charging infrastructure may become more tailored to specific EV models, taking into account their battery technology, charging capabilities, and power requirements. This approach can optimize charging efficiency and ensure compatibility between the charger and the vehicle, maximizing the charging speed and battery health. The establishment of high-power charging hubs at key locations, such as shopping centers, airports, and rest areas, can further enhance the convenience and accessibility of fast charging. These hubs could feature a cluster of high-power charging stations, providing a centralized and efficient charging solution for EV owners. High-power charging infrastructure can also support battery swapping systems, where EV owners can exchange their depleted battery with a fully charged one at designated stations. This approach eliminates the need for long charging times and allows for instant battery replenishment, making it suitable for commercial and fleet applications.

Vehicle-to-Grid (V2G) technology is an innovative concept that allows EVs to not only draw energy from the grid but also to feed excess energy back into the grid when needed. V2G technology enables bidirectional energy flow between EVs and the electrical grid, transforming EVs into mobile energy storage units and facilitating a more flexible and resilient energy system. As V2G technology continues to advance, several developments are expected in the future. The establishment of common standards and protocols for V2G communication and interoperability will be crucial for widespread adoption. Standardization ensures compatibility between different EV models, charging infrastructure, and grid systems, enabling seamless integration and scalability. Integration between V2G technology and advanced grid management systems, such as smart grids and distribution management systems, will enhance the efficiency and coordination of EV charging and discharging. This integration will optimize the use of renewable energy, improve grid stability, and enable more sophisticated grid services.

Wireless charging technology is an emerging trend that has the potential to revolutionize the way EVs are charged in smart cities. Wireless charging eliminates the need for physical cables and connectors, enabling convenient and automated charging experiences. With wireless charging, EVs can be charged simply by parking over a charging pad or driving over a charging lane embedded in the road surface. This technology offers seamless integration of charging infrastructure into existing urban infrastructure, reducing clutter and visual impact. Wireless charging solutions will become more prevalent, especially in public spaces, residential areas, and fleet operations, enhancing the overall user experience and encouraging greater EV adoption. As wireless charging technology continues to evolve, several developments are expected in the future. Wireless charging systems will be able to deliver higher power transfer rates, reducing charging times and increasing the efficiency of charging EVs. The development of dynamic wireless charging systems will enable EVs to charge while on the move. This technology can be integrated into roadways or highways, allowing EVs to replenish their battery levels during long-distance travel, without the need for frequent stops. Furthermore, the establishment of common standards and interoperability between different wireless charging systems will be crucial for widespread adoption. Standardization ensures compatibility between various EV models and charging infrastructure, allowing EV owners to use any wireless charging station, regardless of the vehicle brand.

Vehicle-to-Infrastructure (V2I) communication is a key aspect of integrating EVs into smart cities. It refers to the exchange of information between EVs and the surrounding infrastructure, such as traffic lights, charging stations, parking systems, and grid networks. V2I communication enables seamless interaction between EVs and the urban environment, facilitating efficient and optimized transportation and charging experiences. V2I communication allows EVs to receive real-time traffic information, such as traffic congestion, road conditions, and signal timings, from the infrastructure. With these data, EVs can optimize their routes, avoiding congested areas and selecting the most efficient paths. This not only reduces travel time for EV owners but also helps alleviate traffic congestion and improve overall traffic flow in the city. As V2I communication technology progresses, standardization of V2I communication protocols and connectivity standards will be essential for widespread implementation. Common standards will ensure interoperability between different EV models, infrastructure providers, and cities, enabling seamless communication and integration. The deployment of 5G networks and Cellular Vehicle-to-Everything (C-V2X) technology will enhance V2I communication capabilities. Networks that are 5G offer increased bandwidth, low latency, and high reliability, enabling fast and reliable data exchange between EVs and the infrastructure, facilitating real-time communication and ensuring quick response times. Ongoing advancements in 5G and C-V2X will focus on improving the reliability of V2I communication. This includes developing robust communication protocols, redundancy mechanisms, and ensuring uninterrupted connectivity, even in challenging urban environments.

8. Conclusions

The integration of electric vehicles (EVs) into smart cities presents both challenges and opportunities. While there are hurdles to overcome, successful case studies from around the world demonstrate that the benefits of EV integration are substantial. As we move towards a more sustainable future, the electrification of transportation plays a crucial role in reducing carbon emissions, improving air quality, and enhancing energy efficiency.

The case studies of Amsterdam, San Diego, Shenzhen, Yokohama, and Oslo highlight the diverse approaches and strategies employed by smart cities to overcome challenges and create successful EV integration models. These cities have demonstrated the effectiveness of comprehensive planning, infrastructure development, policy support, and collaboration among stakeholders.

To further accelerate the integration of EVs into smart cities, several key recommendations emerge. First, governments and city planners must prioritize the development of robust charging infrastructure to alleviate range anxiety and promote EV adoption. This includes deploying fast-charging stations in strategic locations, incentivizing private and public entities to invest in charging infrastructure, and leveraging smart grid technologies to manage the increased demand.

Secondly, collaboration and partnerships among different stakeholders are crucial. Governments, utilities, automotive manufacturers, technology companies, and the community must work together to create an integrated ecosystem that supports EV integration. This involves aligning policies, sharing data and resources, and fostering innovation in areas such as vehicle-to-grid (V2G) integration and smart charging solutions.

Thirdly, incentives and supportive policies can play a significant role in accelerating the transition to electric mobility. These can include financial incentives for EV purchases, tax incentives, preferential parking, and dedicated EV lanes. Governments should also consider promoting EV sharing services and electrifying public transportation to reduce private vehicle usage and promote sustainable mobility options.

Furthermore, leveraging smart city technologies and data analytics is essential for optimizing EV integration. Intelligent traffic management systems, integrated energy management systems, and advanced data analytics enable efficient charging infrastructure planning, grid management, and demand-response strategies. By leveraging real-time data, cities can anticipate and respond to EV charging needs, balance energy supply and demand, and optimize charging infrastructure utilization.

Future research on the risks and challenges of electric vehicle integration into smart cities should primarily focus on the development of advanced charging infrastructure and energy management systems. These efforts will aim to optimize the placement and density of charging stations, as well as enable efficient utilization of the power grid through smart charging algorithms and V2G capabilities. Additionally, future research should be directed towards developing energy management systems that can effectively balance the energy demand from EVs with the available supply, considering grid constraints and renewable energy generation. By addressing these areas, future research directions will contribute to the seamless integration of electric vehicles into smart cities, fostering sustainable transportation and enhancing the overall efficiency of urban environments.

In conclusion, the successful integration of EVs into smart cities requires a comprehensive and multi-faceted approach. It involves investment in charging infrastructure, supportive policies, collaboration among stakeholders, and the utilization of smart city technologies. With careful planning, effective governance, and the engagement of citizens, smart cities can realize the full potential of electric mobility, creating sustainable, connected, and livable environments for their residents while contributing to global efforts in combating climate change.

Funding: This research received no external funding.

Data Availability Statement: Not applicable.

Conflicts of Interest: The authors declare no conflict of interest.

References

1. Smarter Planet. Available online: https://www.ibm.com/ibm/history/ibm100/us/en/icons/smarterplanet/ (accessed on 20 May 2023).
2. Taft, D.K. *25 Things You Might Not Know about IBMs Smarter Planet Strategy*; eWeek: Nashville, TN, USA, 2009.
3. IEA. *Global EV Outlook 2023*; IEA: Paris, France, 2023. Available online: https://www.iea.org/reports/global-ev-outlook-2023 (accessed on 22 May 2023).
4. IEA. *Electric Car Registrations and Sales Share in Selected Countries, 2016–2021*; IEA: Paris, France, 2022. Available online: https://www.iea.org/data-and-statistics/charts/electric-car-registrations-and-sales-share-in-selected-countries-2016-2021 (accessed on 22 May 2023).
5. Electric Car Statistics—EV Data [Update: Jan 23] | Heycar. Available online: https://heycar.co.uk/blog/electric-cars-statisticsand-projections (accessed on 24 May 2023).
6. IEA. *Transport, Energy and CO_2*; IEA: Paris, France, 2009. Available online: https://www.iea.org/reports/transport-energy-and-co2 (accessed on 24 May 2023).
7. Shafique, M.; Azam, A.; Rafiq, M.; Luo, X. Life cycle assessment of electric vehicles and internal combustion engine vehicles: A case study of Hong Kong. *Res. Transp. Econ.* **2022**, *91*, 101112. [CrossRef]
8. Abo-Khalil, A.G.; Abdelkareem, M.A.; Sayed, E.T.; Maghrabie, H.M.; Radwan, A.; Rezk, H.; Olabi, A.G. Electric vehicle impact on energy industry, policy, technical barriers, and power systems. *Int. J. Thermofluids* **2022**, *13*, 100134. [CrossRef]
9. Wilken, D.; Oswald, M.; Draheim, P.; Pade, C.; Brand, U.; Vogt, T. Multidimensional assessment of passenger cars: Comparison of electric vehicles with internal combustion engine vehicles. *Procedia CIRP* **2020**, *90*, 291–296. [CrossRef]
10. Albatayneh, A.; Assaf, M.N.; Alterman, D.; Jaradat, M. Comparison of the overall energy efficiency for internal combustion engine vehicles and electric vehicles. *Rigas Teh. Univ. Zinat. Raksti* **2020**, *24*, 669–680. [CrossRef]
11. Verma, S.; Dwivedi, G.; Verma, P. Life cycle assessment of electric vehicles in comparison to combustion engine vehicles: A review. *Mater. Today Proc.* **2022**, *49*, 217–222. [CrossRef]
12. Sanguesa, J.A.; Torres-Sanz, V.; Garrido, P.; Martinez, F.J.; Marquez-Barja, J.M. A review on electric vehicles: Technologies and challenges. *Smart Cities* **2021**, *4*, 372–404. [CrossRef]
13. Mastoi, M.S.; Zhuang, S.; Munir, H.M.; Haris, M.; Hassan, M.; Usman, M.; Bukhari, S.S.H.; Ro, J.S. An in-depth analysis of electric vehicle charging station infrastructure, policy implications, and future trends. *Energy Rep.* **2022**, *8*, 11504–11529. [CrossRef]
14. Rimpas, D.; Kaminaris, S.D.; Aldarraji, I.; Piromalis, D.; Vokas, G.; Papageorgas, P.G.; Tsaramirsis, G. Energy management and storage systems on electric vehicles: A comprehensive review. *Mater. Today Proc.* **2022**, *61*, 813–819. [CrossRef]

15. Yang, C.; Zha, M.; Wang, W.; Liu, K.; Xiang, C. Efficient energy management strategy for hybrid electric vehicles/plug-in hybrid electric vehicles: Review and recent advances under intelligent transportation system. *IET Intell. Transp. Syst.* **2020**, *14*, 702–711. [CrossRef]
16. Ali, E.S.; Hassan, M.B.; Saeed, R.A. Machine learning technologies in internet of vehicles. In *Intelligent Technologies for Internet of Vehicles*; Springer International Publishing: Cham, Switzerland, 2021; pp. 225–252.
17. Mwasilu, F.; Justo, J.J.; Kim, E.K.; Do, T.D.; Jung, J.W. Electric vehicles and smart grid interaction: A review on vehicle to grid and renewable energy sources integration. *Renew. Sustain. Energy Rev.* **2014**, *34*, 501–516. [CrossRef]
18. Potdar, V.; Batool, S.; Krishna, A. Risks and challenges of adopting electric vehicles in smart cities. In *Smart Cities: Development and Governance Frameworks*; Springer: Berlin/Heidelberg, Germany, 2014; pp. 207–240.
19. Chen, T.; Zhang, X.P.; Wang, J.; Li, J.; Wu, C.; Hu, M.; Bian, H. A review on electric vehicle charging infrastructure development in the UK. *J. Mod. Power Syst. Clean Energy* **2020**, *8*, 193–205. [CrossRef]
20. Deb, S.; Pihlatie, M.; Al-Saadi, M. Smart Charging: A Comprehensive Review. In *IEEE Access*; IEEE: Piscaatway, NJ, USA, 2022.
21. Ismail, A.A.; Mbungu, N.T.; Elnady, A.; Bansal, R.C.; Hamid, A.K.; AlShabi, M. Impact of electric vehicles on smart grid and future predictions: A survey. *Int. J. Model. Simul.* **2022**, 1–17. [CrossRef]
22. Deb, S. Machine learning for solving charging infrastructure planning problems: A comprehensive review. *Energies* **2021**, *14*, 7833. [CrossRef]
23. Deb, S. Planning of Sustainable Charging Infrastructure for Smart Cities. In *Flexible Resources for Smart Cities*; Springer: Berlin/Heidelberg, Germany, 2021; pp. 115–132.
24. Petrou, K.; Quiros-Tortos, J.; Ochoa, L.F. Controlling electric vehicle charging points for congestion management of UK LV networks. In Proceedings of the 2015 IEEE Power & Energy Society Innovative Smart Grid Technologies Conference (ISGT), Washington, DC, USA, 17–20 February 2015; pp. 1–5.
25. Ribberink, H.; Darcovich, K.; Pincet, F. Battery life impact of vehicle-to-grid application of electric vehicles. In Proceedings of the 28th International Electric Vehicle Symposium and Exhibition, Goyang, Republic of Korea, 3–6 May 2015; pp. 3–6.
26. Sultan, V.; Aryal, A.; Chang, H.; Kral, J. Integration of EVs into the smart grid: A systematic literature review. *Energy Inform.* **2022**, *5*, 65. [CrossRef]
27. Carryl, C.; Ilyas, M.; Mahgoub, I.; Rathod, M. The PEV security challenges to the smart grid: Analysis of threats and mitigation strategies. In Proceedings of the 2013 International Conference on Connected Vehicles and Expo (ICCVE), Las Vegas, NV, USA, 2–6 December 2013; IEEE: Piscataway, NJ, USA, 2013; pp. 300–305.
28. Reyes, H.I.; Kaabouch, N. Jamming and lost link detection in wireless networks with fuzzy logic. *Int. J. Sci. Eng. Res.* **2013**, *4*, 1–7.
29. Kalinin, M.; Krundyshev, V.; Zegzhda, P. Cybersecurity risk assessment in smart city infrastructures. *Machines* **2021**, *9*, 78. [CrossRef]
30. AlDairi, A. Cyber security attacks on smart cities and associated mobile technologies. *Procedia Comput. Sci.* **2017**, *109*, 1086–1091. [CrossRef]
31. Neshenko, N. Illuminating Cyber Threats for Smart Cities: A Data-Driven Approach for Cyber Attack Detection with Visual Capabilities. Doctoral Dissertation, Florida Atlantic University, Boca Raton, FL, USA, 2021.
32. Alassad, M.; Spann, B.; Al-khateeb, S.; Agarwal, N. Using computational social science techniques to identify coordinated cyber threats to smart city networks. In *Design and Construction of Smart Cities: Toward Sustainable Community*; Springer International Publishing: Berlin/Heidelberg, Germany, 2021; pp. 316–326.
33. Hamid, S.; Bawany, N.Z. Acids: A secure smart city framework and threat model. In Proceedings of the 4th International Conference on Wireless, Intelligent and Distributed Environment for Communication: WIDECOM 2021, Milan, Italy, 11–13 February 2019; Springer International Publishing: Cham, Switzerland; pp. 79–98.
34. Figueiredo, B.J.; Costa RL, D.C.; Santos, L.; Rabadão, C. Cybersecurity and Privacy in Smart Cities for Citizen Welfare. In *Smart Cities, Citizen Welfare, and the Implementation of Sustainable Development Goals, 2022*; IGI Global: Hershey, PN, USA, 2022; pp. 197–221.
35. Wang, P.; Ali, A.; Kelly, W. Data security and threat modeling for smart city infrastructure. In Proceedings of the 2015 International Conference on Cyber Security of Smart Cities, Industrial Control System and Communications (SSIC), Shanghai, China, 5–7 August 2015; IEEE: Piscataway, NJ, USA; pp. 1–6.
36. Basmadjian, R. Communication Vulnerabilities in Electric Mobility HCP Systems: A Semi-Quantitative Analysis. *Smart Cities* **2021**, *4*, 405–428. [CrossRef]
37. Acharya, S.; Dvorkin, Y.; Pandžić, H.; Karri, R. Cybersecurity of smart electric vehicle charging: A power grid perspective. *IEEE Access* **2020**, *8*, 214434–214453. [CrossRef]
38. Sklyar, D. *ChargePoint Home Security Research*; Kaspersky Lab Security Services: Moscow, Russia, 2018.
39. Force, J.T.; Initiative, T. *Guide for Applying the Risk Management Framework to Federal Information Systems*; NIST Special Publication: Gaithersburg, MD, USA, 2010; Volume 800, p. 37.
40. Basnet, M.; Ali, M.H. Exploring Cybersecurity Issues in 5G Enabled Electric Vehicle Charging Station with Deep Learning. *IET Gener. Transm. Distrib.* **2021**, *15*, 3435–3449. [CrossRef]
41. Reeh, D.; Cruz Tapia, F.; Chung, Y.W.; Khaki, B.; Chu, C.; Gadh, R. Vulnerability Analysis and Risk Assessment of EV Charging System under Cyber-Physical Threats. In Proceedings of the 2019 IEEE Transportation Electrification Conference and Expo (ITEC), Detroit, MI, USA, 19–21 June 2019.

42. Johnson, J.; Berg, T.; Anderson, B.; Wright, B. Review of electric vehicle charger cybersecurity vulnerabilities, potential impacts, and defenses. *Energies* **2022**, *15*, 3931. [CrossRef]
43. Falk, R.; Fries, S. Securely connecting electric vehicles to the smart grid. *Int. J. Adv. Internet Technol.* **2013**, *6*, 57–67.
44. Mousavian, S.; Erol-Kantarci, M.; Mouftah, H.T. Cyber-Security and Resiliency of Transportation and Power Systems in Smart Cities. In *Transportation and Power Grid in Smart Cities: Communication Networks and Services*; John & Wiley: Hoboken, NJ, USA, 2018; pp. 507–527.
45. Ntousakis, G.; Ioannidis, S.; Vasilakis, N. Detecting Third-Party Library Problems with Combined Program Analysis. In Proceedings of the 2021 ACM SIGSAC Conference on Computer and Communications Security, Online, 15–19 November 2021; pp. 2429–2431.
46. Vice Article. How a Hacker Controlled Dozens of Teslas Using a Flaw in Third-Party App. 2022. Available online: https://www.vice.com/en/article/akv7z5/how-a-hacker-controlled-dozens-of-teslas-using-a-flaw-in-third-party-app (accessed on 4 May 2023).
47. O'Donnell, L. Volkswagen Cars Open to Remote Hacking, Researchers Warn. 2018. Available online: https://threatpost.com/volkswagen-cars-open-to-remote-hacking-researchers-warn/131571/ (accessed on 15 May 2023).
48. Dibaei, M.; Zheng, X.; Jiang, K.; Abbas, R.; Liu, S.; Zhang, Y.; Yu, S. Attacks and defences on intelligent connected vehicles: A survey. *Digit. Commun. Netw.* **2020**, *6*, 399–421. [CrossRef]
49. Mckinsey. Monetizing Car Data. 2016. Available online: https://www.mckinsey.com/industries/automotive-and-assembly/our-insights/monetizing-car-data (accessed on 15 May 2023).
50. Guo, L.; Ye, J.; Yang, B. Cyberattack detection for electric vehicles using physics-guided machine learning. *IEEE Trans. Transp. Electrif.* **2020**, *7*, 2010–2022. [CrossRef]
51. Aggarwal, C.C.; Aggarwal, C.C. *Data Classification;* Springer International Publishing: Berlin/Heidelberg, Germany, 2015; pp. 285–344.
52. Muhammad, Z.; Anwar, Z.; Saleem, B.; Shahid, J. Emerging Cybersecurity and Privacy Threats to Electric Vehicles and Their Impact on Human and Environmental Sustainability. *Energies* **2023**, *16*, 1113. [CrossRef]
53. De Clercq, D.; Diop, N.F.; Jain, D.; Tan, B.; Wen, Z. Multi-label classification and interactive NLP-based visualization of electric vehicle patent data. *World Pat. Inf.* **2019**, *58*, 101903. [CrossRef]
54. Balan, G.; Arumugam, S.; Muthusamy, S.; Panchal, H.; Kotb, H.; Bajaj, M.; Ghoneim, S.S. An Improved Deep Learning-Based Technique for Driver Detection and Driver Assistance in Electric Vehicles with Better Performance. *Int. Trans. Electr. Energy Syst.* **2022**, *2022*, 8548172. [CrossRef]
55. Ha, S.; Marchetto, D.J.; Dharur, S.; Asensio, O.I. Topic classification of electric vehicle consumer experiences with transformer-based deep learning. *Patterns* **2021**, *2*, 100195. [CrossRef]
56. Mohammed, S.S.; Ahamed, T.I.; Aleem SH, A.; Omar, A.I. Interruptible charge scheduling of plug-in electric vehicle to minimize charging cost using heuristic algorithm. *Electr. Eng.* **2021**, *104*, 1425–1440. [CrossRef]
57. Majhi, R.C.; Ranjitkar, P.; Sheng, M.; Covic, G.A.; Wilson, D.J. A systematic review of charging infrastructure location problem for electric vehicles. *Transp. Rev.* **2021**, *41*, 432–455. [CrossRef]
58. Deb, S.; Tammi, K.; Kalita, K.; Mahanta, P. Review of recent trends in charging infrastructure planning for electric vehicles. *Wiley Interdiscip. Rev. Energy Environ.* **2018**, *7*, e306. [CrossRef]
59. Matanov, N.; Zahov, A. Developments and challenges for electric vehicle charging infrastructure. In Proceedings of the 2020 12th Electrical Engineering Faculty Conference (BulEF), Varna Town, Bulgaria, 9–12 September 2020; IEEE: Piscataway, NJ, USA, 2020; pp. 1–5.
60. Zhang, Y.; Liu, X.; Zhang, T.; Gu, Z. Review of the electric vehicle charging station location problem. In Proceedings of the Dependability in Sensor, Cloud, and Big Data Systems and Applications: 5th International Conference, DependSys 2019, Guangzhou, China, 12–15 November 2019; Proceedings 5; Springer: Singapore, 2019; pp. 435–445.
61. Efthymiou, D.; Chrysostomou, K.; Morfoulaki, M.; Aifantopoulou, G. Electric vehicles charging infrastructure location: A genetic algorithm approach. *Eur. Transp. Res. Rev.* **2017**, *9*, 27. [CrossRef]
62. Savari, G.F.; Sathik, M.J.; Raman, L.A.; El-Shahat, A.; Hasanien, H.M.; Almakhles, D.; Omar, A.I. Assessment of charging technologies, infrastructure and charging station recommendation schemes of electric vehicles: A review. *Ain Shams Eng. J.* **2022**, *14*, 101938. [CrossRef]
63. Fisher, R.; Tang, M.; Le, T.; Yee, D.; White, K. Accelerating beyond early adopters to achieve equitable and widespread electric vehicle use in the san francisco bay area. *World Electr. Veh. J.* **2019**, *11*, 3. [CrossRef]
64. Marwali, M. (Ed.) *Smart Power Grids 2011*; Springer: Berlin/Heidelberg, Germany, 2011.
65. Cossent, R.; Olmos, L.; Gómez, T.; Mateo, C.; Frías, P. Distribution network costs under different penetration levels of distributed generation. *Eur. Trans. Electr. Power* **2011**, *21*, 1869–1888. [CrossRef]
66. Fernandez, L.P.; San Román, T.G.; Cossent, R.; Domingo, C.M.; Frias, P. Assessment of the impact of plug-in electric vehicles on distribution networks. *IEEE Trans. Power Syst.* **2010**, *26*, 206–213. [CrossRef]
67. Méndez, V.H.; Rivier, J.; De La Fuente, J.I.; Gomez, T.; Arceluz, J.; Marin, J.; Madurga, A. Impact of distributed generation on distribution investment deferral. *Int. J. Electr. Power Energy Syst.* **2006**, *28*, 244–252. [CrossRef]
68. Clement-Nyns, K.; Haesen, E.; Driesen, J. The impact of charging plug-in hybrid electric vehicles on a residential distribution grid. *IEEE Trans. Power Syst.* **2009**, *25*, 371–380. [CrossRef]

69. Yong, J.Y.; Ramachandaramurthy, V.K.; Tan, K.M.; Mithulananthan, N. A review on the state-of-the-art technologies of electric vehicle, its impacts and prospects. *Renew. Sustain. Energy Rev.* **2015**, *49*, 365–385. [CrossRef]
70. Wang, Z.; Paranjape, R. An evaluation of electric vehicle penetration under demand response in a multi-agent based simulation. In Proceedings of the 2014 IEEE Electrical Power and Energy Conference, Calgary, AB, Canada, 12–14 November 2014; IEEE: Piscataway, NJ, USA, 2014; pp. 220–225.
71. Wu, D.; Chau, K.T.; Liu, C.; Gao, S.; Li, F. Transient stability analysis of SMES for smart grid with vehicle-to-grid operation. *IEEE Trans. Appl. Supercond.* **2011**, *22*, 5701105.
72. Nguyen, V.L.; Tran-Quoc, T.; Bacha, S. Harmonic distortion mitigation for electric vehicle fast charging systems. In Proceedings of the 2013 IEEE Grenoble Conference, Grenoble, France, 16–20 June 2013; IEEE: Piscataway, NJ, USA, 2013; pp. 1–6.
73. Shareef, H.; Islam, M.M.; Mohamed, A. A review of the stage-of-the-art charging technologies, placement methodologies, and impacts of electric vehicles. *Renew. Sustain. Energy Rev.* **2016**, *64*, 403–420. [CrossRef]
74. Zhou, C.; Wang, H.; Zhou, W.; Qian, K.; Meng, S. Determination of maximum level of EV penetration with consideration of EV charging load and harmonic currents. In *IOP Conference Series: Earth and Environmental Science*; IOP Publishing: Bristol, UK, 2019; Volume 342, No. 1, p. 012010.
75. Melo, N.; Mira, F.; de Almeida, A.; Delgado, J. Integration of PEV in Portuguese distribution grid: Analysis of harmonic current emissions in charging points. In Proceedings of the 11th International Conference on Electrical Power Quality and Utilisation, Lisbon, Portugal, 17–19 October 2011; IEEE: Piscataway, NJ, USA, 2011; pp. 1–6.
76. Monteiro, V.; Gonçalves, H.; Afonso, J.L. Impact of Electric Vehicles on power quality in a Smart Grid context. In Proceedings of the 11th International Conference on Electrical Power Quality and Utilization, Lisbon, Portugal, 17–19 October 2011; IEEE: Piscataway, NJ, USA, 2011; pp. 1–6.
77. Hilshey, A.D.; Rezaei, P.; Hines, P.D.; Frolik, J. Electric vehicle charging: Transformer impacts and smart, decentralized solutions. In Proceedings of the 2012 IEEE Power and Energy Society General Meeting 2012, San Diego, CA, USA, 22–26 July 2012; IEEE: Piscataway, NJ, USA, 2012; pp. 1–8.
78. Gray, M.K.; Morsi, W.G. Power quality assessment in distribution systems embedded with plug-in hybrid and battery electric vehicles. *IEEE Trans. Power Syst.* **2014**, *30*, 663–671. [CrossRef]
79. Nour, M.; Chaves-Ávila, J.P.; Magdy, G.; Sánchez-Miralles, Á. Review of positive and negative impacts of electric vehicles charging on electric power systems. *Energies* **2020**, *13*, 4675. [CrossRef]
80. Das, H.S.; Rahman, M.M.; Li, S.; Tan, C.W. Electric vehicles standards, charging infrastructure, and impact on grid integration: A technological review. *Renew. Sustain. Energy Rev.* **2020**, *120*, 109618. [CrossRef]
81. Bräunl, T. *EV Charging Standards*; University of Western Australia: Perth, Australia, 2012; pp. 1–5.
82. Photovoltaics, D.G.; Storage, E. IEEE standard for interconnection and interoperability of distributed energy resources with associated electric power systems interfaces. *IEEE Std.* **2018**, *1547*, 1547–2018.
83. Miller, C.R. *Illustrated Guide to the National Electrical Code*; Cengage Learning: Boston, MA, USA, 2014.
84. National Electrical Code. Available online: https://www.nfpa.org/NEC (accessed on 30 May 2023).
85. SAE Standards 2019. Available online: https://www.sae.org/standards/ (accessed on 30 May 2023).
86. CHAdeMO Protocol Development (2018). Available online: https://www.chademo.com/activities/protocol-development/ (accessed on 30 May 2023).
87. Oron, A. *Top 10 Countries in the Global EV Revolution: 2018 Edition*; Inside EVs: Miami, FL, USA, 2018.
88. Tseng, H.K.; Wu, J.S.; Liu, X. Affordability of electric vehicles for a sustainable transport system: An economic and environmental analysis. *Energy Policy* **2013**, *61*, 441–447. [CrossRef]
89. Wu, G.; Inderbitzin, A.; Bening, C. Total cost of ownership of electric vehicles compared to conventional vehicles: A probabilistic analysis and projection across market segments. *Energy Policy* **2015**, *80*, 196–214. [CrossRef]
90. Randall, T. *Here's How Electric Cars Will Cause the Next Oil Crisis*; Bloomberg: New York, NY, USA, 2016.
91. Glitman, K.; Farnsworth, D.; Hildermeier, J. The role of electric vehicles in a decarbonized economy: Supporting a reliable, affordable and efficient electric system. *Electr. J.* **2019**, *32*, 106623. [CrossRef]
92. König, A.; Nicoletti, L.; Schröder, D.; Wolff, S.; Waclaw, A.; Lienkamp, M. An overview of parameter and cost for battery electric vehicles. *World Electr. Veh. J.* **2021**, *12*, 21. [CrossRef]
93. Waclaw, A.; Aloise, T.; Lienkamp, M. Charging Infrastructure Design for Commercial Company Sites with Battery Electric Vehicles: A Case Study of a Bavarian Bakery. In Proceedings of the 2020 Fifteenth International Conference on Ecological Vehicles and Renewable Energies (EVER), Monte-Carlo, Monaco, 10–12 September 2020; IEEE: Piscataway, NJ, USA, 2020; pp. 1–10.
94. Li, J.; Du, Z.; Ruther, R.E.; An, S.J.; David, L.A.; Hays, K.; Wood, D.L. Toward low-cost, high-energy density, and high-power density lithium-ion batteries. *JOM* **2017**, *69*, 1484–1496. [CrossRef]
95. Thomas, C.S. How green are electric vehicles? *Int. J. Hydrogen Energy* **2012**, *37*, 6053–6062. [CrossRef]
96. Huang, B.; Pan, Z.; Su, X.; An, L. Recycling of lithium-ion batteries: Recent advances and perspectives. *J. Power Sources* **2018**, *399*, 274–286. [CrossRef]
97. Baum, Z.J.; Bird, R.E.; Yu, X.; Ma, J. Lithium-ion battery recycling—Overview of techniques and trends. *ACS Energy Lett.* **2022**, *7*, 712–719. [CrossRef]
98. Hossain, M.S.; Kumar, L.; Islam, M.M.; Selvaraj, J. A comprehensive review on the integration of electric vehicles for sustainable development. *J. Adv. Transp.* **2022**, *2022*, 3868388. [CrossRef]

99. Gardiner, J. The Rise of Electric Cars Could Leave us with a Big Battery Waste Problem. *Guardian*, 10 August 2017. Available online: https://www.theguardian.com/sustainable-business/2017/aug/10/electric-cars-big-battery-waste-problem-lithium-recycling(accessed on 15 May 2023).
100. Casals, L.C.; García, B.A.; Aguesse, F.; Iturrondobeitia, A. Second life of electric vehicle batteries: Relation between materials degradation and environmental impact. *Int. J. Life Cycle Assess.* **2017**, *22*, 82–93. [CrossRef]
101. Marra, F.; Yang, G.Y.; Træholt, C.; Larsen, E.; Rasmussen, C.N.; You, S. Demand profile study of battery electric vehicle under different charging options. In Proceedings of the 2012 IEEE Power and Energy Society General Meeting, San Diego, CA, USA, 22–26 July 2012; IEEE: Piscataway, NJ, USA, 2012; pp. 1–7.
102. Kostopoulos, E.D.; Spyropoulos, G.C.; Kaldellis, J.K. Real-world study for the optimal charging of electric vehicles. *Energy Rep.* **2020**, *6*, 418–426. [CrossRef]
103. Wager, G.; Whale, J.; Braunl, T. Driving electric vehicles at highway speeds: The effect of higher driving speeds on energy consumption and driving range for electric vehicles in Australia. *Renew. Sustain. Energy Rev.* **2016**, *63*, 158–165. [CrossRef]
104. Avci, B.; Girotra, K.; Netessine, S. Electric vehicles with a battery switching station: Adoption and environmental impact. *Manag. Sci.* **2015**, *61*, 772–794. [CrossRef]
105. Petrasch, R.J.; Petrasch, R.R. Data Integration and Interoperability: Towards a Model-Driven and Pattern-Oriented Approach. *Modelling* **2022**, *3*, 105–126. [CrossRef]
106. Tomasz, G. Integration flows modeling in the context of architectural views. *IEEE Access* **2023**, *11*, 35220–35231.

energies

MDPI

Article

Power Cloud Framework for Prosumer Aggregation to Unlock End-User Flexibility

Giovanni Brusco [1], Daniele Menniti [2], Anna Pinnarelli [2], Nicola Sorrentino [2] and Pasquale Vizza [2,*]

1 Creta Energie Speciali Srl, 87036 Rende, Italy; giovanni.brusco@cretaes.com
2 Department of Mechanical, Energy and Management Engineering, University of Calabria, 87036 Rende, Italy; daniele.menniti@unical.it (D.M.); anna.pinnarelli@unical.it (A.P.); nicola.sorrentino@unical.it (N.S.)
* Correspondence: pasquale.vizza@unical.it

Abstract: The behind-the-meter technologies integrating "all-in-one" photovoltaic plants, storage systems, and other technological solutions can transform consumers into active prosumages to both reduce their energy costs and provide flexibility to the grid. To exploit those flexibility services, it is necessary to manage the end-users in an aggregated form. End-user aggregation is currently becoming a suitable solution to manage energy flows to obtain environmental, economic, and social benefits. In this scope, the paper presents an algorithm to opportunely manage the energy flows inside this aggregation operating in a Power Cloud framework. The algorithm schedules the energy flows that the end-user storage systems must exchange inside the aggregation to maximize the use of renewable sources, provide grid flexibility services, and simultaneously provide balancing services. The algorithm is organized into three different steps: the day-ahead step, the real-time step, and the balancing one. Some simulation results are illustrated to demonstrate the effectiveness of the proposed algorithm.

Keywords: power cloud; energy aggregation; energy management

1. Introduction

The widespread presence of both prosumers and prosumages (a prosumage is a prosumer integrating a storage system) allows the implementation of several solutions to reduce the cost of energy and to provide grid flexibility services in the energy transition scenario.

There are some studies in the literature that provide analysis concerning the participation of end-user aggregation in providing services to the transmission or distribution grids. In [1], the authors introduce the concept of a distributed resource aggregator, which is an active part of the distribution system that enables small resources to participate in the electricity market, providing ancillary services. In this framework, the aggregator covers the role of a system coordinator. In [2,3], the aggregation of electric vehicles (EVs) to provide ancillary services in the presence of V2G (vehicle-to-grid) charging stations is considered. In particular, in [2], real-time EV charging controllers allowing participation in the ancillary services markets have been implemented. One of the parameters utilized is the charging efficiency. In [3], the EV aggregations provide a secondary frequency response considering the EV user's preferences.

The management of the aggregation to reduce congestion and provide flexibility using distributed energy resources is also an important issue. In [4], a heuristic dispatching approach is used. In [5], distributed resources are involved in providing voltage support and optimizing real-time operations in the distribution and transmission networks.

The aggregation can provide services to the grid both using storage systems and by opportunely managing the loads in real-time or scheduling them in advance. In [6–8], home appliances are considered flexible resources to provide services to the grid if they are required; algorithms to manage them in real-time for a 24-hour time horizon are also implemented. This operating mode can lead to modifying the users' habits or, in any case, can require the user's interaction, often leading to the impossibility of providing flexible services with the possibility of penalties for the aggregator, as in [9]. On the contrary, if energy storage systems (ESSs) are used, several features can be carried out without requiring users' interaction [10,11].

In [12], a review of the different management models of storage systems is carried out. It is highlighted how, among the different dispatching domains, between financial and technical, the financial one is the most requested and used by users who require greater profits. On the contrary, the technical ones are most requested by network operators. In the model that will be discussed in this paper, a technical approach has been considered, but without neglecting the financial one, considering the earning possibilities for the users themselves that will interact with the grid operator. Different from other methods, this one considers both users and grid necessities.

In [13], an energy management system for the minimization of the daily cost of energy and the maximization of self-consumption for a community microgrid is discussed. It is shown how the participation of a community microgrid allows it to have significant economic benefits. Such benefits are also increased using a peer-to-peer (P2P) mechanism.

In [14], a two-stage approach is proposed to manage a community, allowing separate energy management operations and economic aspects; they show that in the French framework, the user's bill savings belonging to the community are 10% more convenient compared to the users not belonging to the community.

In [15], an examination of the possible aggregation of users and how they can be managed is performed. Among the different models, the flexibility of aggregation has been introduced. Since market access is difficult for small consumers, the implementation of aggregations allows them to access the market and, at the same time, provide flexible services to the grid.

In [16], an energy management approach is presented. The main characteristic is that it is based both on a minimization cost algorithm and, at the same time, integrates a demand-side approach.

In [17], an energy management strategy that uses fuzzy logic to allocate electricity to hydrogen storage and electricity storage is proposed. It is applied to a near-zero-energy community. The proposed multi-objective optimization strategy allows for a percentage greater than 80% of renewable energy. In [18], an aggregation of users is considered, and an approach to managing the internal microgrids is presented. In particular, the internal aspects, from different points of view—market, environmental, and economic—are examined. In this energy management strategy, a dispatchable biomass plant is considered to increase flexibility.

To provide flexibility services, it is necessary to forecast the available flexibility. The purpose of [19] is to predict the flexibility of a local energy community (LEC), so the use of power consumption of controllable loads is useful to this aim. Such available flexibility was predicted using a particular artificial neural network (ANN).

1.1. Contribution of the Paper

This paper, starting from the concept of Power Cloud [20] and its related advantages, proposes an algorithm to opportunely manage the energy flows inside an end-user aggregation from day-ahead to real-time. The Power Cloud concept is a solution to facilitate the integration of distributed renewable energy systems with new environmentally friendly and smart enabling technologies for final end-user active participation. Specifically, distributed flexibility resources, such as storage systems, are managed to maximize the use of RES and provide transmission system operator (TSO) and/or distribution system operator (DSO) ancillary services. At this scope, EVs may also be easily introduced into the algorithm as other flexibility resources, like those considered by the authors in [21].

The proposed algorithm is organized in three steps: the day-ahead step, the real-time step, and the balancing one. The first one determines the charge/discharge of storage systems present in the aggregation to maximize the renewable energy prosumers/prosumages self-consumption using forecasted hourly production and consumption. A reference power profile is determined to exchange with the grid. The second one determines in real-time how to modify that profile if flexibility services are required by the TSO/DSO, and in the last step, the third one, to compensate for errors in forecasting production and/or consumption. Differently from [22,23], where particular bidding and optimization strategies are proposed to work in real-time in the reserve market, the proposed algorithm operates in a deterministic way under the existing electricity market rules.

The implemented algorithm is divided into steps that can operate separately, and one of its advantages is that it does not strictly need field communication among and with end-users to operate. Indeed, if some communication problems occur, it keeps working, still guaranteeing acceptable results; only the balancing action needs measurements from the field. Another advantage is that the algorithm can be implemented on a management platform with relatively low computational efforts if the platform that enables the exchange of data, mainly the charge/discharge of the storage systems between the aggregator and end-users, is blockchain-based in order to assure transparency, immutability, and security [20].

1.2. Structure of the Paper

This paper is structured in four sections. After a summary of the principal issue to be faced, Section 2 describes some aspects concerning end-user aggregation. Sections 3 and 4 describe, respectively, the day-ahead and real-time steps of the proposed algorithm, with a specific focus on the balancing one. In Section 5, a case study and the related simulation results are illustrated and discussed.

2. Technical and Economic Issues of Power Cloud Management Model

The end-user aggregations are new energy management models to produce, consume, and share energy in the Power Cloud framework as proposed in [20]. The end-users virtually operate within a geographical local perimeter to reduce the energy exchange outside the aggregation and maximize its economic, environmental, and social benefits.

To allow them to achieve those benefits, the implementation of the proposed algorithm must consider the electricity and energy market framework and its rules.

For most of the structure of the electricity markets in Europe, it is necessary to define in advance the energy schedule for both injected and absorbed power, the so-defined day-ahead schedule of the exchanged power. Specifically, in the presence of renewable energy sources, it is necessary to schedule their production.

At the same time, considering that the ESSs are the most important resources to provide grid flexibility services [20], it is necessary to define in advance the flexibility schedule that they can offer. It is equally important to have a day-ahead schedule of the power that the ESSs may exchange to maximize the use of the energy produced by renewable sources.

To this purpose, two issues must be faced: the first is the use of accurate consumption and production forecasting algorithms to evaluate a reliable power profile schedule and avoid power imbalances between the scheduled and real power profiles; the second is to have reliable measures from the field in terms of the power produced by all renewable energy sources, absorbed by the loads, exchanged with the grid and with the ESS, and the state of charge (SOC) of the ESSs.

For the first issue, there are several studies that correlate the positive effect of forecasting models with imbalance charges [21]. In this paper, the forecasting algorithm described in [24] is used to forecast PV power production. Moreover, the balancing step is introduced to compensate for possible forecasting errors that may occur and to avoid charges for not having fulfilled the undertaken commitments.

For the second issue, although the above-mentioned measures do not intervene in the day-ahead step, they become necessary in the balancing step. For this reason, the use of a smart meter with a small measurement time range (from minutes up to a few seconds) is assumed.

3. Power Cloud Day Ahead Energy Management Algorithm

The aim of the proposed algorithm is the optimal use of renewable energy source generation inside an end-user aggregation in a Power Cloud framework, with the possibility of providing grid flexibility services. The management algorithm consists of a three-step algorithm. The first step is the day-ahead step (DA), which receives as input the consumption and production forecasts, as well as the forecasts of the ESS SOC. It provides the scheduling of power and the availability of flexibility services, sharing both the surplus and deficit of energy among the end-users equipped with the ESSs. Starting from the DA results, the real-time algorithm (RTA) is processed. It considers the flexibility requests sent by the grid operator, and it operates for each hour using the results of the first step, changing the scheduled power profile. In this step, the flexibility services, determined by the availability of ESS, are offered, processed, and executed. In the end, based on the real measures of the exchanged power and of the storage system parameters, the third step, the balancing step, operates in real time to reduce any imbalance between the scheduled power profile and the real one.

3.1. Day-Ahead Step

The day-ahead step operations are illustrated in Figure 1. It consists of many functions that are carried out, one in sequence to the next, starting from the forecasting of production and consumption for the single users, aggregating them until the exchange of battery profiles and the availability to provide services to the grid are calculated. Such operations are represented by different equations that are reported below.

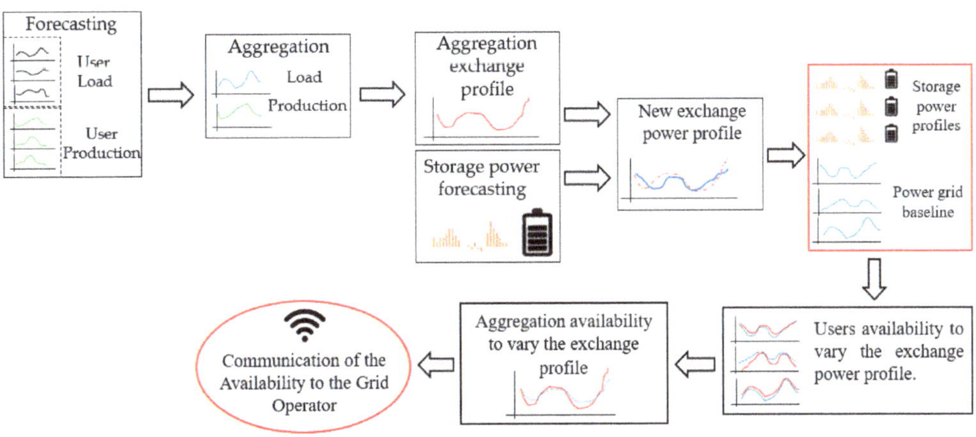

Figure 1. Day-ahead step flow chart.

Let us consider an aggregation of n end-users: n_p is the number of end-users equipped with a PV plant, n_c is the number of consumers, and n_s is the number of end-users equipped with an ESS.

Starting from n_p hourly energy production forecasts $\left(E^{f,u}_{p,h}\right)$ and n_c hourly energy consumption forecasts $\left(E^{f,u}_{c,h}\right)$, the aggregated hourly energy consumption and production forecasts, respectively, $E^{f}_{c,h}$and $E^{f}_{p,h}$, are obtained.

$$E^{f}_{c,h} = \sum_{u=1}^{n_c} E^{f,u}_{c,h} \tag{1}$$

$$E^{f}_{p,h} = \sum_{u=1}^{n_c} E^{f,u}_{p,h} \tag{2}$$

The exchange energy profile with the grid for each end-user ($E^{f,u}_{ex,h}$) and for the aggregation ($E^{f}_{ex,h}$) can be obtained as the difference between production and consumption.

$$E^{f,u}_{ex,h} = E^{f,u}_{p,h} - E^{f,u}_{c,h} \tag{3}$$

$$E^{f}_{ex,h} = E^{f}_{p,h} - E^{f}_{c,h} \tag{4}$$

Using the rated capacity of each ESS and their *SOC* forecast for the last hour of the day before ($SoC^{f,u}_{24}$), the energy that can be stored in the ESS for each end-user ($E^{f,u}_{av,ch}$) and for the aggregation ($E^{f}_{av,ch}$) is determined.

$$E^{f,u}_{av,ch} = C^{u} \cdot \left(1 - SoC^{f,u}_{24}\right) \tag{5}$$

where C^{u} is the capacity for each *ESS*.

The hourly energy surplus ($E^{f,u}_{sur,h}$) and deficit ($E^{f,u}_{def,h}$) for the day ahead are determined for each end-user.

$$E^{f,u}_{sur,h} = E^{f,u}_{p,h} - E^{f,u}_{c,h} \qquad if\ \left(E^{f,u}_{p,h} - E^{f,u}_{c,h}\right) > 0 \tag{6}$$

$$E^{f,u}_{def,h} = E^{f,u}_{c,h} - E^{f,u}_{p,h} \qquad if\ \left(E^{f,u}_{p,h} - E^{f,u}_{c,h}\right) < 0 \tag{7}$$

At the same time, the daily energy surplus $\left(E^{f,u}_{sur,d}\right)$ for each end-user and for the aggregation is obtained $\left(E^{f}_{sur,d}\right)$.

$$E^{f,u}_{sur,d} = \sum_{h}\left(E^{f,u}_{p,h} - E^{f,u}_{c,h}\right) \quad if\ \left(E^{f,u}_{p,h} - E^{f,u}_{c,h}\right) > 0 \tag{8}$$

$$E^{f}_{sur,d} = \sum_{u=1}^{n} E^{f,u}_{sur,d} \tag{9}$$

Similarly, the energy deficit for each end-user ($E^{f,u}_{def,d}$) and for the aggregation ($E^{f}_{def,d}$) can be determined.

$$E^{f,u}_{def,d} = \sum_{h}\left(E^{f,u}_{c,h} - E^{f,u}_{p,h}\right) \quad if\ \left(E^{f,u}_{p,h} - E^{f,u}_{c,h}\right) < 0 \tag{10}$$

$$E^{f}_{def,d} = \sum_{u=1}^{n} E^{f,u}_{def,d} \tag{11}$$

Once such variables have been determined, at this point, it is necessary to calculate opportune distribution coefficients to maximize the amount of energy to be shared among the n end-users. These distribution coefficients (F^{u}_{d}) are determined starting from $E^{f,u}_{av,ch}$, referred to each end-user equipped with an ESS. Defining $E^{f}_{av,ch}$ the overall available charging energy for the aggregation, F^{u}_{d} is determined as the ratio between $E^{f,u}_{av,ch}$ and $E^{f}_{av,ch}$.

$$E^{f}_{av,ch} = \sum_{u=1}^{n} E^{f,u}_{av,ch} \tag{12}$$

$$F_d^u = \frac{E_{av,ch}^{f,u}}{E_{av,ch}^{f}} \quad (13)$$

An example of F_d^u calculation is reported in Figure 2. Five different end-users equipped with ESS are considered. The ESSs have different *SOCs* and capacities. The values of F_d^u are determined and reported.

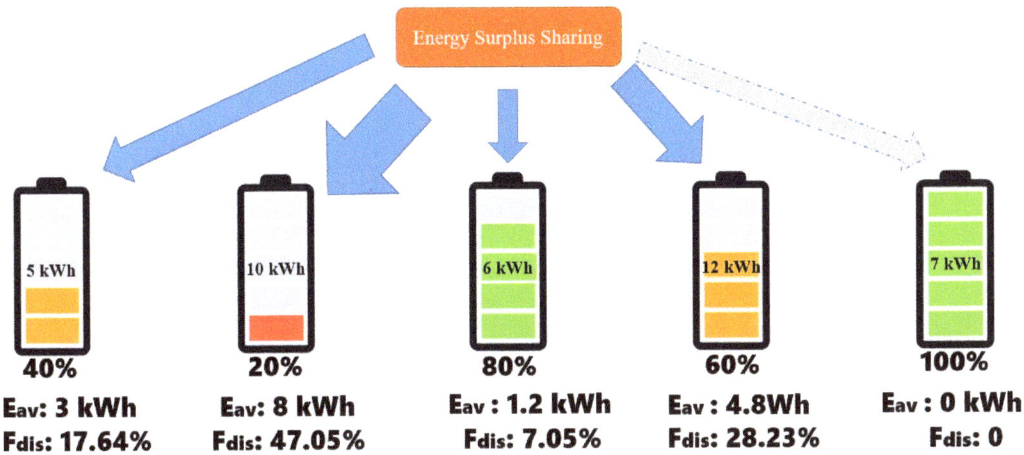

Figure 2. An example of distribution coefficients (F_d^u) calculations.

Such distribution coefficients are used to determine the energy that each end-user u has to store during the following day ($E_{crg,d}^{f,u}$); it will be function both of $E_{sur,d}^{f}$ and of $E_{av,ch}^{f,u}$. It is defined as:

$$E_{crg,d}^{f,u} = \min\left(F_d^u \cdot E_{sur,d}^{f},\ E_{av,ch}^{f,u}\right) \quad \forall\, u \in aggregation \quad (14)$$

Once $E_{crg,d}^{f,u}$ has been calculated, the power profiles that each end-user has to exchange with the ESS are available. The energy profile for the ESS ($E_{bat,ch,h}^{f,u}$) is obtained considering the distribution coefficient F_d^u in (11) and the aggregation hourly energy surplus ($E_{sur,h}^{f}$) if it exists.

$$E_{sur,h}^{f} = \sum_u \left(E_{p,h}^{f,u} - E_{c,h}^{f,u}\right) \quad (15)$$

$$\text{where, } if\ E_{sur,h}^{f},\ E_{crg,d,res}^{f,u} > 0$$

$$E_{bat,ch,h}^{f,u} = \min\left(E_{sur,h}^{f} \cdot F_d^u,\ E_{crg,d,res}^{f,u}\right) \quad (16)$$

where $E_{crg,d,res}^{f,u}$ is calculated iteratively as follows:

$$E_{crg,d,res}^{f,u} = E_{crg,d}^{f,u}\ for\ h = 1 \quad (17)$$

$$E_{crg,d,res}^{f,u} = E_{crg,d,res}^{f,u} - E_{bat,ch,h}^{f,u}\ \ for\ h > 1 \quad (18)$$

If the hourly aggregate energy production profile is lower than the aggregate consumption energy profile, an energy deficit occurs ($E_{def,h}^{f} > 0$). Thus, the energy that the ESS must supply is calculated ($E_{bat,ds,h}^{f,u}$) as in (15). Only the load of the end-users equipped with the ESS can be supplied by that ESS. The $E_{bat,ds,h}^{f,u}$ is determined as the minimum between $E_{def,h}^{f,u}$

and the residual energy in the ESS that is useful to supply the load ($E^{f,u}_{dis,res,h}$). Naturally, it can be defined only if the $SOC^{f,u}_{h}$ of the ESS is greater than the minimum admissible value of SOC (SOC_{min}).

$$E^{f,u}_{bat,ds,h} = \min\left(E^{f,u}_{def,h}, E^{f,u}_{dis,res,h}\right) \tag{19}$$

$$\text{where, } if \left(E^{f,u}_{c,h} - E^{f,u}_{p,h}\right) > 0$$

$$E^{f,u}_{def,h} = E^{f,u}_{c,h} - E^{f,u}_{p,h} \tag{20}$$

$$\text{and } E^{f}_{def,h} = \sum_{u=1}^{n} E^{f,u}_{def,h} \tag{21}$$

$$E^{f,u}_{dis,res,h} = \left(SOC^{f,u}_{h} - SOC_{min}\right) \cdot C^{u} \tag{22}$$

$$SOC^{f,u}_{h+1} = SOC^{f,u}_{h} + \left(E^{f,u}_{bat,ch,h} - E^{f,u}_{bat,ds,h}\right) / C^{u} \tag{23}$$

The baseline grid power exchange profile ($E^{f,u}_{grid,h}$), with an hourly time step, can be defined for each end-user and for the entire aggregation.

$$E^{f,u}_{grid,h} = E^{f,u}_{c,h} - E^{f,u}_{p,h} + E^{f,u}_{bat,ch,h} - E^{f,u}_{bat,ds,h} \tag{24}$$

At this point, the flexibility and availability of the single end-user must be estimated. At this scope, a fundamental assumption is that only the reduction of the baseline can be carried out, so the hours where $E^{f}_{def,h}$ is greater than zero are considered. The number of hours when $E^{f}_{def,h} > 0$ is calculated (n_{def}). It is used to determine the hourly energy amount that end-users make available to modify their power profile. Indeed, the available capacity ($E^{f,u}_{av,var}$) to vary the power profile is determined as the difference between $SOC^{f,u}_{24}$, and the SOC under which it is not possible to modify the profile ($SOC_{min,var}$).

The flexibility availability of the single end-user profile $E^{f,u}_{av,var,h}$ is so determined by comparing the load profile of the single end-user and the ratio between $E^{f,u}_{av,var}$ and n_{def}.

The aggregation availability $E^{f}_{av,var,h}$ can be determined as the sum of the availability of the single end-user profiles.

$$E^{f,u}_{av,var} = \left(SOC^{f,u}_{24} - SOC_{min,var}\right) \cdot C^{u} \tag{25}$$

$$E^{f,u}_{av,var,h} = \min\left(E^{f,u}_{c,h}, \frac{E^{f,u}_{av,var}}{n_{def}}\right) \tag{26}$$

$$E^{f}_{av,var,h} = \sum_{u=1}^{n} E^{f,u}_{av,var,h} \tag{27}$$

After this step, DA can be considered concluded, so both $E^{f,u}_{grid,h}$ and $E^{f,u}_{av,var,h}$ are evaluated and can be communicated to the grid operator, representing the interface with the aggregator.

3.2. The Real-Time Algorithm (RTA) Step

The real-time step operation is illustrated in Figure 3.

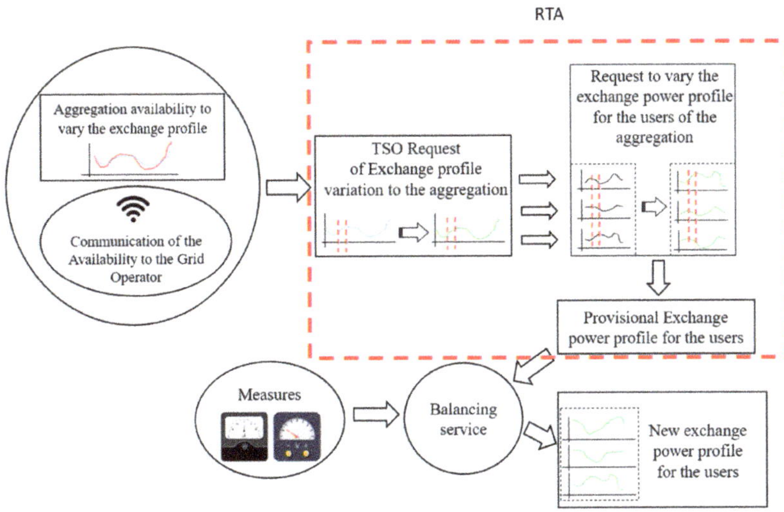

Figure 3. Real-time step flow chart.

The results of the DA are used to operate the real-time step (RT). Starting from $E^{f}_{av,var,h}$, the grid operator can send a flexibility request to the aggregation. Such a request, considering the end-users' flexibility and availability and the end-user distribution coefficients ($F^{u}_{d,var}$), is distributed among the n end-users. The new ESS and grid exchange power profiles ($E^{f,u}_{bat,h,var}$, $E^{f,u}_{grid,h,var}$) are so calculated for each end-user as follows:

$$F^{u}_{d,var} = \frac{E^{f,u}_{av,var}}{E^{f}_{av,var}} \tag{28}$$

$$E^{f,u}_{bat,h,var} = E^{f,u}_{bat,ch,h} - E^{f,u}_{bat,ds,h} - E^{f,u}_{av,var,h} \tag{29}$$

At the end of the process, there is a check to avoid overcharging ($SOC > 100\%$) and overdischarging (less than $SOC = 5\%$) of each ESS.

3.3. Balancing Step (BS)

The previous steps were performed using only the production and consumption forecasts as input. Therefore, it is possible that forecast errors occur, causing corresponding power imbalances and, consequently, significant imbalances in costs. For this issue, the balancing step becomes useful (Figure 4).

The BS considers, first of all, the configuration data (n, n_p, n_c, n_s) for the aggregation, the measured and forecasted grid power (P^{m}_{grid}, $P^{f}_{grid,h,var}$), the ESS energy ($E^{f,u}_{bat,h}$), the ESS capacity of the aggregation (C^{agg}). The SOC for the entire aggregation (SOC^{agg}) is determined as the weighted average of the end-users ESSs SOC, considering the capacity as an element to provide such weight. The applied criterion is to store the energy starting from the ESS that has a measured SOC less than the forecasted one and vice versa when a deficit of energy occurs, as described in [25]. First, it can be observed that for the end-users belonging to the aggregation, it is possible to significantly reduce imbalances, thanks also to self-dispatching.

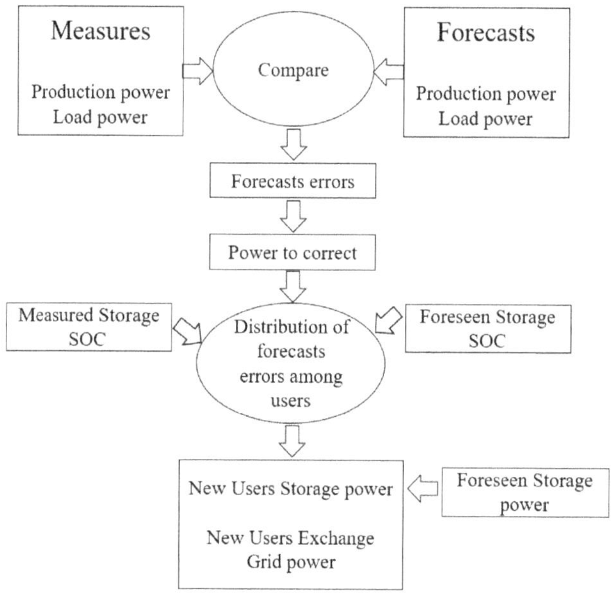

Figure 4. Balancing step flow chart.

In the balancing step, three main scenarios 'S' are analyzed:

- S1, where the power exchange with the grid is equal to the scheduled one (output of the DA step);
- S2, where the real grid exchange power profile is less than the scheduled one;
- S3, where the real grid exchange power profile is greater than the scheduled one.

In the case of S1, no operation is required; the schedule is operated.

In cases S2 and S3, it is necessary to operate to reduce as much as possible the error between the scheduled grid exchange power profile and the real one.

In the case of S2, two different events can occur: in the first event, the actual SOC^{agg} is greater than the forecasted value; in the second one, it is lower. In this second event, no correction is necessary because a possible operation of the ESS could invalidate the following time steps.

In the first event, instead, the variation in the power that the ESSs have to supply is determined (P^{agg}_{batt}).

$$P^{agg}_{batt} = P^{f}_{bat,h,var} + \min\left(\begin{array}{c} \left(P^{f}_{grid,h,var} - P^{m}_{grid}\right), \\ \left(SOC^{agg} - SOC^{f}_{h,var}\right) \cdot \frac{C^{agg}}{100} \end{array} \right) \quad (30)$$

where $P^{f}_{grid,h,var}$ and $P^{f}_{bat,h,var}$ are determined, respectively, from $E^{f,u}_{grid,h}$ and $E^{f,u}_{bat,h,var}$ considering the constant power during the hour; $SOC^{f}_{h,var}$ is the forecasted weighted average SOC for the ESSs of the aggregation.

Similarly, the BS can operate in S3. In this case, the SOC of the aggregation, SOC^{agg}, is evaluated: if it is equal to 100%, no operation is necessary; if it is less than 100%, the variation in the power that the ESSs have to supply is determined (P^{agg}_{batt}) as follows:

$$P^{agg}_{batt} = P^{f}_{bat,h,var} - \min\left(\begin{array}{c} \left(P^{m}_{grid} - P^{f}_{grid,h,var}\right), \\ \left(100 - SOC^{agg}\right) \cdot \frac{C^{agg}}{100} \end{array} \right) \quad (31)$$

At the end of the BS, the variation in the power that the ESSs have to supply P_{batt}^{agg} is distributed among the end-users of the aggregation. The power P_{batt}^{agg} can be distributed according to different criteria. The criteria used in the paper is to sort the end-user and distribute according to the difference between the real and the forecasted SOC in a proportional way, prioritizing the ESS that is less charged.

4. Case Study Description and Results

To test the proposed algorithm, a case study has been considered. It consists of 20 residential end-users located in the south of Italy, where 15 of them (from User A to User O) are provided with a PV plant, and 10 (from User A to User J) are also equipped with an ESS. The configuration of each end-user in terms of load, generation, and storage capacity is summarized in Table 1. The end-users are really monitored using a smart meter installed for each end-user.

Table 1. End-user configuration data.

End-Users	Load Contracted Power [kW]	Generation Contracted Power [kW]	Storage Capacity [kWh]
User A	3	3	5
User B	3	2	3
User C	4	3	6
User D	6	4	3
User E	3	3	6
User F	3	2	4
User G	3	3	3
User H	6	3	6
User I	10	6	12
User J	6	5	8
User K	3	3	-
User L	3	2	-
User M	15	10	-
User N	3	2	-
User O	6	3	-
User P	3	-	-
User Q	3	-	-
User R	6	-	-
User S	3	-	-
User T	15	-	-

The forecasted and instantaneous power data have been used for testing the algorithm, considering the month of March 2018.

4.1. DA Test Results

To carry out the test, the $SOC_{min,load}$ has been set to 50%, and the $SOC_{min,var}$ has been set to 15%.

First, to show how the algorithm operates, a single day is considered. The load and production aggregated power profiles are reported in Table 2. The energy surplus $E_{sur,d}^{f}$ is equal to 223.5 kWh, while the deficit $E_{def,d}^{f}$ is equal to 231.7 kWh.

Table 2. Aggregated load and production profile.

Time [h]	1	2	3	4	5	6	7	8	9	10	11	12	13	14	15	16	17	18	19	20	21	22	23	24
E_c^f[kWh]	7.68	9.37	7.32	5.57	5.39	7.94	10.40	13.67	12.28	14.65	15.45	16.13	13.43	11.52	12.32	10.00	14.34	16.76	19.76	22.13	27.70	17.61	17.76	12.17
E_p^f[kWh]	0.00	0.00	0.00	0.00	0.00	0.00	0.36	9.08	23.85	34.47	41.37	44.06	45.66	41.67	35.74	24.37	10.61	1.88	0.00	0.00	0.00	0.00	0.00	0.00

The parameter $SOC_{24}^{f,u}$ is equal to 50% for each ESS of the aggregation. For the considered day, $E_{av,ch}^{f}$ is equal to 43.5 kWh, determined using Equation (8). In this case, $E_{av,ch}^{f}$ limits the power that can be stored because $E_{av,ch}^{f} < E_{sur,d}^{f}$.

The distribution coefficients F_{d}^{u} are determined (Table 3) and used to obtain $E_{av,ch}^{f,u}$. Therefore, the energy profiles that the ESSs can exchange to be charged and to supply the load are scheduled.

Table 3. The distribution coefficients and the ESS energy to store.

User	A	B	C	D	E	F	G	H	I	J
F_{d}^{u} [%]	9.2	12.6	10.3	6.9	12.6	8.0	6.9	10.3	12.6	10.3
$E_{av.ch}^{f.u}$ [kWh]	4	5.5	4.5	3	5.5	3.5	3	4.5	5.5	4.5

The forecast ESS energy profiles for every end-user are obtained (Figure 5). It is possible to observe that generally, the ESSs are charged in the first part of the day when a surplus exists, and they are discharged in the evening.

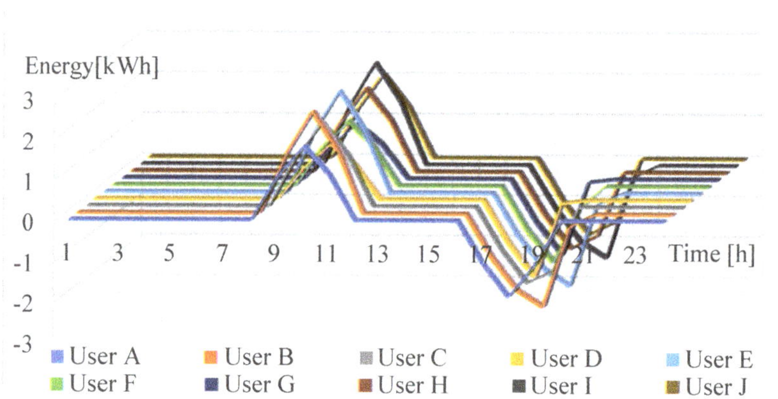

Figure 5. Day-ahead ESS forecasted profiles.

Once the day-ahead scheduling has been carried out, the availability of flexibility for the next day can be planned.

First $E_{av,var}^{f,u}$ is determined to obtain the hourly energy profile for each user $E_{av,var,h}^{f,u}$. Then, the availability to provide flexible services to the grid for the aggregation $E_{av,var,h}^{f}$ is determined using Equation (23). Such variables are reported in Tables 4 and 5. Starting from such results, it is possible to proceed with the RTA step. It is assumed that there are grid operator flexibility requests for the aggregation (see Table 6). These requests must be distributed among the end-users according to $E_{av,var}^{f,u}$. Then, using Equation (29), $E_{bat,h,var}^{f,u}$ is obtained (see Table 7).

Table 4. Overall flexibility and availability.

User	A	B	C	D	E	F	G	H	I	J
$E_{av.var}^{f.u}$ [kWh]	2.8	3.70	3.15	1.94	3.77	2.14	2.1	1.20	3.62	3.05

Table 5. Hourly aggregated flexibility and availability.

Hour	1	2	3	4	5	6	7	8	9	10	11	12	13	14	15	16	17	18	19	20	21	22	23	24
$E^{f}_{av,var,h}$ [kWh]	1.86	1.65	1.63	1.72	1.61	1.59	1.61	1.82	0	0	0	0	0	0	0	0	1.65	1.65	1.73	1.90	1.72	1.73	1.74	1.89

Table 6. Grid operator flexibility requests.

Hour	1	2	3	4	5	6	7	8	9	10	11	12	13	14	15	16	17	18	19	20	21	22	23	24
Flex Request [kWh]	1.85	0	0	1.72	0	0	1.61	0	0	0	0	0	0	0	0	0	0	1.64	0	0	0	1.72	1.73	1.88

Table 7. Real-time forecasted battery profile.

	1	2	3	4	5	6	7	8	9	10	11	12	13	14	15	16	17	18	19	20	21	22	23	24
A	−0.18	0	0	−0.18	0	0	−0.18	0	1.06	1.82	1.11	0	0	0	0	0	−1.06	−2.04	−1.07	0	0	−0.18	−0.18	−0.18
B	−0.24	0	0	−0.24	0	0	−0.22	0	1.46	2.51	1.53	0	0	0	0	0	−1.06	−2.08	−2.28	−0.29	0	−0.24	−0.24	−0.24
C	−0.20	0	0	−0.20	0	0	−0.20	0	1.20	2.05	1.25	0	0	0	0	0	−1.06	−2.07	−1.57	0	0	−0.20	−0.20	−0.20
D	−0.09	0	0	−0.10	0	0	−0.13	0	0.80	1.37	0.83	0	0	0	0	0	−1.06	−2.00	−0.07	0	0	−0.13	−0.13	−0.13
E	−0.24	0	0	−0.24	0	0	−0.17	0	1.46	2.51	1.53	0	0	0	0	0	−1.06	−2.11	−2.28	−0.29	0	−0.24	−0.24	−0.24
F	−0.15	0	0	−0.10	0	0	−0.11	0	0.93	1.59	0.97	0	0	0	0	0	−1.06	−2.02	−0.57	0	0	−0.15	−0.15	−0.15
G	−0.13	0	0	−0.13	0	0	−0.13	0	0.80	1.37	0.83	0	0	0	0	0	−1.06	−2.00	−0.07	0	0	−0.13	−0.13	−0.13
H	−0.20	0	0	−0.18	0	0	−0.04	0	1.20	2.05	1.25	0	0	0	0	0	−1.06	−1.90	−1.57	0	0	−0.02	−0.03	−0.18
I	−0.24	0	0	−0.19	0	0	−0.24	0	1.46	2.51	1.53	0	0	0	0	0	−1.06	−2.11	−2.28	−0.29	0	−0.24	−0.24	−0.24
J	−0.20	0	0	−0.16	0	0	−0.20	0	1.20	2.05	1.25	0	0	0	0	0	−1.06	−2.01	−1.57	0	0	−0.20	−0.20	−0.20

In Figure 6, for end-user A, the comparison between the ESS power profile scheduled (output of the DA step) and the ESS power profile evaluated in the RTA step is shown. The difference is limited, and the two profiles are comparable.

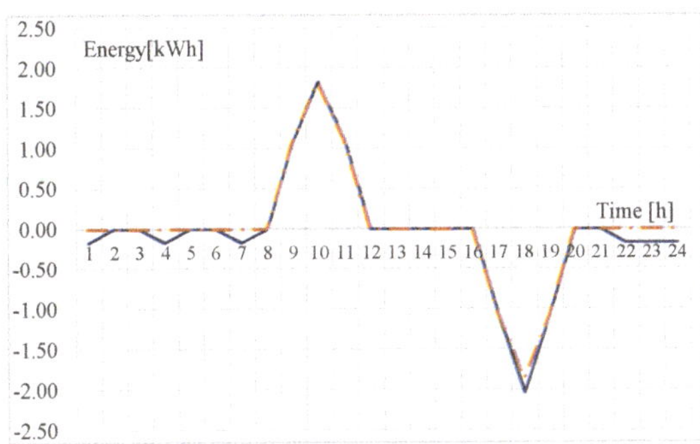

Figure 6. End-user A—comparison between the day-ahead ESS power profile (orange) and the real-time ESS power profile (blue) with a grid operator flexibility request.

In the end, the forecasts of the ESSs SOC are determined; their variation depends only on possible forecasting errors.

4.2. BS Results

At the end of the RTA, the BS starts. Firstly, it compares the hourly production and consumption forecasts with the measures (real power values) every 10 min. Starting from the results of DA, the $SOC^{f,agg}_{h,var}$ and the SOC^{agg} must be determined.

If an error between the forecasted and measured power occurs, it is necessary to verify if such error can be covered by the ESSs.

When an excess of power exists, the BS operates with the aim of storing the excess energy as far as possible, as expressed in Equation (27), while when a deficit occurs, Equation (26) is used.

Once the contribution to the balancing step has been determined, it will be possible to calculate the power that the ESSs have to exchange after such a correction. Thanks to the balancing step, it is possible to significantly reduce the imbalances. Based on the accuracy of the forecasts and the type of day, it is possible to cover more than 60% of the imbalances that would have occurred if the algorithm had not been used.

4.3. Long-Term Performance Analysis

After the results for a single day have been presented to understand the operation of the algorithm, an overall analysis of the results for the considered month is carried out.

The results are discussed, considering three cases:

- Case A, which is the base case described before;
- Case B, where $SOC_{min,load}$ is decreased to 40%;
- Case C, where the capacity of the ESSs is increased (50% more than the base case).

4.3.1. Case A

For the considered period, the energy surplus obtained for the entire aggregation is equal to 3629 kWh, while the energy deficit is equal to 6328 kWh.

In the first step of the algorithm, in the day ahead, the energy surplus adsorbed by the ESSs is equal to 1373 kWh, while the energy supplied by the ESSs is equal to 1110 kWh.

The energy stored in the ESSs is about equal to 30% of the aggregation energy surplus; this is due to the limit of the overall ESS capacity and to the possibility of scheduling the ESS in this step until a deep of discharge (DoD) of 50% to supply the load.

On the other hand, the overall availability provided to carry out flexibility services to the grid is equal to 600 kWh, while the grid operator flexibility services actually requested and supplied (unless forecast errors) are equal to 227 kWh, which is 38% of the communicated availability for flexibility services (Figure 7).

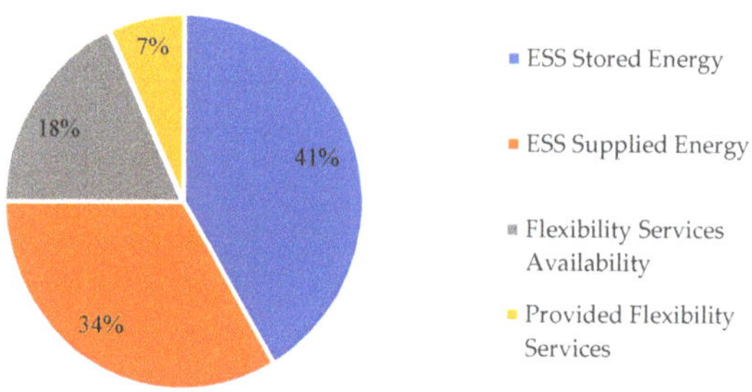

Figure 7. Use of stored energy in supplying loads and providing services—Case A.

In summary, the energy used to carry out services to the grid is about 16.5% of the stored energy surplus; the remaining part, about 83.5%, is scheduled to supply the loads.

Such values obviously depend both on the overall capacity of the ESSs and on the limits imposed on the possibility of being able to discharge the storage system under the defined DoD.

4.3.2. Case B

If $SOC_{min,load}$ is decreased (setting it equal to 40%), it can be observed that the energy surplus absorbed in this case by the ESSs is equal to 1512 kWh, while that supplied to the loads is equal to 1318 kWh. The energy stored is approximately equal to 42% of the energy surplus from the aggregation. On the other hand, an overall availability to carry out grid operator flexibility services equal to 433 kWh is provided, while the services requested and provided (unless forecast errors) are equal to 165 kWh, equal to 38% of the availability, which is similar to the previous case (Figure 8). In this case, only 10% of the stored energy surplus is used to provide flexibility.

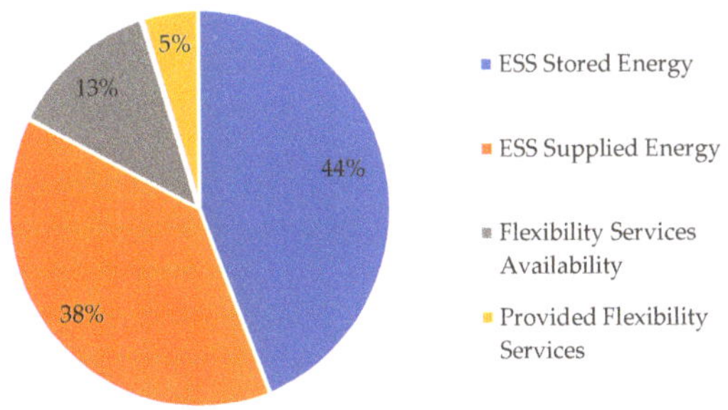

Figure 8. Use of stored energy in supplying loads and providing services—Case B.

It is possible to observe that after changing $SOC_{min,load}$, the energy stored and discharged from the ESSs (supplying the load and providing flexibility services) is greater, compared to the previous case.

4.3.3. Case C

The overall capacity of the ESSs is increased to 50% while maintaining the $SOC_{min,load}$ equal to 50%. The energy surplus absorbed in this case by the ESSs is equal to 1926 kWh, while the supplied energy is equal to 1547 kWh. The energy stored in the ESSs is equal to 53% of the surplus of the aggregation (Figure 9).

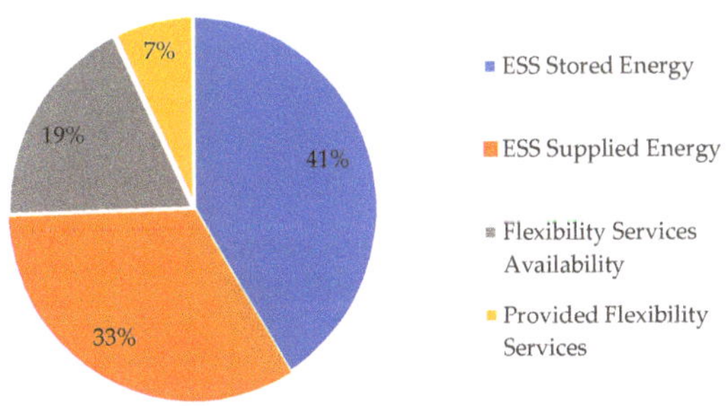

Figure 9. Use of stored energy in supplying loads and providing services—Case C.

The overall availability to provide grid operator flexibility services is equal to 861 kWh, while the services requested and provided (except for forecast errors) are equal to 325 kWh, or approximately 38% of the available resources.

4.3.4. Summary

To obtain a complete picture, the results obtained in the three different cases are summarized below (Table 8).

Table 8. Long-term analysis performance.

	Case A **Base Case**	**Case B** $SOC_{min,load}$ **= 40%**	**Case C** **ESSs Capacity Increased**
ESSs Stored Energy	1373 kWh	1512 kW	1926 kWh
ESSs Supplied energy	1110 kWh	1318 kWh	1547 kWh
Flexibility Services Availability	600 kWh	433 kWh	861 kWh
Provided Flexibility Services	227 kWh	165 kWh	325 kWh

Starting from the base case, it is possible to observe how the possibility to supply the loads until a SOC of 40% (CASE B) allows both to supply and store more energy with the

ESSs; naturally, it also increases the ESSs capacity, but not proportionally. On the contrary, reducing the minimum SOC to supply the loads to 40% does not allow for more flexibility services, while increasing the ESS capacity makes it possible to increase the amount of energy for flexibility services.

To find a better solution to maximize both energy used to supply the loads and to provide services to the grid, it becomes necessary to combine the correct parameters used, depending on the PV production and the loads.

5. Conclusions

In this paper, an algorithm to opportunely manage the energy flows inside an aggregation operating following the Power Cloud concept has been presented. In this framework, different kinds of users (prosumers, consumers, and prosumages) connected via a public distribution network are aggregated. The main objective was to verify the possibilities of exploiting the flexibility of behind-the-meter technologies integrating "all-in-one" photovoltaic plants, storage systems, and other technological solutions in order to maximize benefits to end-users.

The algorithm schedules the energy that the end-user ESSs must exchange inside the aggregation to maximize the use of renewable sources, provide grid flexibility services, and simultaneously provide balancing services. The algorithm is organized into three different steps: the day-ahead step, the real-time step, and the balancing one.

The first step determines the charge/discharge of storage systems present in the aggregation to maximize renewable energy self-consumption using forecasted hourly production and consumption. A reference power profile is determined to exchange with the grid.

The second one determines in real-time how to modify that profile if flexibility services are required by the TSO or DSO, and the third one compensates for errors in forecasting production and/or consumption.

A test has been implemented to illustrate a real aggregation of 20 end-users, demonstrating the effectiveness of the proposed algorithm in reducing by about 60% the power imbalances that can occur if forecast errors and/or grid flexibility requests exist.

One of the principal characteristics of the proposed algorithm to be highlighted is that it can manage the charge/discharge phases of the ESSs present on the user side in a blind manner for the final user. So, an adequate level of aggregate self-consumption and the ancillary services required by the power system operator are met without any change in user habits. The energy storage systems are used to increase self-consumption and provide real-time grid operator flexibility services.

The main drawback of the proposed algorithm is related to considering some variables (especially power) as deterministic ones while some kind of uncertainty exists, especially in the day-ahead step.

In the future work of the authors, (i) the above-mentioned uncertainty will be considered in order to avoid errors deriving from deterministic assumptions; (ii) a possible improvement in the benefit of the proposed algorithm will be investigated considering the use of an opportune ICT platform, respecting the concept of Power Cloud, able to exchange data among aggregators and end-users in order to increase the end-users advantage from self-consumption and flexibility services delivered to the system without changing their power consumption behavior and habits.

Author Contributions: Conceptualization, P.V.; methodology, P.V., N.S. and A.P.; software, D.M. and P.V.; validation, A.P. and G.B.; formal analysis, A.P. and N.S.; investigation, P.V. and D.M.; resources, N.S.; data curation, G.B.; writing—original draft preparation, P.V.; writing—review and editing, A.P., G.B. and N.S.; visualization, P.V. and D.M.; supervision, A.P.; project administration, N.S.; funding acquisition, D.M. and A.P. All authors have read and agreed to the published version of the manuscript.

Funding: This work was financed by the Italian Ministry of Economic Development (MISE) and the Ministry of Education, University and Research (MIUR) through the National Operational Program for Business and Competitiveness 2014–2020, PON F/050159/01-03/X32, and by the European Union's Horizon 2020 research and innovation program under grant agreement n° 864283.

Data Availability Statement: Not applicable.

Conflicts of Interest: The authors declare no conflict of interest.

Abbreviations

EV	Electric vehicle	$E^{f,u}_{def,d}$	Aggregated daily energy deficit
EC	Energy Community	$E^{f}_{def,d}$	Aggregated daily energy deficit
ESS	Energy storage System	$E^{f}_{sur,h}$	Aggregated hourly energy surplus
P2P	Peer-to-peer	$E^{f,u}_{bat,ch,h}$	Charging Energy profile for the ESS
LEC	Local energy community	$E^{f,u}_{crg,d}$	Energy that each user has to store daily
ANN	Artificial neural network	F^{u}_{d}	Surplus distribution factor
SOC	State of charge	$E^{f,u}_{crg,d,res}$	Daily residual energy that each user has to store
DAA	Day-ahead algorithm	$E^{f,u}_{bat,ds,h}$	Hourly energy that the user equipped with ESS has to supply
RTA	Real-time algorithm	$E^{f,u}_{dis,res,h}$	Hourly residual energy in the ESS is useful to supply the load
TSO	Transmission system operator	$SOC_{min,load}$	SOC under that the ESS discharge cannot be scheduled
n_p	Number of producers	$E^{f,u}_{grid,h}$	Hourly profile to be exchanged with the grid (baseline)
n_c	Number of consumers	$E^{f,u}_{av,var}$	Daily available capacity to vary the profile
n_s	Number of end-users equipped with ESS	$SOC_{min,var}$	SOC, under which it is not possible to modify the profile
$E^{f,u}_{p,h}$	Hourly production forecasts for each user	$E^{f,u}_{av,var,h}$	Hourly flexibility availability of the single end-user
$E^{f,u}_{c,h}$	Hourly load forecasts for each user	n_{def}	Number of hours where a deficit exists
$E^{f}_{c,h}$	Aggregated hourly consumption forecast	$E^{f}_{av,var,h}$	End-users aggregation availability
$E^{f}_{p,h}$	Aggregated hourly production forecast	$F^{u}_{d,var}$	User availability and the distribution factor
$E^{f,u}_{ex,h}$	Exchange power profile with the grid for each user	$E^{f,u}_{bat,h,var}$	New ESS power profiles
$E^{f}_{ex,h}$	Aggregated exchange power profile with the grid	$E^{f,u}_{grid,h,var}$	New grid power profiles
$SoC^{f,u}_{24}$	SOC forecast for the last hour of the day before	C^{agg}	Aggregated ESS capacity
$E^{f,u}_{av,ch}$	Energy that can be stored in the ESSs for each user	SOC^{agg}	SOC for the entire aggregation
$E^{f}_{av,ch}$	Aggregated energy that can be stored in the ESSs	P^{m}_{grid}	Measured grid power
C^{u}	Capacity for each ESS	$P^{f}_{grid,h,var}$	Forecasted grid power
$E^{f,u}_{sur,h}$	Hourly energy surplus for each user	P^{agg}_{batt}	Aggregated power variation that ESSs have to supply
$E^{f,u}_{def,h}$	Hourly energy deficit for each user	$P^{f}_{bat,h,var}$	Forecasted ESS power
$E^{f,u}_{sur,d}$	Daily energy surplus for each user	$SOC^{f,agg}_{h,var}$	Forecasted weighted average SOC for aggregation ESSs
$E^{f}_{sur,d}$	Aggregated daily energy surplus	V2G	Vehicle to Grid

References

1. Lesniak, A.; Chudy, D.; Dzikowski, R. Modelling of Distributed Resource Aggregation for the Provision of Ancillary Services. *Energies* **2020**, *13*, 4598. [CrossRef]
2. Wenzel, G.; Negrete-Pincetic, M.; Olivares, D.E.; MacDonald, J.; Callaway, D.S. Real-time charging strategies for an electric vehicle aggregator to provide ancillary services. *IEEE Trans. Smart Grid* **2017**, *9*, 5141–5151. [CrossRef]
3. Clairand, J.M. Participation of electric vehicle aggregators in ancillary services considering users' preferences. *Sustainability* **2020**, *12*, 8. [CrossRef]
4. Edmunds, C.; Elders, I.; Galloway, S.; Stephen, B.; Postnikov, A.; Varga, L.; Hu, Y.; Kipouros, T. Congestion management with aggregated delivery of flexibility using distributed energy resources. In Proceedings of the 2020 6th IEEE International Energy Conference (ENERGYCon), Tunis, Tunisia, 28 September–1 October 2020; pp. 616–620.

5. Karagiannopoulos, S.; Mylonas, C.; Aristidou, P.; Hug, G. Active distribution grids providing voltage support: The swiss case. *IEEE Trans. Smart Grid* **2020**, *12*, 268–278. [CrossRef]
6. Lezama, F.; Soares, J.; Canizes, B.; Vale, Z. Flexibility management model of home appliances to support DSO requests in smart grids. *Sustain. Cities Soc.* **2020**, *55*, 102048. [CrossRef]
7. Lipari, G.; Del Rosario, G.; Corchero, C.; Ponci, F.; Monti, A. A real-time commercial aggregator for distributed energy resources flexibility management. *Sustain. Energy Grids Netw.* **2018**, *15*, 63–75. [CrossRef]
8. Henríquez, R.; Wenzel, G.; Olivares, D.E.; Negrete-Pincetic, M. Participation of demand response aggregators in electricity markets: Optimal portfolio management. *IEEE Trans. Smart Grid* **2017**, *9*, 4861–4871. [CrossRef]
9. Roos, A.; Ottesen, S.Ø.; Bolkesjø, T.F. Modeling consumer flexibility of an aggregator participating in the wholesale power market and the regulation capacity market. *Energy Procedia* **2014**, *58*, 79–86. [CrossRef]
10. Kumar, G.V.; Palanisamy, K. A Review of Energy Storage Participation for Ancillary Services in a Microgrid Environment. *Inventions* **2020**, *5*, 63. [CrossRef]
11. Agbonaye, O.; Keatley, P.; Huang, Y.; Bani-Mustafa, M.; Ademulegun, O.O.; Hewitt, N. Value of demand flexibility for providing ancillary services: A case for social housing in the Irish DS3 market. *Util. Policy* **2020**, *67*, 101130. [CrossRef]
12. Yang, Y.; Bremner, S.; Menictas, C.; Kay, M. Modelling and optimal energy management for battery energy storage systems in renewable energy systems: A review. *Renew. Sustain. Energy Rev.* **2022**, *167*, 112671. [CrossRef]
13. Elkazaz, M.H.; Sumner, M.; Pholboon, S.; Thomas, D. A Peer to Peer Energy Management System for Community Microgrids in The Presence of Renewable Energy and Storage Systems. In Proceedings of the 10th International Conference on Power Electronics, Machines and Drives (PEMD 2020), Online, 15–17 December 2020.
14. Mustika, A.D.; Rigo-Mariani, R.; Debusschere, V.; Pachurka, A. A two-stage management strategy for the optimal operation and billing in an energy community with collective self-consumption. *Appl. Energy* **2022**, *310*, 118484. [CrossRef]
15. Reis, I.F.; Gonçalves, I.; Lopes, M.A.; Antunes, C.H. Business models for energy communities: A review of key issues and trends. *Renew. Sustain. Energy Rev.* **2021**, *144*, 111013. [CrossRef]
16. Karimi, H.; Jadid, S. Multi-layer energy management of smart integrated-energy microgrid systems considering generation and demand-side flexibility. *Appl. Energy* **2023**, *339*, 120984. [CrossRef]
17. Fan, G.; Liu, Z.; Liu, X.; Shi, Y.; Wu, D.; Guo, J.; Zhang, S.; Xinyan, Y.; Zhang, Y. Energy management strategies and multi-objective optimization of a near-zero energy community energy supply system combined with hybrid energy storage. *Sustain. Cities Soc.* **2022**, *83*, 103970. [CrossRef]
18. Tomin, N.; Shakirov, V.; Kozlov, A.; Sidorov, D.; Kurbatsky, V.; Rehtanz, C.; Lora, E.E. Design and optimal energy management of community microgrids with flexible renewable energy sources. *Renew. Energy* **2022**, *183*, 903–921. [CrossRef]
19. Firoozi, H.; Khajeh, H.; Laaksonen, H. Flexibility Forecast at Local Energy Community Level. In Proceedings of the 2021 IEEE PES Innovative Smart Grid Technologies Europe (ISGT Europe), Espoo, Finland, 18–21 October 2021; pp. 1–5.
20. Mendicino, S.; Menniti, D.; Palumbo, F.; Pinnarelli, A.; Sorrentino, N.; Viggiano, L.; Vizza, P. An IT platform for the management of a Power Cloud community leveraging IoT, data ingestion, data analytics and blockchain notarization. In Proceedings of the 2021 IEEE PES Innovative Smart Grid Technologies Europe (ISGT Europe), Espoo, Finland, 18–21 October 2021; pp. 1–6.
21. Menniti, D.; Pinnarelli, A.; Sorrentino, N.; Vizza, P.; Brusco, G.; Barone, G.; Marano, G. Techno economic analysis of electric vehicle grid integration aimed to provide network flexibility services in Italian regulatory framework. *Energies* **2022**, *15*, 2355. [CrossRef]
22. Iria, J.; Scott, P.; Attarha, A.; Gordon, D.; Franklin, E. MV-LV network-secure bidding optimisation of an aggregator of prosumers in real-time energy and reserve markets. *Energy* **2022**, *242*, 122962. [CrossRef]
23. Sun, G.; Shen, S.; Chen, S.; Zhou, Y.; Wei, Z. Bidding strategy for a prosumer aggregator with stochastic renewable energy production in energy and reserve markets. *Renew. Energy* **2022**, *191*, 278–290. [CrossRef]
24. Burgio, A.; Brusco, G.; Menniti, D.; Pinnarelli, A.; Sorrentino, N.; Vizza, P. Economic evaluation in using storage to reduce imbalance costs of renewable sources power plants. In Proceedings of the 2017 14th International Conference on the European Energy Market (EEM), Dresden, Germany, 6–9 June 2017; pp. 1–6.
25. Menniti, D.; Pinnarelli, A.; Sorrentino, N.; Vizza, P.; Brusco, G.; Barone, G. How to Provide Dispatching Services in a Smart Community Microgrid to Reduce Imbalance Charges. In Proceedings of the 2018 15th International Conference on the European Energy Market (EEM), Lodz, Poland, 27–29 June 2018; pp. 1–5.

energies

MDPI

Review

Leveraging Artificial Intelligence to Bolster the Energy Sector in Smart Cities: A Literature Review

José de Jesús Camacho [1], Bernabé Aguirre [1], Pedro Ponce [1,*], Brian Anthony [2] and Arturo Molina [1]

1 Institute of Advanced Materials for Sustainable Manufacturing, Tecnológico de Monterrey, Monterrey 64849, Mexico; a00837152@tec.mx (J.d.J.C.); a01338617@tec.mx (B.A.); armolina@tec.mx (A.M.)
2 Department of Mechanical Engineering, Massachusetts Institute of Technology, 77 Massachusetts Avenue, Cambridge, MA 02139, USA; banthony@mit.edu
* Correspondence: pedro.ponce@itesm.mx

Abstract: As Smart Cities development grows, deploying advanced technologies, such as the Internet of Things (IoT), Cyber–Physical Systems, and particularly, Artificial Intelligence (AI), becomes imperative for efficiently managing energy resources. These technologies serve to coalesce elements of the energy life cycle. By integrating smart infrastructures, including renewable energy, electric vehicles, and smart grids, AI emerges as a keystone, improving various urban processes. Using the PRISMA (Preferred Reporting Items for Systematic Reviews and Meta-Analyses) and the Scopus database, this study meticulously reviews the existing literature, focusing on AI technologies in four principal energy domains: generation, transmission, distribution, and consumption. Additionally, this paper shows the technological gaps when AI is implemented in Smart Cities. A total of 122 peer-reviewed articles are analyzed, and the findings indicate that AI technologies have led to remarkable advancements in each domain. For example, AI algorithms have been employed in energy generation to optimize resource allocation and predictive maintenance, especially in renewable energy. The role of AI in anomaly detection and grid stabilization is significant in transmission and distribution. Therefore, the review outlines trends, high-impact articles, and emerging keyword clusters, offering a comprehensive analytical lens through which the multifaceted applications of AI in Smart City energy sectors can be evaluated. The objective is to provide an extensive analytical framework that outlines the AI techniques currently deployed and elucidates their connected implications for sustainable development in urban energy. This synthesis is aimed at policymakers, urban planners, and researchers interested in leveraging the transformative potential of AI to advance the sustainability and efficiency of Smart City initiatives in the energy sector.

Keywords: smart cities; energy sector; artificial intelligence

1. Introduction

The concept of Smart Cities can be interpreted in many different ways, which highlights the importance of having a universally accepted definition. An example of this is highlighted in [1], where the concept of "smartness" in the context of Smart Cities is addressed as the unification of sustainability objectives, which ensures that technology integration serves a purpose beyond mere automation and actively engages users in achieving environmental goals. However, it is also recognized that the definition of Smart Cities is flexible and can be adapted to suit specific situations and contextual factors [2]. The integration of communication and information technologies in Smart Cities offers a wide range of potential benefits. The primary objective of Smart Cities is to effectively manage resources and achieve efficient energy consumption while facilitating seamless data communication to ensure smooth Smart City operations [3]. To cope with increasing power demands while protecting citizens from the detrimental effects of Greenhouse Gases (GHGs) emissions, it becomes crucial to monitor and manage energy in an efficient manner throughout its entire life cycle, which encompasses generation, transmission, distribution, and consumption processes. By adopting comprehensive energy management strategies, Smart Cities can play a pivotal role in promoting sustainability and mitigating environmental impacts [4]. Indeed, the integration of smart energy management is vital to further enhance the efficiency of Smart Cities. This integration can be achieved through the incorporation of Artificial Intelligence (AI) techniques. AI has the potential to significantly contribute to the performance and integration of alternative energy sources within the Smart City infrastructure. By leveraging AI algorithms, smart energy management systems can intelligently analyze large amounts of data from various sources, including renewable energy generation, energy consumption patterns, weather conditions, and demand forecasts. This enables the optimization of energy distribution and consumption, leading to more efficient utilization of resources and reduced reliance on traditional fossil fuel-based energy sources. AI also plays a crucial role in enhancing the integration of renewable energy into the existing energy grid. With its ability to adapt and learn from real-time data, AI can dynamically adjust energy distribution and storage strategies, ensuring the seamless integration of fluctuating renewable energy sources like solar and wind power. Furthermore, AI-driven predictive analytics can aid in anticipating energy demand fluctuations and identifying potential areas for energy savings and optimization. This empowers Smart Cities to proactively manage energy resources, reduce waste, and minimize carbon emissions, contributing to a more sustainable and eco-friendly urban environment. The incorporation of AI techniques into smart energy management systems holds significant promise for improving alternative energy performance and integration in Smart Cities, fostering a more sustainable and energy-efficient future [5]. Various AI algorithms are currently under study and review for diverse applications in industries that have shown increasing interest in them. This includes the implementation of the Smart City concept [2].

One of the definitions of AI can be found in [6]. In this context, the concept of Artificial Intelligence (AI) is initially defined as the capacity of an entity to function effectively in response to its environment. Therefore, AI involves the development of machines endowed with this characteristic, enabling them to perform functions within their environment, effectively mechanizing human thought processes.

In the domain of energy systems, AI finds several valuable applications. In particular, it can be employed for energy management to achieve savings, enable control functions, and optimize the forecasting of alternative energy generation. AI also facilitates accurate demand forecasting, consumption prediction, and the efficient monitoring of energy grids, contributing to enhanced operational efficiency and sustainability in the context of Smart Cities and beyond [7]. Therefore, the use of AI methods is an important tool to be used to improve different stages of the energy cycle in Smart Cities. AI has some important features that can be used in the energy sector, as shown below:

1. Learning: This aspect of AI relies on data collection and analysis to create algorithms that facilitate the development of efficiency-promoting processes. By processing large data sets, AI systems can identify patterns, trends, and correlations, enabling them to learn from past experiences and make informed decisions.
2. Cognition: The cognitive capabilities of AI help to recognize similarities and patterns in previous processes. Drawing on past experience and knowledge, AI can interpret complex data, recognize trends, and generate valuable insights, improving decision-making processes.
3. Actions: AI is capable of making automatic decisions within specified time frames. This involves the real-time processing of data to generate responses and take actions based on predetermined rules or learned behaviors. Such automation can streamline various tasks, leading to increased efficiency and precision. These fundamental AI functionalities play a crucial role in a wide array of industries, including energy management in Smart Cities, predictive analytics, natural language processing, robotics, and more. As AI continues to advance, its potential for transformative impact across diverse fields remains substantial [8].

To recognize the main elements in which Artificial Intelligence helps Smart Cities in the generation, transmission, distribution, and consumption of energy, the following questions are formulated for answering through the literature review:

- How can the implementation of AI methods address crucial challenges in the energy sector?
- What AI methodologies are currently being employed in Smart Cities to address challenges in the energy sector?
- What advantages does the application of AI in the energy sector offer that conventional methods cannot achieve?
- What are the latest trends in applying AI in Smart Cities to improve the energy sector?

This paper presents a review of the literature conducted using the Scopus search engine employing the PRISMA (Preferred Reporting Items for Systematic Reviews and Meta-Analyses) methodology. PRISMA primarily focuses on the reporting of reviews evaluating the effects of interventions but can also be used as a basis for reporting systematic reviews with objectives other than evaluating interventions [9]. This methodology is a guideline for writing clear, detailed reports on systematic reviews and meta-analyses, ensuring transparency and reliability. The guidelines include a Title and Abstract, which should be clear and concise, summarizing the reviewer's goals, methods, results, and conclusions. The Introduction section explains the rationale and objectives of the review. The Methods section describes the criteria for selecting studies and the data collection and analysis methods. In the Results section, the study selection process is illustrated (often with a flow chart), together with the characteristics of the studies and the main findings. The Discussion section interprets the results, considering the strengths and weaknesses of the evidence, and discusses the generalizability and applicability of the findings. PRISMA also assists authors in ensuring that they include all essential information and improve the transparency of their review or analysis. This research focuses on the utilization of Artificial Intelligence (AI) in the energy sector for Smart Cities. This includes the use of AI for control, the integration of renewable energy, smart grids, energy production, and the forecasting of energy consumption, as well as other AI applications. As the main results, statistical information is presented, such as the annual scientific production, most relevant sources, most cited articles, most relevant words, and keyword co-occurrence. These results can help in visualizing the trend of publications for each category, identifying areas of opportunity for publications, identifying the main concepts that appear for each category, highlighting the potential of using AI in the energy sector, and showing some case studies where it has been successfully implemented.

This document is divided into four main categories: energy generation, transmission, distribution, and consumption. Statistical data are presented about the annual publications, the most cited articles, and principal keywords. In addition, a keyword co-occurrence network is presented for each section in order to recognize the main concepts of each category and the relationship between the principal topics.

2. Methodology

This research article employs the Preferred Reporting Items for Systematic Reviews and Meta-Analyses (PRISMA 2020) as the proposed methodology. PRISMA 2020 offers effective reporting guidance for systematic literature reviews. The PRISMA method can be summarized using the chart presented in Figure 1. The initial phase involves the identification and screening of relevant papers to be included in the research. The exploration of articles was conducted by performing a comprehensive search on Scopus using specific terms related to "Smart Cities". The search targeted these terms within the title, abstract, and keywords of the articles. In addition, the following terms are used: "**AND** Energy **AND** Artificial Intelligence **OR** AI" (*Smart Cities-Energy-Artificial Intelligence*). The articles had to meet certain eligibility criteria, including being of the document type "journal" and published in English within the last ten years prior to conducting the research (from 2013 to 2023). Subsequently, the base search was repeated by incorporating additional terms,

Generation, *Transmission*, *Distribution* and *Consumption*, including one term at a time after the term *Energy* of the base term. The potential articles identified during the initial search were subjected to a rigorous full-text review, and those that met all the specified criteria were included for further analysis. The subsequent phase involved data extraction and synthesis, which required retrieving relevant information using a standardized approach, including statistical meta-analysis. The search results derived from Scopus, with the base term, were exported and subjected to statistical analysis using the "Blibliometrix" software version 4.1.4. Additionally, a keyword co-occurrence network (KCN) was generated using the "VOSviewer" software version 1.6.19.

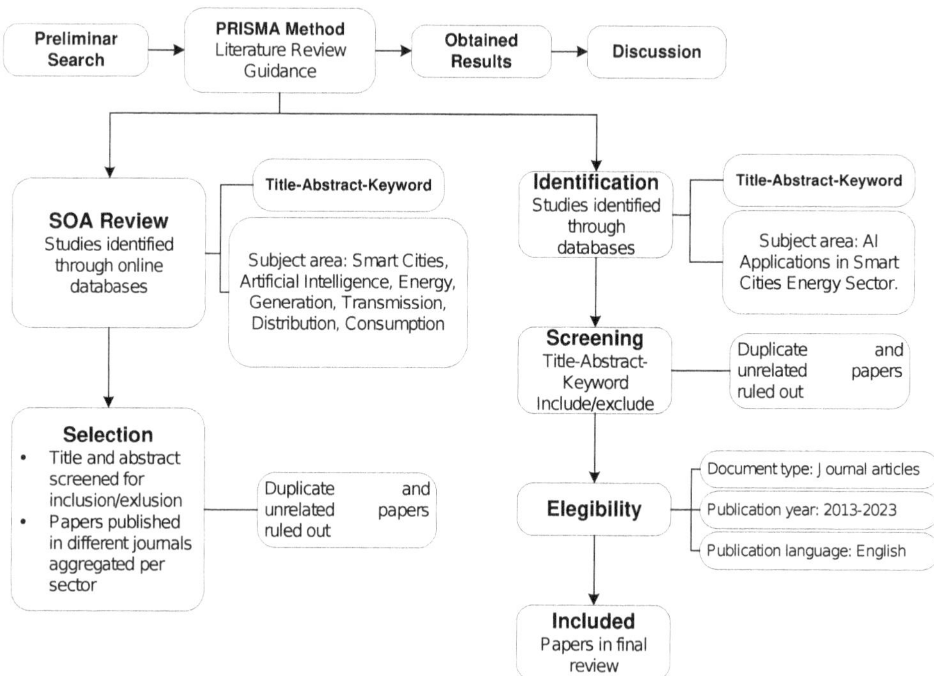

Figure 1. PRISMA method flowchart for paper inclusion/exclusion.

3. Results

3.1. Literature Review and Content Analysis

Using the methodology described in the previous section, the following results were obtained for the base terms. One way to easily analyze the results is by having a keyword co-occurrence network. Figure 2 shows a map made with the VOSviewer software; in this map, concepts such as IoT, energy efficiency, machine learning and smart grids are presented. A separate map was created for each category: energy generation, transmission, distribution and consumption. The purpose was to identify which category showed a higher prevalence of concepts in the overall map.

The following subsections analyze the statistical results of the search by adding a term each time, these terms being *Generation*, *Transmission*, *Distribution* and *Consumption*, in the *Smart City Energy* sector.

3.1.1. Energy Generation

The initial search was conducted on Scopus once more, incorporating the term "Generation" into the query. The search parameters were configured as follows: "Smart Cities" AND "Energy" AND "Generation" AND "Artificial Intelligence" OR "AI" (Smart Cities-Energy-

Generation-Artificial Intelligence). This yielded a total of 45 journal publications published since 2013. For this search, 35 sources were examined, involving 200 different authors.

Figure 3 illustrates the yearly scientific output on this subject. Given the relatively limited number of articles yielded by the initial search, the screening process primarily relied on keywords within titles and abstracts. However, a notable average annual growth rate of 42.62% was observed. The year 2023 has the highest publication count, totaling 12 publications. Notably, this observation was made at the midpoint of the year, suggesting the likelihood of additional publications during this period. This potential for further contributions could continue the upward trajectory in the research output.

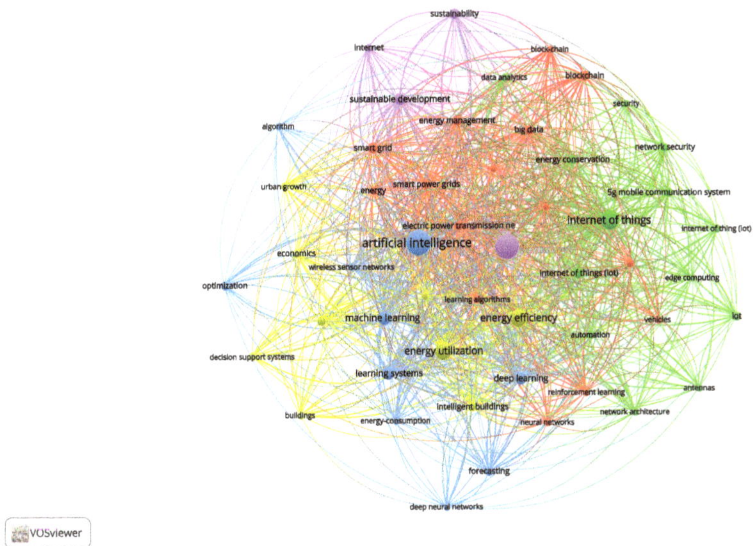

Figure 2. Keyword co-occurrence network for *Smart Cities-Artificial Intelligence-Energy*.

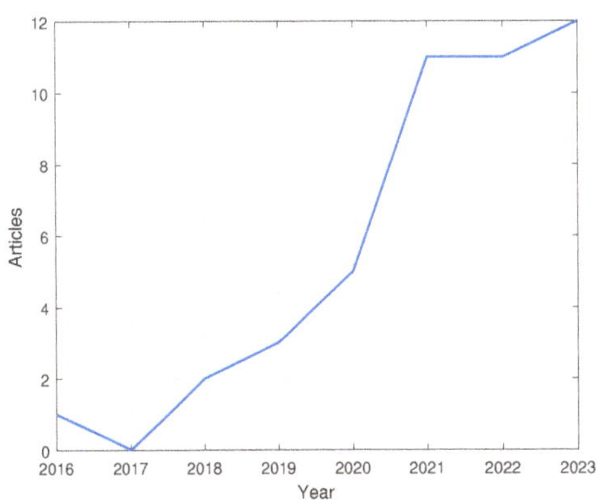

Figure 3. Annual scientific production (articles per year) for *Smart Cities-Artificial Intelligence-Energy-Generation*.

These publications involve authors from a variety of countries, the most prominent nationalities being China (38), India (20), Canada (12), the UK (12), Saudi Arabia (11), Iran (10), South Korea (10), the USA (10), Spain (9), and Japan (7). Additionally, when considering the number of citations received by articles originating from different countries, the leading nations are as follows: USA (423), China (127), South Korea (71), Romania (66), Turkey (36), India (29), South Africa (25), the United Kingdom (18), Qatar (14), and Bulgaria (6).

Table 1 shows the most relevant journals for paper publishing, exhibiting the number of articles published by each one of the journals, ranking in first place "Sustainable Cities and Society" with a total of 4 articles.

Table 1. Most relevant sources for *Smart Cities-Energy-Generation-Artificial Intelligence.*

Rank	Publisher	No. of Articles
1	*Sustainable Cities and Society*	4
2	*IEEE Access*	3
3	*Sensors*	3
4	*Applied Sciences*	2
5	*IET Smart Cities*	2

Table 2 shows the most cited articles worldwide for *Smart Cities-Artificial Intelligence-Energy Generation*. The most cited article is that of Ghadmai et al. with 75 citations, published in 2021 for *Sustainable Cities and Society*.

Table 2. Most globally cited articles published for *Smart Cities-Energy-Generation-Artificial Intelligence.*

Authors, Year	Title	Citations	Source
Ghadami N. et al. [2021] [10]	Implementation of solar energy in Smart Cities using an integration of artificial neural network, photovoltaic system, and classical Delphi methods	75	*Sustainable Cities and Society*
Serban AC. et al. [2020] [11]	Artificial Intelligence for smart renewable energy sector in Europe—smart energy infrastructures for next-generation Smart Cities	66	*IEEE Access*
Azzaoui AE, [2020] [12]	Block5GIntell: Blockchain for AI-enabled 5G networks	55	*IEEE Access*
Lee YL, [2021] [13]	6G massive radio access networks: Key applications, requirements and challenges	52	*IEEE Open Journal of Vehicular Technology*
Zhang, N., [2016] [14]	Semantic framework of Internet of Things for Smart Cities: Case studies	47	*Sensors*

The keyword search for *Smart Cities-Artificial Intelligence-Energy-Generation* yielded the following most common keywords: *Smart Cities* (23), *Artificial Intelligence* (19), *Internet of Things* (16), *Energy Utilization* (14), *Mobile Communication Systems* (11), *Energy Efficiency* (10), *Deep Learning* (9), *Learning Systems* (8), *Decision Making* (7), and *Economics* (7) as shown in Figure 4.

To analyze the word co-concurrency, VOSviewer software was utilized to generate a KCN (keyword co-concurrence network). Figure 5 shows the map obtained from the file with the Scopus search containing the data from articles previously referenced. This map or KCN shows that words like *sustainable development, decision making, and energy* are closely related to *Artificial Intelligence* and *energy generation*. While bigger nodes show the frequency of occurrence of each word as can be seen on the KCN, the biggest nodes (not containing *Smart Cities*) are *Internet of Things, learning systems, deep learning, mobile communications systems, energy utilization* and *energy efficiency*, which have the greatest occurrence, which coincides with the previous keyword analysis.

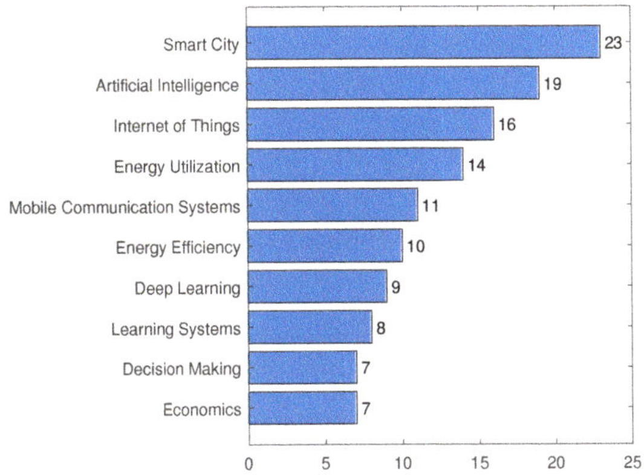

Figure 4. Most relevant words for *Smart Cities-Energy-Consumption-Artificial Intelligence.*

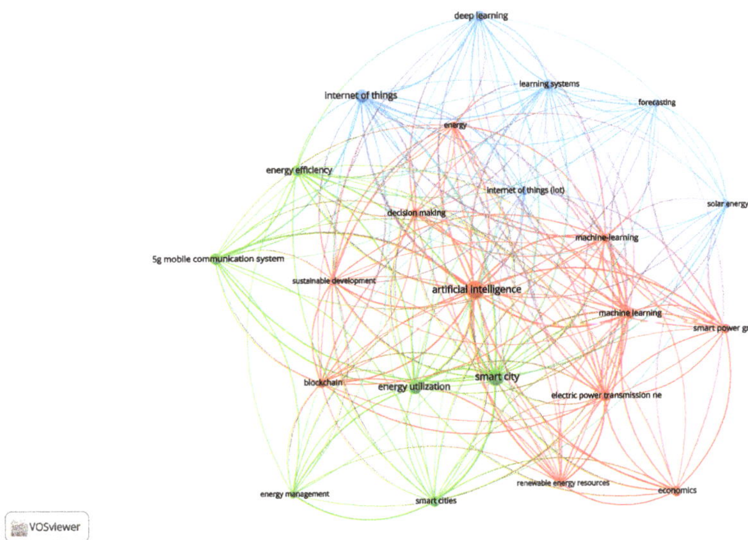

Figure 5. Keyword co-occurrence network for *Smart Cities-Artificial Intelligence-Energy-Generation.*

3.1.2. Renewable Energy Sources within Smart Cities

To highlight the significance of renewable sources in previous years and their prospective impact in the coming years, a subsection was created under the category of Power Generation. This subsection is dedicated to underscoring the contribution of renewable energy sources. A search was performed on Scopus to identify the currently relevant topics and potential future trends in the use of two of the most widely adopted renewable energy sources, namely solar and wind energy, within the context of Smart Cities. The search query employed was as follows: "Smart Cities **AND** Energy **AND** Generation **AND** Solar **AND** Wind" (*Smart Cities-Energy-Generation-Solar-Wind*). For this search, the selected timeframe ranged from 2013 to 2023, and only journal publications in the English language were considered eligible. This search yielded a total of 47 articles, yet only 23 met the specified criteria outlined above. The average annual growth rate for this search is calculated at 1.84% from the start of the chosen period to its conclusion; this trend can be attributed to the publications between 2013 and 2016 showing a declining pattern. However, in subsequent years, there has been a consistent and continuous upward trend in the number of publications. Figure 6 illustrates the annual scientific production, revealing a significant increase in article output from 2013 to 2023. This underscores the recent upward trajectory in the field.

Indeed, it is crucial to recognize that at the time of composing this paper, there may still be forthcoming publications for 2023, rendering the final count for articles in this year subject to potential change. Concerning the search results, the volume of articles published on this topic emphasizes "Energy" as the most prominent source, boasting six articles. "Renewable Energy" closely follows with four articles, and "Energies" holds the third position with three articles, establishing these as the top three journals with the highest number of publications. Furthermore, Table 3 shows other notable sources identified in this search section.

Among the total of 47 published articles, the production per country indicates that India is the largest contributor to this renewable energy section, with 27 authors originating in this country. Saudi Arabia and China are closely behind, with 24 and 20 authors, respectively. Additionally, significant contributions are observed from authors hailing from Italy, with 18 of them. Furthermore, several countries have actively participated in this research area, including South Korea (12), the USA (8), Iran (6), Singapore (6), Canada (5), and the United Arab Emirates (5). This international involvement underscores the global significance and collaborative efforts in advancing knowledge related to renewable energy generation sources for applications in Smart Cities.

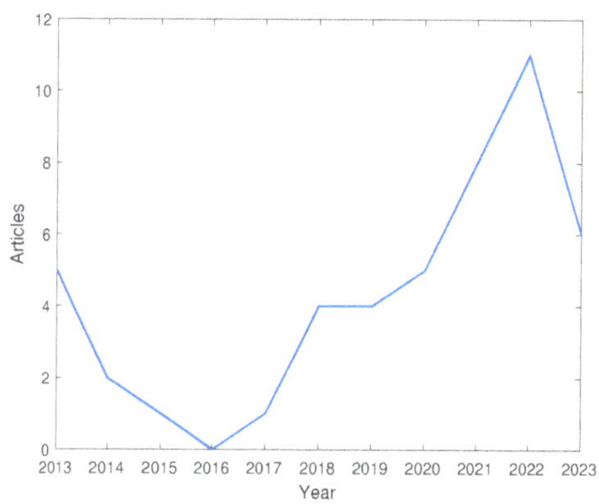

Figure 6. Annual scientific production for *Smart Cities-Energy-Generation-Solar-Wind*.

Table 3. Most relevant sources for *Smart Cities-Energy-Generation-Solar-Wind*.

Rank	Publisher	No. of Articles
1	*Energy*	6
2	*Renewable Energy*	4
3	*Energies*	3
4	*Energy Reports*	3
5	*Applied Energy*	2

Table 4 shows the articles from the search with the most citations; ref. [15] is the one that has the most citations.

From this search made, Figure 7 shows the most relevant words obtained; Renewable Energies (42) and Solar Power Generation (28) are the ones that appear the most in the search. Going deeper into the relevant words, it can also be seen that Smart City (16), Wind Power (16), and Smart Grid (15) also appear frequently in the articles. Economic Analysis (8) and Electric Power Transmission Networks (8) are also important concepts for the introduction of alternative energy sources, as they allow us to know the feasibility of an energy project and energy transportation, respectively. Sustainable Development (7) is also an important concept mentioned, as one of the main goals of these energy sources is to reach this sustainable development in Smart Cities and reduce fossil fuel dependency. Other relevant words are Power Generation (7) and Energy Storage (6).

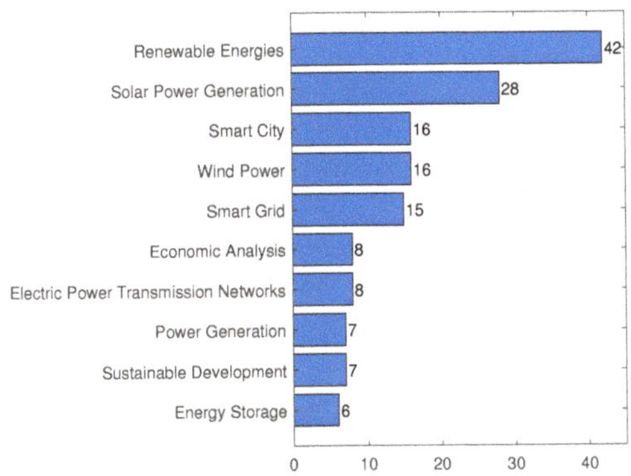

Figure 7. Most relevant words for *Smart Cities-Energy-Generation-Solar-Wind*.

Table 4. Most globally cited articles published for *Smart Cities-Energy-Generation-Solar-Wind*.

Authors, Year	Title	Citations	Source
Ramli Mam et al. [2018] [15]	Optimal sizing of PV/wind/diesel hybrid microgrid system using multi-objective self-adaptive differential evolution algorithm	345	*Renewable Energy*
Yang D. et al. [2013] [16]	Solar irradiance forecasting using spatial–temporal covariance structures and time-forward kriging	121	*Renewable Energy*
Oldenbroek V, et al. [2017] [17]	Fuel cell electric vehicle as a power plant: Fully renewable integrated transport and energy system design and analysis for Smart City areas	76	*International Journal of Hydrogen Energy*
De Luca G. et al. [2018] [18]	A renewable energy system for a nearly zero greenhouse city: Case study of a small city in southern Italy	51	*Energy*
Soliman Ms et al. [2021] [11]	Supervisory energy management of a hybrid battery/PV/tidal/wind sources integrated into a DC-microgrid energy storage system	42	*Energy Reports*

A KCN was obtained using VOSviewer by introducing the information obtained from the Scopus search. Figure 8 illustrates the KCN focusing on solar power generation and its interconnected nodes, including renewable energy, hybrid systems and electric power. Another important concept depicted in the KCN is wind power, highlighting the frequent association between smart grids, Smart Cities, and wind energy generation. This diagram serves as a valuable tool to analyze crucial topics related to the integration of renewable energies into Smart Cities. One notable relationship showcased in this diagram is the energy storage and photovoltaic systems. It exemplifies the essential role of storage devices in supporting the operation of photovoltaic systems. Additionally, the diagram features an economic analysis component closely tied to solar power generation. This economic analysis can be linked to the feasibility assessment of solar photovoltaic (PV) systems from a financial perspective. This type of graphical representation effectively illustrates associations between commonly co-searched keywords. An examination of data extracted from Bibliometrix reveals that topics such as energy utilization, economics, and greenhouse gases are currently capturing the attention of scientists as indicated by their relevance and ongoing development. These subjects are intricately linked to the global issue of climate change and represent a pressing global concern. Furthermore, emerging themes highlighted in the diagram include forecasting, deep learning, solar radiation, and weather stations. This aligns with the consensus that the use of renewable energy sources is key to achieving the Sustainable Development Goals. The prominence of Artificial Intelligence underscores its increasing relevance and growth in the current year. Overall, this KCN offers a comprehensive overview of the interconnections and areas of interest within renewable energy and Smart City development. It serves as a valuable resource for understanding the evolving landscape of sustainable energy solutions and their critical role in addressing global challenges such as climate change.

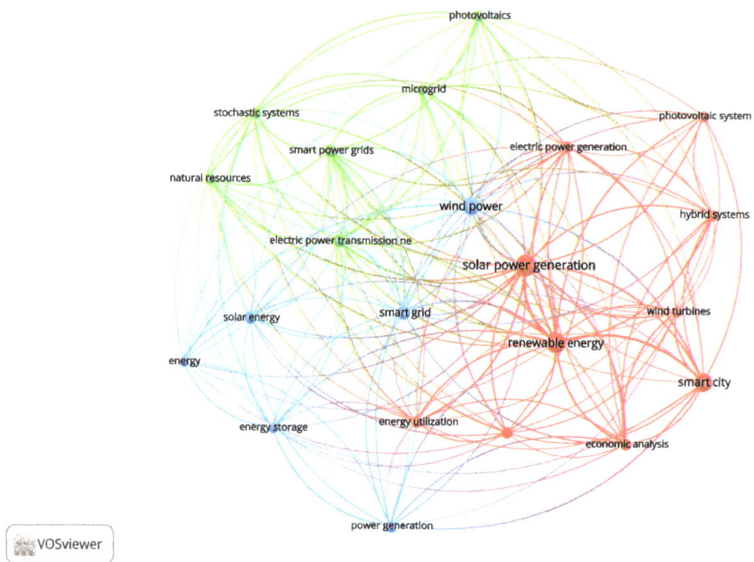

Figure 8. Keyword co-occurrence network for *Smart Cities-Energy-Generation-Solar-Wind*.

3.1.3. Energy Transmission

For the *Transmission* section, a Scopus search was performed by adding the aforementioned term to the base structure, resulting in "Smart Cities **AND** Energy **AND** Transmission **AND** Artificial Intelligence **OR** AI" (*Smart Cities-Energy-Transmission-Artificial Intelligence*. It has a total of 31 articles, with an average annual growth rate of 14 percent from the year 2013 to 2023. Figure 9 shows the Annual Scientific production. It is essential to note that the analysis excludes articles published from 2013 to 2015 for this specific topic or area, as there were no publications during that period. Therefore, the graph and statistics presented focus only on the years following 2015.

As is evident from the analysis, there is a clear tendency to increase the number of publications over time. However, it is noteworthy that there is a notable lack of publications specifically focused on energy transmission. It is essential to acknowledge that this paper's analysis was conducted before the end of 2023, leaving room for additional publications during the remaining part of the year. Therefore, the number of publications for 2023 may still witness growth, potentially shedding more light on the area of energy transmission in the context of Smart Cities. The most relevant sources are presented in Table 5, *IEEE Access*, *Sustainable Cities and Society*, *Wireless Communications and Mobile Computing* being the three sources with the most publications in this field, each having three publications.

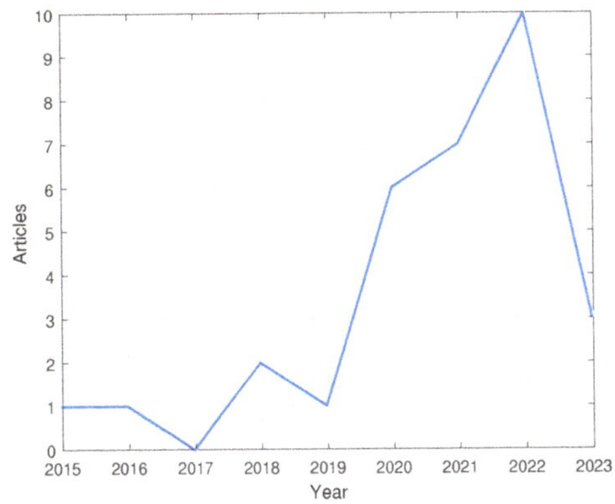

Figure 9. Annual scientific production of energy transmission keyword co-occurrence network for *Smart Cities-Artificial Intelligence-Energy-Transmission*.

Table 5. Most relevant sources for *Smart Cities-Artificial Intelligence-Energy-Transmission*.

Rank	Publisher	No. of Articles
1	*IEEE Access*	3
2	*Sustainable Cities and Society*	3
3	*Wireless Communications and Mobile Computing*	3
4	*Sustainability*	2
5	*Computer Communications*	1

Indeed, it is crucial to highlight that China exhibits the highest number of authorships, with 29 articles featuring at least one Chinese author. Following closely are Germany and Saudi Arabia, each with 11 publications. South Korea and India also demonstrate a significant presence in the research, with 10 and 9 articles published by authors of their respective nationalities, respectively. Furthermore, countries such as France, Lebanon, the United Kingdom, and Malaysia, among others, have contributed to the research in the selected category, although to a lesser extent, each having eight articles. The diverse international participation underscores the global interest and participation in this subject matter.

Table 6 presents the articles with the most citations on this research. The work presented in [3] is the most cited, with 258 citations.

Table 6. Most globally cited articles published for *Smart Cities-Energy-Transmission-Artificial Intelligence*.

Authors, Year	Title	Citations	Source
Ullah N. et al. [2020] [3]	Applications of Artificial Intelligence and Machine learning in Smart Cities	258	*Computer Communications*
Serrano W. et al. [2018] [4]	Digital systems in Smart Cities and infrastructure	66	*Smart Cities*
Aguilar J. et al. [2021] [19]	A systematic literature review on the use of Artificial Intelligence in energy self-management in smart buildings.	53	*Renewable and Sustainable Energy Reviews*
Sharma H, [2021] [20]	Machine learning in wireless sensor networks for Smart Cities: a survey.	52	*Electronics*
Khan N., [2021] [21]	DB-Net: A novel dilated CNN-based multi-step forecasting model for power consumption in integrated local energy systems.	50	*International Journal of Electrical Power*

The most relevant words for *Smart Cities-Artificial Intelligence-Energy-Transmission* are presented using Figure 10, having *electric power transmission networks* (17) as the category with the highest occurrence. Other important concepts can be identified as *smart power grids* (12) and *Internet of Things* (11), which are part of AI and Smart City, which are part of the base structure.

The map obtained from *VOSviewer* with the information from the Scopus search is presented in Figure 11. A relationship between *Smart City* and *Artificial Intelligence* is shown; while *Smart City* is closely related to *Internet of Things* and *Energy Utilization*, on the other hand, AI is related to *Smart Power Grids*, and this word to *Electric Power Transmission networks*. These are interesting relationships, as smart power grids have AI applications in electric power transmission networks.

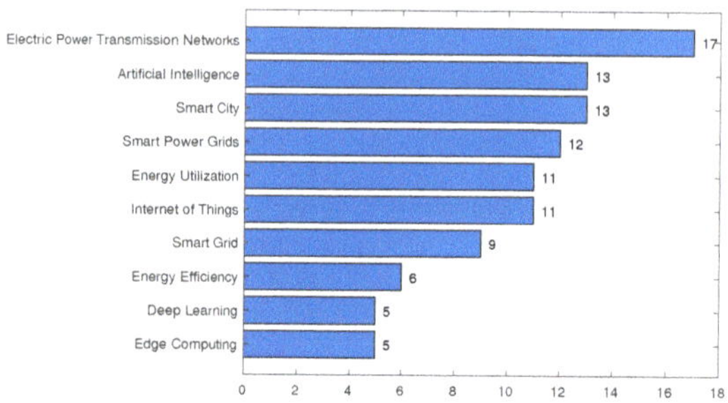

Figure 10. Most relevant words for *Smart Cities-Artificial Intelligence-Energy-Transmission.*

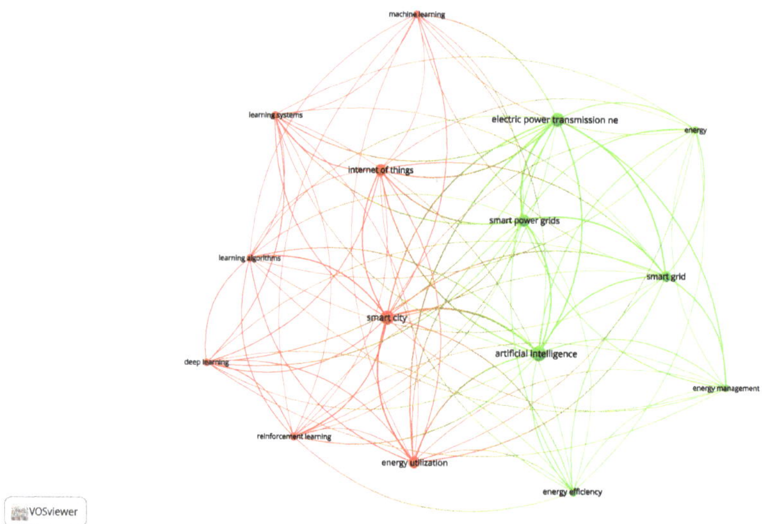

Figure 11. Keyword co-occurrence network for *Smart Cities-Artificial Intelligence-Energy-Transmission.*

3.1.4. Energy Distribution

This time, a search on Scopus for the base structure was performed, adding *Distribution*, resulting in "Smart Cities **AND** Energy **AND** Distribution **AND** Artificial Intelligence **OR** AI" (*Smart Cities-Energy-Distribution-Artificial Intelligence.* A total of 21 articles were found, with an average annual growth rate of 10. 41% from 2013 to 2023. Figure 12 shows the annual scientific production; similarly to the *Energy Transmission* subsection, *Energy Distribution* has no reported articles found by Scopus from 2013 to 2016. For this reason, the previously mentioned years are not considered in Figure 12.

For the *Energy Distribution* topic, an intermittent pattern in the number of publications is identified, with notable fluctuations over the years. Specifically, there was an increase in the number of publications in 2019 and 2021, indicating a surge in research activity during those periods. However, a decrease in publication frequency was observed in 2020 and 2022. Additionally, as mentioned earlier, the year 2023 is subject to the constraint of limited publications in the area of energy transmission and distribution. This intermittent trend in research output underscores the dynamic nature of the field and suggests that further research is required to determine the factors influencing these fluctuations.

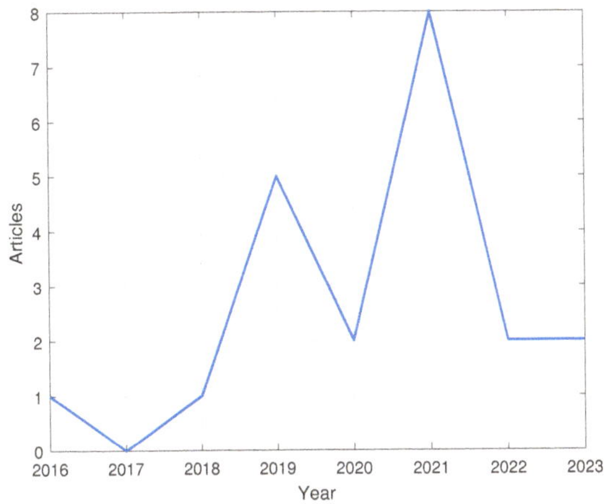

Figure 12. Annual scientific production of energy transmission keyword co-occurrence network for *Smart Cities-Artificial Intelligence-Energy-Distribution*.

After analyzing the number of articles published regarding this topic, the most relevant source was *Sustainability* with four articles, followed by *Applied Sciences* with three. The most relevant sources of this section search are presented in Table 7.

Table 7. Most relevant sources for *Smart Cities-Artificial Intelligence-Energy-Distribution*.

Rank	Publisher	No. of Articles
1	*Sustainability*	4
2	*Applied Sciences*	3
3	*Energy and Buildings*	2
4	*Energy*	1
5	*Future Generation Computer Systems*	1

Upon analyzing production by country, China emerges as the leading contributor in the distribution section, with 18 articles authored by nationals. Following behind are Spain and Lebanon with eight and six articles, respectively. Noteworthy contributions are also observed from India, Japan, and Saudi Arabia, each with six articles, among several other countries that have actively participated in this area of research. The diverse international participation emphasizes the global interest and collaboration in advancing knowledge related to energy distribution within Smart Cities.

In Table 8 are presented the articles with the most citations of this research, with [22] being the most cited, having a total of 182 citations.

The most relevant words obtained for the distribution can be seen in Figure 13, being *Artificial Intelligence* (7) and *Smart Cities* (6). As they are part of the base search, some of the relevant words were *Energy Efficiency* (5) and *Data Analytics* (4). Energy efficiency is an important concept to consider for distribution networks since most articles seek to minimize energy losses to obtain better efficiency. For data analytics, it is important to monitor the distribution of energy, as AI could help to perform this task. Other relevant words are *Economics* (4) and *Deep Learning* (4).

A KCN was obtained using VOSviewer by feeding the information obtained from the Scopus search; Figure 14 shows this KCN. The KCN shows that the main topic is *Artificial Intelligence* which is connected to several nodes, such as *Learning Systems*, *Sustainability*, *Machine Learning*, and *Energy Efficiency*. This last keyword is connected to the second-highest co-occurrence word, which is *Smart Cities*, which shows that AI and Smart Cities are often paired with energy efficiency.

Table 8. Most globally cited articles published for *Smart Cities-Artificial Intelligence-Energy-Distribution*.

Authors, Year	Title	Citations	Source
Le L.T. et al. [2019] [22]	A comparative study of PSO-ANN, GA-ANN, ICA-ANN, and ABC-ANN in estimating the heating load of buildings' energy efficiency for Smart City planning	182	*Applied Sciences*
Idowu S. et al. [2016] [23]	Applied machine learning: Forecasting heat load in district heating system	152	*Energy and Buildings*
Zhou Z. et al. [2019] [24]	Blockchain and computational intelligence inspired incentive-compatible demand response in internet of electric vehicles	91	*IEEE Transactions on Emerging Topics in Computational Intelligence*
Le L.T, [2019] [25]	Estimating the heating load of buildings for Smart City planning using a novel Artificial Intelligence technique PSO-XGBoost.	71	*Applied Sciences*
Ingwersen, P., [2018] [26]	Smart City research 1990–2016.	36	*Scientometrics*

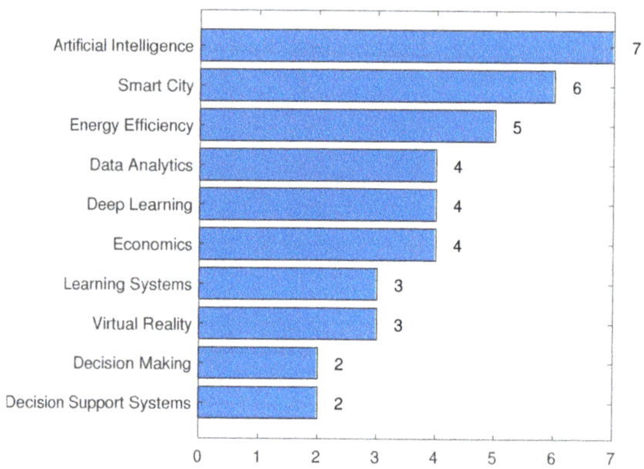

Figure 13. Most relevant words for *Smart Cities-Artificial Intelligence-Energy-Distribution*.

3.1.5. Energy Consumption

A search was conducted on Scopus for articles containing the terms "Smart Cities **AND** Energy **AND** Consumption **AND** Artificial Intelligence **OR** AI" (*Smart Cities-Energy-Consumption-Artificial Intelligence*). The results of the previously mentioned search show a total of 103 documents published from 2013 to 2023, having been published by 60 sources with a total of 432 authors. Figure 15 shows the number of articles published each year since 2014 (there were no publications during 2013); an interesting pattern can be observed in the publication trends. In 2015, there was a significant drop in the number of publications, reaching zero. However, in the subsequent years, there was a steady increase in the number of articles published, indicating a rising trend. Although 2023 shows a lower production number so far, it is essential to consider that the year has not concluded, leaving room for additional articles to be published, potentially continuing the upward trend from previous years.

The average annual growth rate for publications containing the specified terms stands at 35.11%, demonstrating a notable increase in research activity over time. Furthermore, the average number of citations per article is 27.54, highlighting the impact and relevance of the research output in this field. These metrics emphasize the significance and growing interest in the subject of AI applications in the energy sector within Smart Cities.

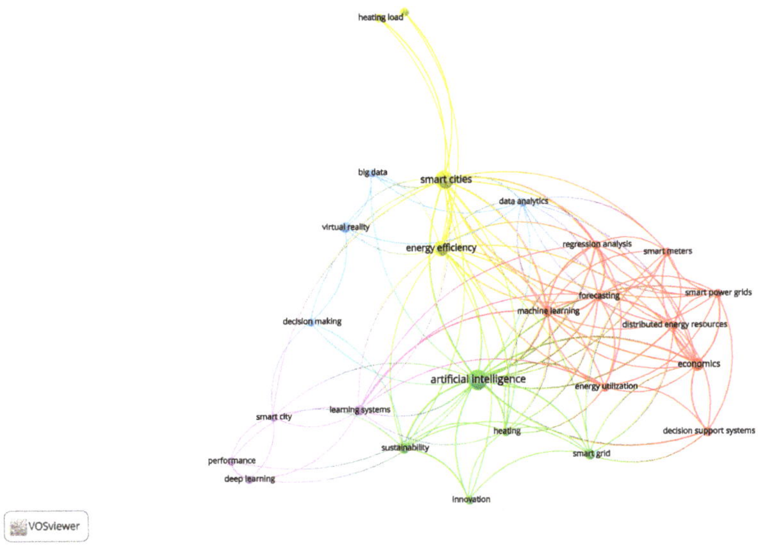

Figure 14. Keyword co-occurrence network for *Smart Cities-Artificial Intelligence-Energy-Distribution*.

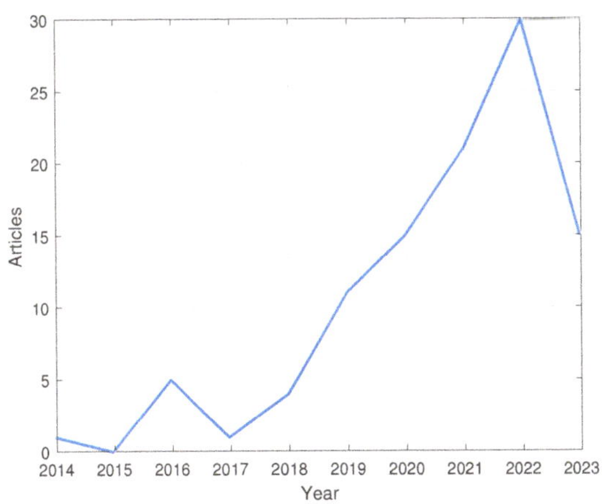

Figure 15. Annual scientific production (articles per year) for *Smart Cities-Artificial Intelligence-Energy-Consumption*.

The most relevant sources are presented in Table 9, IEEE Access (nine publications), Sustainable Cities and Society (with nine articles) and Sustainability (with six papers) are the three sources with the most publications in this field.

The publications encompass authors from various countries, with the most prevalent nationalities among the authors being China (109), the USA (32), India (31), Saudi Arabia (30), Spain (27), the United Kingdom (23), South Korea (21), Italy (11), Japan (11), and Canada (10), which comprise the top ten contributors in this category. In particular, South Korea stands out as the country with the highest number of citations, with an accumulated total of 148 citations for its contributions in this area. This indicates the significant impact and recognition of research conducted by authors from South Korea in the field of AI applications in the energy sector within Smart Cities. The wide array of contributing nations reflects the global interest and collaboration in advancing knowledge in this domain.

Table 9. Most relevant sources for *Smart Cities-Artificial Intelligence-Energy-Consumption*.

Rank	Publisher	No. of Articles
1	*IEEE Access*	9
2	*Sustainable Cities and Society*	9
3	*Sustainability*	6
4	*Energies*	5
5	*Sensors*	5

Table 10 shows the most globally cited articles for the period comprehended from 2014 to 2023 based on the search *Smart Cities-Artificial Intelligence-Energy-Consumption*. The article with the most citations is [27] with a total of 423 citations, which was published in *Applied Energy* in the year 2018.

Table 10. Most globally cited articles published for *Smart Cities-Artificial Intelligence-Energy-Consumption*.

Authors, Year	Title	Citations	Source
Rahman A., [2018] [27] *	Predicting electricity consumption for commercial and residential buildings using deep Recurrent Neural Networks	423	*Applied Energies*
Ullah, Z., [2020] [28]	Applications of Artificial Intelligence and machine learning in Smart Cities	258	*Computer Communications*
Alsamhi, S. [2019] [29]	Survey on collaborative smart drones and Internet of Things for improving smartness of Smart Cities	195	*IEEE Access*
Vázquez-Canteli, J. R., [2019] [30]	Fusing TensorFlow with building energy simulation for intelligent energy management in Smart Cities	113	*Sustainable Cities and Society*

* ref. [27] also appears in the search for energy generation but the topic better fits energy consumption.

The search for keywords related to *Smart Cities-Artificial Intelligence-Energy-Consumption* provided the following results: *Energy utilization* (73), *Artificial Intelligence* (56), *Smart City* (52), *Energy Efficiency* (30), *Internet Of Things* (25), *Learning Systems* (18), *Deep Learning* (15), *Sustainable Development* (14), *Decision Making* (10) and *Economics* occupy the top ten keywords results for this search.

As in all previous subsections, a KCN was performed using VOSviewer software, allowing to visualize the most frequent occurrence of a word and how often words are searched together. As previously mentioned, it can be corroborated using Figure 16. The analysis of the keyword frequency of occurrence reveals that terms such as "energy utilization", "energy efficiency", "Artificial Intelligence ", "learning systems", and "deep learning" are among the most frequently mentioned keywords. In the visual representation of the data, the larger nodes correspond to terms with the highest frequency of occurrence. On the other hand, terms with more links represent keywords with a greater co-occurrence with other terms. This graphical representation provides valuable insights into the prominence and relationships among the key concepts in the research domain of AI applications in the energy sector within Smart Cities.

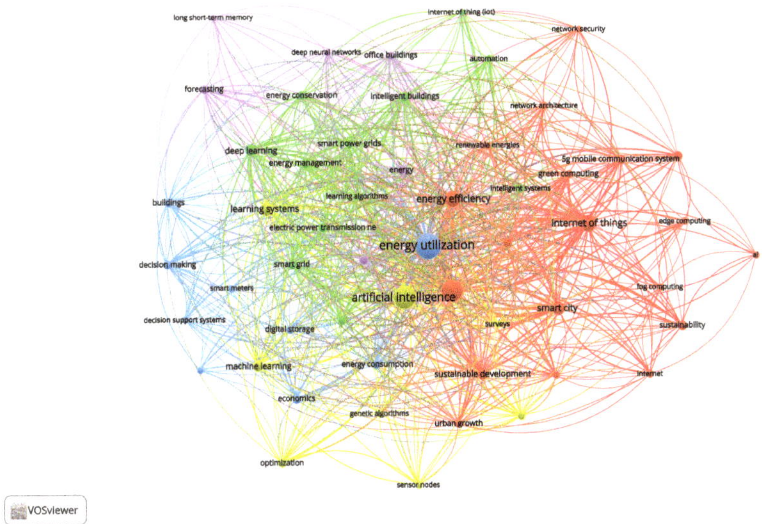

Figure 16. Keyword co-occurrence network for *Smart Cities-Artificial Intelligence-Energy-Consumption*.

4. Discussion

This section delves into a more detailed analysis of the selected papers that met all the specified eligibility criteria. Using the PRISMA methodology, this paper aims to provide a comprehensive overview of the existing literature on AI applications in the energy sector within Smart Cities. This systematic review allows to identify common patterns, consistencies, and trends in each category. By doing so, the reader can gain valuable insight into the advancements and developments within these research areas. The following subsections will present the findings for each of the topics in a structured and coherent manner.

Table 11 shows the articles obtained after screening and will be referred to in the following sections. The table presents the rich landscape of research in Smart Cities and energy applications, showcasing the evolution of methodologies and technologies over time. The diversity of the methods with which researchers have approached the areas described previously varies from algorithms of machine learning, fuzzy logic, optimization, and structural analyses; this diversity indicates the interdisciplinary nature of the research performed in the Smart Cities context and the application that AI has in the same area. Future research could focus on validating and standardizing approaches to ensure the compatibility and reproducibility of results in terms of real-world implementation. Possible areas of exploration include the scalability of the proposed solutions, addressing challenges in deployment and integration into existing urban infrastructure. Articles from the early years (2018–2020) showcase a keen interest in topics such as wind characterization, digital systems for transmission networks, and microgrid management. As emerging technologies continue to evolve, IoT technologies are becoming increasingly popular. This trend is evident in the table, where a search for the frequency of the term in the "Method" column shows a total of 13 occurrences, starting from the year 2020. Climate change is a pressing issue globally, as evidenced by the increasing number of papers on renewable energy being integrated into Smart Cities. It is highly recommended to continue research on renewable energy sources since this topic is still in its nascent stages and remains critical for achieving long-term energy goals and the Sustainable Development Goals (SDGs). The articles related to renewables focus on wind turbines, solar energy, hydrogen production, and renewable energy in high greenhouse-gas-emitting sectors. Methods include structural and modal analyses, fuzzy methods, social and political analysis, and optimization algorithms.

Table 11. Articles that fulfilled all eligibility criteria.

Area of Application	Author(s)	Year	Category	Method	Source
Generation	Miyasawa A. et al. [31]	2023	Demand forecasting	Smart metering, nonparametric regression models.	*IET Smart Cities*
Generation	Shafiullah M. et al. [32]	2023	Energy Systems	Artificial Intelligence, IoT	*Smart Cities*
Generation	Wu Z. et al. [33]	2023	Wind power forecasting	Probabilistic Physics-informed AI for completing dataset caused by ocassional shutdwon	*CMES—Computer Modeling in Engineering and Sciences*
Transmission	Fakhar A. et al. [34]	2023	Smart Grids with Renewable Energy	Cloud Computing, IoT, Blockchain	*International Journal of Green Energy*
Consumption	Bayer D. and Pruckner M. [35]	2023	Energy Systems in Buildings	Digital Twin	*Energy Informatics*
Consumption	Alymani M. et al. [36]	2023	Forecasting energy consumption	Stacked Autoencoder (SAE), Deep Neural Network (DNN), Bidirectional Long Short-Term Memory (BiLSTM)	*Sustainable Energy Technologies and Assessments*
Consumption	Al-Habaibeh A. et al. [37]	2023	Estimate crowds in cities	Internet of Things (IoT)	*Ain Shams Engineering Journal*
Consumption	Selvaraj R. et al. [38]	2023	Energy consumption management	Artifitial Intelligence Technique for Monitoring Systems in Smart Buildings (AIMS-SB)	*Sustainable Energy Technologies and Assessments*
Consumption	Feng Y. et al. [39]	2023	Energy Saving	Reinforcement Learning	*IEEE Access*
Consumption	Jiang R. et al. [40]	2023	Demand Prediction	Deep-chain echo state network (DCESN)	*IEEE Transactions on Industrial Informatics*
Consumption	AlHajri I. et al. [41]	2023	Urban Planning	Long-Short Term memory networks	*Energy*
Renewables	Fantin Irudaya Raj E. et al. [42]	2023	Wind turbines in smart cities	Structural, modal, and harmonic analyses performed using ANSYS	*MRS Energy and Sustainability*
Generation	Khan N. et al. [43]	2023	Power Generation Forecasting	Multi-Head Attention (MHA)-based deep Autoencoder(AE) with Extreme Gradient Boosting (XBG) algorithm	*IEEE Internet of Things Journal*
Renewables	Ulpiani G. et al. [44]	2023	Renewable energy in high GHG's emitting sectors in cities	Social and Political analysis	*Renewable and Sustainable Energy Reviews*
Renewables	Kedir N. et al. [45]	2023	Solar/PV Systems	Fuzzy Hybrid Methods	*Energies*
Consumption	Icaza-Alvarez D. et al. [46]	2023	Estimation on the power demand	Energy Plan tool	*Energy Reports*
Generation	Moon J. et al. [47]	2022	Electrical load forecasting	Explainable Electrical Load Forecasting (XELF) Methodology	*Sustainable Energy Technologies and Assessments*
Transmission	Said D. [48]	2022	Demand-Side Management	Big data, Blockchain, Machine learning (ML), IoT	*IEEE Engineering Management Review*
Generation	Heidari A. et al. [49]	2022	Smart cities power and energy management	Convolutional Neural Network (CNN) Long Short-Term Memory (LSTM)	*Sustainable Cities and Society*
Generation	Chang E.-C. et al. [50]	2022	PV tracking and control	Finite Time Terminal Attractor (FFTA)	*Wireless Communications and Mobile Computing*
Transmission	Liu Z. et al. [51]	2022	Smart Power Grids, Power Systems	Machine learning (ML)	*Energy Reports*
Transmission	Khosrojerdi F. et al. [52]	2022	Smart Grid	Artificial Intelligence Analytics (AIA)	*International Journal of Energy Sector Management*

Table 11. *Cont.*

Area of Application	Author(s)	Year	Category	Method	Source
Consumption	Chavhan S. et al. [53]	2022	Energy-efficient transport	AI-IoT System	*ACM Transactions on Internet Technology*
Consumption	Al-Hawawreh M. et al. [54]	2022	Smart decision making	Deep Reinforcement Learning (DRL)	*IEEE Sensors Journal*
Consumption	Singh S. et al. [55]	2022	Clustering for Wireless Sensors Networks	Improved gray wolf optimization (IGWO)	*Sensors*
Consumption	Huang J. et al. [56]	2022	Building energy forecast	Three ML algorithms (SVR, XGBoost, and LSTM)	*Applied Sciences*
Consumption	Mohamed H. et al. [57]	2022	Reduce energy consumption	TOPSIS fuzzy	*Electronics*
Consumption	Ren Y. et al. [58]	2022	Data management in energy consumption	Quantum-inspired Reinforcement Learning (QRL)	*IEEE Transactions on Green Communications and Networking*
Consumption	Murthy Nimmagadda S. and Harish K.S. [59]	2022	Building smart cities	Internet of Things (IoT), Connectivity, Cloud computing and AI	*Multimedia Tools and Applications*
Consumption	Islam N. et al. [60]	2022	Data management	Secured protocol with collaborative learning for IoT using AI techniques	*Sustainability*
Consumption	Zamponi M.E. and Barbierato E. [61]	2022	Forecast energy consumption	Different AI algorithms	*Smart Cities*
Consumption	Naveed Q.N. et al. [62]	2022	Transportation data management	Improved phase timing optimization (IPTO)	*Sensors*
Consumption	Akkad M.Z. et al. [63]	2022	Energy consumption and emissions	IoT, Smart bins, multi-percentage sensors	*Designs*
Consumption	Garlik B. [64]	2022	Energy consumption reduction in buildings	Artificial Intelligence with IoT	*Applied Sciences*
Consumption	Saba D. et al. [65]	2022	Smart home electricity management	Decision-making tool (IRRHEM)	*Applied Sciences*
Consumption	Zaimen K. et al. [66]	2022	Wireless Sensor Networks	Generic algorithm, particle swarm optimization, flower pollination, and ant colony optimization	*IEEE Access*
Consumption	Serrano W. [67]	2022	Smart Buildings	Neural networks with deep learning structure	*Neural Computing and Applications*
Renewables	Li J. et al. [68]	2022	Hydrogen production and conversion	Fuzzy Methods	*Sustainable Cities and Society*
Renewables	Doosti R. et al. [69]	2022	Industrial Building renewable energy	CPLEX Solver	*IET Smart Cities*
Renewables	Vyas M. et al. [70]	2022	Urban Space Utilization	Use of PV Trees on urban areas	*Renewable Energy*
Renewables	AlHammadi A. et al. [71]	2022	Hybrid renewable systems for vehicle charging	Hybrid Optimization of Multiple Energy Resources	*Energies*
Renewables	Balabel A. et al. [72]	2022	Solar energy in building sectors	Solatube technology analysis	*Alexandria Engineering Journal*
Renewables	Nuvvula R.S.S. et al. [73]	2022	Optimal configuration of PV and wind conversion system	Particle Swarm Optimization	*Sustainable Energy Technologies and Assessments*
Renewables	Ponce P. et al. [74]	2022	Solar/PV Systems	Fuzzy TOPSIS	*Energies*
Generation	Konhäuser W. [75]	2021	Local energy generation	Blockchain Technology (BCT)	*Wireless Personal Communications*
Generation	Pérez-Romero Á. et al. [76]	2021	Operation and Maintenance of PV plants	Five AI-based models	*Applied Sciences*

Table 11. *Cont.*

Area of Application	Author(s)	Year	Category	Method	Source
Generation	Zhou H. et al. [77]	2021	PV energy generation forecasting	Hybrid Deep Learning	*Wireless Communications and Mobile Computing*
Generation	Saini G.S. et al. [78]	2021	Resource Management	Fuzzy Logic	*Recent Advances in Computer Science and Communications*
Transmission	Antonopoulos I. et al. [79]	2021	Smart Grid Smart City (SGSC) project	Energy Demand Response Modeling	*Energy and AI*
Distribution	Wang K. et al. [2]	2021	Decision making in Smart Cities	IoT and Artificial Intelligence	*Sustainability*
Distribution	Calamaro N. et al. [80]	2021	Energy losses detection	Energy fraud detection algorithm	*Sustainability*
Consumption	Manman L. et al. [81]	2021	Energy efficiency	Distributed Artificial Intelligence (DAI)	*Sustainable Cities and Society*
Consumption	Li J. et al. [82]	2021	Decision Support Systems	Internet of Things (IoT)	*Computers and Industrial Engineering*
Consumption	Ghadami N. et al. [10]	2021	Forecast energy consumption	Artificial Neural Network (ANN)	*Sustainable Cities and Society*
Consumption	Cirella G.T. et al. [83]	2021	Smart Electricity	Different AI algorithms	*Energies*
Consumption	Hu Y.-C. et al. [84]	2021	Energy decomposition in smart meter	Neuro-fuzzy classifier	*Processes*
Consumption	Mahmood D. et al. [85]	2021	Energy Management	Demand Side Management (DSM)	*International Journal of Advanced and Applied Sciences*
Consumption	Wang X. et al. [86]	2021	Energy consumption of ac-grid system	Fuzzy Logic	*Journal of Intelligent and Fuzzy Systems*
Generation	Kanase-Patil A.B. et al. [87]	2020	Power generation in Smart City	Different AI algorithms	*Environmental Technology Reviews*
Generation	Serban A.C. and Lytras M.D. [11]	2020	Renewable Energy in Smart Cities	Different AI algorithms	*IEEE Access*
Generation	Jiang Y. et al. [88]	2020	Improvement of Urban Development	IoT	*IEEE Access*
Transmission	Cheng Y. et al. [89]	2020	Demand Forecasting	Neural Network	*IET Smart Cities*
Transmission	Ullah Z. et al. [3]	2020	Energy Efficiency of Smart Grids	Machine learning (ML), Deep Reinforcement Learning	*Computer Communications*
Distribution	Loose N. et al. [90]	2020	Smart grids, energy networks	Unified energy agent	*Sustainability*
Distribution	Fattahi J. et al. [91]	2020	Financial Resources	Distributed Energy Management System (DERMS)	*Sustainable Cities and Society*
Consumption	Sharma S. [92]	2020	Infrastructure	Smart vs Intelligent comparison	*International Journal of Advanced Research in Engineering and Technology*
Consumption	Marinakis V. et al. [93]	2020	Energy consumption reduction	Novel framework with reward schemes	*Sensors*
Consumption	Ullah A. et al. [28]	2020	Energy consumption prediction	Clustering based analysis	*Sensors*
Consumption	Shah A.S. et al. [94]	2020	Energy consumption in smart buildings	Bat algorithm, fuzzy logic	*IEEE Access*
Consumption	Azzaoui A.E. et al. [12]	2020	Energy saving	Blockchain and AI	*IEEE Access*
Consumption	Guo Y. et al. [95]	2020	Minimizing cost of energy consumption	Ant colony optimization (ACO)	*IEEE Access*
Consumption	Zhuang H. et al. [96]	2020	Building energy management	ANN and Fuzzy Logic Controller	*Environmental Modeling and Software*

Table 11. *Cont.*

Area of Application	Author(s)	Year	Category	Method	Source
Renewables	Algieri A. et. al. [97]	2020	Biofuel Rankine Cycle combined with Renewable sources	Multivariable optimization	*Energies*
Renewables	Kumar D. [98]	2020	Urban energy system	Simulation of hybrid urban renewable energy systems	*Energy Exploration and Exploitation*
Generation	Aghajani D. et al. [99]	2019	Wind Turbines Monitoring	Geographic information system and RETSCREEN software	*International Journal of Environmental Science and Technology*
Generation	Oun A. et al. [100]	2019	Energy rationalization of a steel plant	AI, Smart Metering (SM), Automated Decision Making (ADM)	*International Journal of Advanced Computer Science and Applications*
Consumption	Salehi H. et al. [101]	2019	Structural Health Monitoring Energy Consumption	Data mining with pattern recognition, an innovative probabilistic approach, and machine learning	*Expert Systems with Applications*
Consumption	Dong Y. et al. [102]	2019	Energy consumption management	Fairness cooperation algorithm (FCA)	*IEEE Internet of Things Journal*
Consumption	Krayem A. et al. [103]	2019	Energy consumption prediction	Archetypal classification	*Energy and Buildings*
Consumption	Marin-Perez R. et al. [104]	2019	Energy consumption improvement	PLUG-N-HARVEST architecture	*Sensors*
Consumption	Vázquez-Canteli J.R. et al. [30]	2019	Energy saving and demand response	Advanced machine learning algorithms	*Sustainable Cities and Society*
Consumption	Aymen F. and Mahmoudi' C. [105]	2019	Electric Vehicles energy	Support vector classification	*Energies*
Renewables	Khoury D. and Keyrouz F. [106]	2019	Wind and Solar power forecasting	Convolutional Neural Network (CNN)	*WSEAS Transactions on Power Systems*
Consumption	Alhussein M. et al. [107]	2019	Microgrid energy management	Deep learning model	*Energies*
Consumption	Risso C. [108]	2019	Energy demand control	Combinatorial optimization dispatch models	*Revista Facultad de Ingeniería*
Generation	Njuguna Matheri A. et al. [109]	2018	Waste quantification for biofuel	Simple Multi Attribute Rating (SMART) of Multiple Criteria Decision Analysis (MCDA)	*Renewable and Sustainable Energy Reviews*
Transmission	Serrano W. [4]	2018	Digital Systems, Transmission Networks	Digital as a Service (DaaS), IoT, Blockchain, Virtual Reality	*Smart Cities*
Consumption	Chui K.T. et al. [110]	2018	Electricity consumption	Hybrid genetic algorithm support vector machine kernel learning approach (GA-SVM-MKL)	*Energies*
Consumption	Rahman A. et al. [27]	2018	Medium-long term energy predictions	Recurrent Neural Network (RNN)	*Applied Energy*
Renewables	Ramli M.A.M. et al. [27]	2018	Microgrid Systems	Multi-Objective Self-Adaptive Different Evolution (MOSaDE) algorithm	*Renewable Energy*
Renewables	Ashfaq A. and Ianakiev A. [111]	2018	Technical and cost design to decarbonize the heating network	Large Scale heat pump and thermal heat storage	*Energy*
Generation	Laiola E. and Giungato P. [112]	2018	Wind characterization	Statistical methods used for meteorological and economic data	*Journal of Cleaner Production*
Renewables	Oldebroek V. et al. [17]	2017	Fully Renewable integrated transport	An energy balance and cost analysis	*International Journal of Hydrogen Energy*

Table 11. *Cont.*

Area of Application	Author(s)	Year	Category	Method	Source
Transmission	Rekik M. et al. [113]	2016	Smart Grid	Ant Colony Optimization (ACO)	*Sustainable Cities and Society*
Consumption	De Paz J.F. et al. [114]	2016	Optimization of energy consumption and cost	Artificial neural networks (ANN), multi-agent systems (MAS)	*Information Sciences*
Consumption	Peña M. et al. [115]	2016	Energy inefficiencies detection in smart buildings.	Data mining	*Expert Systems with Applications*
Consumption	Huang J. et al. [116]	2016	Energy consumption of train	Decision tree, data mining	*IEEE Transactions on Computers*
Consumption	Sędziwy A. and Kotulski L. [117]	2016	Lighting system energy consumption	Dynamic street lighting control	*Energies*
Transmission	Lützenberger M. et al. [118]	2015	Energy in Smart Cities	Distributed Artificial Intelligence Laboratory	*Journal of Ambient Intelligence and Humanized Computing*
Consumption	Fernández C. et al. [119]	2014	Energy consumption modeling	Automated vacuum waste collection (AVWC)	*Sustainability*
Consumption	Yang D. et al. [16]	2013	Solar irradiance forecasting	Time-Forward kriging	*Renewable Energy*
Generation	Park C.J. et al. [120]	2013	Forecasting of renewable energy	Spatio-Temporal correlation	*International Journal of Multimedia and Ubiquitous Engineering*

4.1. Energy Generation

The Scopus search yielded a total of 46 articles as a result of the initial exploration. After filtering the search to focus solely on articles with a primary emphasis on energy generation, a total of 17 articles were identified that met all eligibility criteria and were not duplicated in other sections. These 17 articles will form the basis of our analysis and in-depth examination of the advancements and research trends related to energy generation in the context of AI applications in Smart Cities.

For the renewable energy section, a total of 48 articles were obtained, but after filtering these articles, choosing just the ones that were mostly related to renewable energy generation in Smart Cities, a total of 24 articles were obtained. These articles are added to the energy generation section in order to have a more complete review.

Energy generation forecasting, particularly of alternative sources, emerged as one of the main topics identified in this section. The selected articles delve into the application of AI methodologies for predicting energy generation from renewable and sustainable sources. This area of research is of significant importance in the context of Smart Cities, as it contributes to enhancing the integration and utilization of alternative energy sources to meet the growing energy demands sustainably. The analysis of these articles will provide valuable insights into the advancements and challenges in accurately forecasting energy generation from renewable sources, enabling better energy planning and management in Smart City environments. In [31], the authors developed a method based on non-parametric regression models that forecasts the demand and generation of energy with information provided by smart meters. Another application for forecasting purposes can be reviewed in [33], where a physics-informed AI is applied that forecasts wind power generation, with information on a wind farm in China and ML methods. Reference [45] shows an extensive review of Fuzzy Hybrid Methods, future possible challenges, and opportunities in this sector. This study shows that combining fuzzy logic systems can enhance the efficiency of solar energy applications, and the approach outlined in this research can be applied to investigate various renewable energy sources.

Another application found was the Maximum PowerPoint Tracking of photovoltaic panels, which is the point where solar PV panels produce the maximum energy possible, and the tracking of this point increases their energy efficiency; in [50], a Finite-Time Terminal Attractor (FFTA) is combined with Gradient Particle Swarm Optimization (GPSO) to track MPPT of a solar PV system. Another application in solar energy can be found in [10], in which an Artificial Neural Network (ANN) is applied to create a decision-making tool based on the generation and consumption of solar PV systems that can aid decision makers in creating strategies towards energy generation; these strategies include reducing costs and/or maximizing solar energy generation.

Hence, a trend that can be noticed is that AI has been applied for energy generation forecasting and also to increase the energy efficiency from PV systems, which increases the power generation. AI helps to manage large amounts of data to predict these forecasting models and also manages generation data to track the MPPT for PV panels. In addition, AI has been applied in decision-making tools that consider generation and consumption to provide alternatives that can bring benefits in terms of economic and energy efficiency.

In addition, the study presented in [121] primarily concentrates on employing fuzzy decision-making tools in supply chain management rather than tackling energy issues directly. The paper mainly emphasizes developing and applying dynamic spherical fuzzy aggregation operators (AOs) for multi-period decision making (MPDM) within supply chain management. This concentration differs significantly from typical areas of interest in energy-related studies in Smart Cities, which generally include energy generation, distribution, consumption, efficiency, renewable sources, and sustainable energy practices. Furthermore, the technical approach, involving mathematical models and algorithms for decision making, is designed to manage uncertainty and imprecision in these processes. Although these methodologies benefit their respective fields, they do not contribute directly to advancing energy systems or technologies in Smart Cities, particularly those utilizing AI in the energy sector. The case studies and examples provided in the paper focus on supply chain management, an essential aspect of urban systems. However, they do not cover critical elements of energy management in Smart Cities, such as energy generation, conservation, or optimization strategies, often involving smart grids and the integration of renewable energy.

The paper [122] shows a unique approach, focusing on and contributing to energy management and sustainability. It searches the expansion of renewable energy, exploring its applications and open research problems. It emphasizes the evolution, theoretical underpinnings, and practical applications of various renewable energy sources, including solar, wind, bioenergy, hydraulic energy, and waste-to-energy conversions. Conducting an extensive literature review, it covers developments from 2010 to 2022 and proposes innovative ideas and cost-effective models for implementing renewable energy across different sectors.

In contrast, this review paper is oriented towards integrating Artificial Intelligence (AI) in the energy sector within Smart Cities. It employs the PRISMA methodology and the Scopus database to review the literature, systematically analyzing 122 peer-reviewed articles. This review underscores the pivotal role of AI technologies in improving urban energy processes across four key domains: generation, transmission, distribution, and consumption. It identifies trends, high-impact articles, and emerging keyword clusters, providing a comprehensive analytical framework on the multifaceted applications of AI in the energy sectors of Smart Cities.

While the study in [122] is aimed primarily at researchers and practitioners in the field of renewable energy, offering innovative solutions and pinpointing open areas for future research, this review paper is geared towards policymakers, urban planners, and researchers. It emphasizes the use of AI to achieve sustainable and efficient energy management in Smart Cities. Although both articles contribute to the larger conversation on energy management and sustainability, they have distinct focal points: [122] centers on the various applications of renewable energy, while this review paper focuses on the transformative impact of AI in managing energy resources in the context of Smart Cities.

4.2. Energy Transmission

After conducting a thorough search using the method described above, a total of 31 articles were obtained for this section. The filters applied considered both energy and transmission; however, it is worth noting that some of the results were centered on IoT networks and data transmission (not energy transmission). To ensure precision, only articles that focused on energy transmission in their abstract were considered, resulting in a total of 10 relevant articles.

In [34], the authors elaborate on various concepts related to energy transmission lines in Smart Cities, particularly with the introduction of smart grids. Smart grids are advanced systems that integrate automation, data transmission, and energy monitoring at each stage of the energy supply chain, from generation to final consumption. This integration facilitates seamless communication and coordination between different components of the energy transmission network within a Smart City. This paper emphasizes that concepts such as the Internet of Things (IoT) and cloud computing play a vital role in enhancing energy transmission. These technologies enable efficient data management, as they can handle vast amounts of information exchanged between various energy transmission networks within a Smart City. By leveraging IoT and cloud computing, smart grids optimize the management and distribution of energy resources, contributing to the achievement of Smart Cities' objectives. The implementation of smart grids in energy transmission is pivotal for effectively carrying energy from its source to the end-users consumption. The integration of these advanced systems enables greater energy efficiency, sustainability, and alignment with the goals of Smart City development. Smart grids integration is also mentioned in [3,52,79], so there is a close connection between smart grids and energy transmission for AI applications in Smart Cities.

In [89], it is stated that in order to have an appropriate electric transmission, an anomaly detection system is required to avoid power losses. In this article, an AI method called PowerNet, which is based on neural networks, is proposed to detect anomalies for electricity theft detection in the smart grids, so the AI could make a contribution to automatically detect these problems and report them. Even though the main contribution of this article is regarding energy forecasting, the proposed AI method is explored in this article, and its application for transmission is reviewed.

The work presented in [48] mentions several methodologies for the demand side management (DSM) application in power grids, which is again mentioned for energy transmission and the application of IOT, but also introduces the SM, which is a mixture between software and hardware that helps to give real-time data, with valuable precision, about the energy consumption, which can be useful for monitoring purposes and home area networks (HANs), which are utilized to connect electric devices at home.

4.3. Energy Distribution

Using the methodology this time for *Energy Distribution*, a total of 21 relevant articles were obtained. Among these, some discussed the heating load in buildings, some others focused on electric vehicles, which briefly mentioned the Smart City concept. To narrow down the selection of the articles, the criterion used was to include articles whose abstract prominently features energy distribution as one of the main topics. Upon careful review, it was found that only four papers either focused solely on energy distribution or made significant contributions in this area. These selected articles will be subject to a thorough examination to identify their key findings, contributions, and emerging trends in the field of energy distribution within Smart Cities. Although the number of articles in this category may be relatively small, their significance lies in the valuable insights they provide regarding energy distribution strategies and advancements in the context of smart cities.

The work presented in [2] shows Artificial Neural Networks (ANNs) and machine learning algorithms like Support Vector Machine (SVM) to train energy prediction models that could contribute to obtaining the amount of energy consumption of buildings in Smart Cities, helping distribution systems to be more efficient.

From the literature review, in [90], a multi-agent-based simulation of the distribution networks of a city in Germany reached the objective of the article of successfully simulating the dynamic model of the city energy network, and this simulation was compared to professional simulation tools, which had a relative error lower than 0.0000084%. Another important approach to the application of AI in energy distribution is present in [80], which explains the development of an algorithm to solve energy fraud detection to minimize energy loss in the electricity grid. This algorithm used Convolutional Neural Networks (CNNs) and Robotic Process Automation (RPA) to detect precise fraud identification in electricity networks, the main objective of which was to separate electricity fraud from the many other anomalies that could be presented in the network. The last article reviewed in this section is the work presented in [91], which is considered an energy internet architecture to reach an economic mechanism for clean energy management. This work is not completely focused on an energy distribution approach, but the work could contribute to clean energy integration, which is an important part of Smart Cities. Indeed, a noticeable trend can be observed in the area of energy distribution. This trend revolves around three key aspects: simulating distribution networks, detecting faults in distribution networks, and predicting energy.

4.4. Energy Consumption

The search conducted for the base terms followed by energy consumption resulted in a total of 103 articles. After the screening, 49 articles of this set fulfilled the eligibility criteria, and said articles are listed in Table 11 under *Consumption* Area of Application. Common topics among the reviewed articles were forecasting, data mining, economy, energy management, user profiling, behavior modeling, electric vehicles, and computing.

A paper that alludes to forecasting found during this research is [35], which proposes a smart meter time series from generation sources, the electric grid, and localized buildings to attain a *Digital Twin* of the entire system with the added benefit of allowing geospatial information to be fed to the *twin*. The results show that when geospatial data are not available, a 7% overestimation of the grid level is performed during the summer days. A different approach to forecasting the energy consumption of individual residential households is the one proposed in [89]; this case study presents a neural network architecture named PowerNet, which can incorporate historical supply and demand, weather data, and date information to forecast energy consumption. Some of the machine learning algorithms used by *PowerNet* are Gradient Boosting Tree (GBT), Support Vector Regression (SVR), Random Forest (RF), and Gated Recurrent Unit (GRU). Additionally, PowerNet can also be used for anomaly detection in power systems. The use of Recurrent Neural Networks is proposed in [27] to forecast the consumption of commercial and residential buildings in medium- to long-term time horizons (greater than a week). The model proposed in this article demonstrates the capability to predict unknown transient responses, making it promising for forecasting hourly electricity consumption as well. In [74], the Fuzzy-TOPSIS decision-making method is used to evaluate the selection of Mexican manufacturing companies for installing solar panels. Additionally, the S4 framework is implemented to improve the decision-making process by Fuzzy-TOPSIS.

One of the articles that focuses on building energy management to minimize power consumption is [67], This paper examines the application of deep learning structures. Initially, sensory neurons are spread throughout the smart building, collecting data from the environment. Subsequently, a reinforcement learning algorithm is employed to predict values and trends, thereby aiding the building managers with the decision-making task. The proposal called *iBuilding* is validated using a public research data set, demonstrating that Artificial Intelligence within smart buildings allows the real-time monitoring and accurate predictions of its variables.

Previously mentioned examples show a growing trend of the use of AI in the forecasting of the energy consumption sector due to the availability of more information on weather conditions and domestic appliances' energy consumption. It is possible to use the data gathered to train neural networks and models to have more precision. The historical patterns can be studied.

4.5. Research Gaps

As detailed in this literature review, AI applications in the energy sector reveal the current state of research and uncover several technological gaps that present opportunities for future innovation. The role of AI in forecasting alternative sources like wind and solar is well established in energy generation. However, a technological gap remains in the accurate energy output prediction under varying environmental conditions. Additionally, while AI has made strides in the operation and monitoring of these sources, there is a need for more advanced AI systems that can dynamically adapt to changing energy demands and integrate seamlessly with traditional energy grids. This gap is particularly evident in the context of hybrid energy systems in Smart Cities, where a diverse array of energy sources must be efficiently managed. In transmission and distribution, AI has significantly contributed to smart grid management and energy loss detection. However, a notable gap exists between the long-distance transmission efficiency and the integration of distributed energy resources. Current AI models often struggle with the complexity of large-scale, interconnected networks. There is a need for more sophisticated AI algorithms capable of optimizing energy flow across massive and varied geographical areas. On the consumption side, AI has been instrumental in reducing energy usage, forecasting demand, and managing building energy systems. Yet, a technological gap is evident in personalizing energy consumption strategies. AI systems that can learn and adapt to individual user behaviors and preferences in real time are in progress. Such systems could significantly enhance energy efficiency at the consumer level. While AI applications in forecasting and EV charging are advancing in the renewable energy sector, there is a gap in the comprehensive integration of various renewable sources. Research heavily focuses on solar, wind, and hydrogen, but other sources like hydropower, geothermal, and bio-energy are less explored. AI applications in these areas are not as developed, indicating a gap in diversifying renewable energy sources within Smart Cities. In addition, integrating AI systems is a cross-cutting technological gap across all these sectors. Figure 17 shows a brief description in this section. While AI is being applied in siloed aspects of the energy sector, there is a lack of holistic, interconnected AI frameworks that can manage the energy life-cycle from generation to consumption in a unified manner. Such integration is crucial for the development of truly smart, efficient cities. Finally, while AI has made significant inroads in the energy sector, our review highlights several technological gaps. Addressing these gaps requires focused research and development efforts, aiming for more sophisticated, adaptable, and integrated AI solutions that can meet the complex demands of modern energy systems, particularly in the context of Smart Cities.

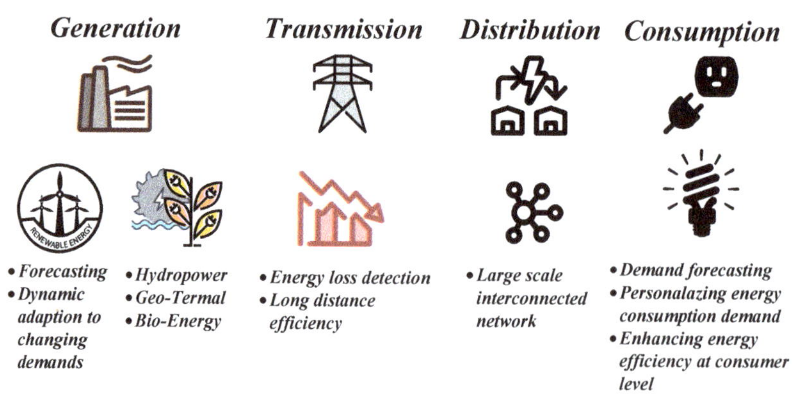

Figure 17. Illustrative description on research gaps section.

5. Conclusions

This paper contributes to understanding the integration and application of Artificial Intelligence (AI) within the energy sector, especially in Smart Cities. The study systematically delineates the existing literature into four key domains, energy generation, transmission, distribution, and consumption, meticulously reviewing a compendium of 122 articles sourced from Scopus. A salient feature of this paper is its temporal analysis, revealing an exponential surge in publication growth over the years. In energy generation alone, publications have increased from a single article in 2016 to 45 in 2023, representing a staggering annual growth rate of 42.62 percent. Comparable ascendant trajectories are discerned in other domains, such as energy transmission and distribution. Furthermore, the paper sheds light on renewable energy, focusing on the integration of solar and wind energy in Smart Cities, outlined in the Energy Generation section. A crucial aspect of this research lies in keyword co-occurrence analysis, enabling the identification of predominant AI techniques in current practice. Concepts such as artificial intelligence, economics, energy efficiency, data analytics, renewable energy, and deep learning have emerged as integral to the formulation of intelligent energy solutions. This comprehensive analysis provides a robust framework for deciphering the multifarious AI techniques and their application. From machine learning algorithms, such as Ant Colony Optimization and neural networks, to the nascent trends in smart grids, these innovations are pivotal in spearheading advancements in energy forecasting, consumption prediction, and intelligent control systems. The aim of this paper is to present a profound and up-to-the-minute overview of the influential role of AI in the energy sector of Smart Cities. It meticulously tracks the evolution and impact of research in this field and pinpoints the foremost technologies and methodologies expected to shape forthcoming innovations. This paper serves as a repository of information on current trends and a compass pointing toward the future trajectories in AI-mediated intelligent energy solutions within the progressive ecosystems of Smart Cities.

Author Contributions: Conceptualization, J.d.J.C., P.P., B.A. (Bernabé Aguirre), B.A. (Brian Anthony) and A.M.; methodology, J.d.J.C., B.A. (Bernabé Aguirre), P.P., B.A. (Brian Anthony) and A.M.; validation, J.d.J.C., B.A. (Bernabé Aguirre), P.P., B.A. (Brian Anthony) and A.M.; formal analysis, J.d.J.C., B.A. (Bernabé Aguirre), P.P., B.A. (Brian Anthony) and A.M.; investigation, J.d.J.C., P.P., B.A. (Bernabé Aguirre) and A.M.; data curation, J.d.J.C., P.P. and A.M.; writing—original draft preparation, J.d.J.C., B.A. (Bernabé Aguirre), P.P., B.A. (Brian Anthony) and A.M.; visualization, J.d.J.C. and B.A. (Bernabé Aguirre); supervision, P.P., B.A. (Brian Anthony) and A.M. All authors have read and agreed to the published version of the manuscript.

Funding: This research received support from the Institute of Advanced Materials for Sustainable Manufacturing at Tecnologico de Monterrey.

Institutional Review Board Statement: Not applicable.

Data Availability Statement: Data are contained within the article.

Conflicts of Interest: The authors declare no conflicts of interest.

Abbreviations

The following abbreviations are used in this manuscript:

AI	Artificial Intelligence
ACO	Ant Colony Optimization
ADM	Automatic Decision Making
AIA	Artificial Intelligence Analytics
AIMS-SB	Artificial Intelligence Technique for Monitoring Systems in Smart Buildings
ANN	Artificial Neural Network
AVWC	Automated Vacuum Waste Collection
BCT	Blockchain Technology
BiKSTM	Bidirectional Long Short-Term Memory
CNN	Convolutional Neural Network

DaaS	Device as a Service
DAI	Distributed Artificial Intelligence
DCESN	Deep-Chain Echo State Network
DCNN	Dilated Convolutional Neural Network
DERMS	Distributed Energy Management System
DNN	Deep Neural Network
DRL	Deep Reinforcement Learning
DSM	Demand Side Management
FCA	Fairness Cooperation Algorithm
FFTA	Finite-Time Terminal Attractor
GA	Genetic Algorithm
GHG	Greenhouse gas
GPSO	Gradient Particle Swarm Optimization
IGWO	Improved Gray Wolf Optimization
IOT	Internet of Things
IPTO	Improved Phase Timing Optimization
LSTM	Long Short-Term Memory
MAS	Multi Agent Systems
MCDA	Multiple Criteria Decision Analysis
MKL	Machine Kernel Learning
ML	Machine Learning
MPPT	Maximum Power Point Tracking
PV	Photovoltaic
QRL	Quantum-Inspired Reinforcement Learning
RNN	Recurrent Neural Network
SAE	Stacked Autoencoder
SM	Smart Metering
SMART	Simple Multi-Attribute Rating
SVM	Support Vector Machine
XELF	Explainable Electrical Load Forecasting

References

1. Llanez-Caballero, I.; Ibarra, L.; Peña-Quintal, A.; Catzín-Contreras, G.; Ponce, P.; Molina, A.; Ramirez-Mendoza, R. The "Smart" concept from an electrical sustainability viewpoint. *Energies* **2023**, *16*, 3072. [CrossRef]
2. Wang, K.; Zhao, Y.; Gangadhari, R.K.; Li, Z. Analyzing the adoption challenges of the Internet of things (Iot) and artificial intelligence (ai) for smart cities in china. *Sustainability* **2021**, *13*, 10983. [CrossRef]
3. Ullah, Z.; Al-Turjman, F.; Mostarda, L.; Gagliardi, R. Applications of artificial intelligence and machine learning in smart cities. *Comput. Commun.* **2020**, *154*, 313–323. [CrossRef]
4. Serrano, W. Digital systems in smart city and infrastructure: Digital as a service. *Smart Cities* **2018**, *1*, 134–154. [CrossRef]
5. Moghaddam, S.M.S.H.; Dashtdar, M.; Jafari, H. AI applications in smart cities' energy systems automation. *Repa Proc. Ser.* **2022**, *3*, 1–5. [CrossRef]
6. Mehmood, M.U.; Chun, D.; Han, H.; Jeon, G.; Chen, K. A review of the applications of artificial intelligence and big data to buildings for energy-efficiency and a comfortable indoor living environment. *Energy Build.* **2019**, *202*, 109383. [CrossRef]
7. Ahmad, T.; Zhu, H.; Zhang, D.; Tariq, R.; Bassam, A.; Ullah, F.; AlGhamdi, A.S.; Alshamrani, S.S. Energetics Systems and artificial intelligence: Applications of industry 4.0. *Energy Rep.* **2022**, *8*, 334–361. [CrossRef]
8. Li, G.; Jin, Y.; Akram, M.; Chen, X.; Ji, J. Application of bio-inspired algorithms in maximum power point tracking for PV systems under partial shading conditions—A review. *Renew. Sustain. Energy Rev.* **2018**, *81*, 840–873. [CrossRef]
9. Page, M.J.; McKenzie, J.E.; Bossuyt, P.M.; Boutron, I.; Hoffmann, T.C.; Mulrow, C.D.; Shamseer, L.; Tetzlaff, J.M.; Akl, E.A.; Brennan, S.E.; et al. Declaración PRISMA 2020: Una guía actualizada para la publicación de revisiones sistemáticas. *Rev. Espa Nola De Cardiol.* **2021**, *74*, 790–799. [CrossRef]
10. Ghadami, N.; Gheibi, M.; Kian, Z.; Faramarz, M.G.; Naghedi, R.; Eftekhari, M.; Fathollahi-Fard, A.M.; Dulebenets, M.A.; Tian, G. Implementation of solar energy in smart cities using an integration of artificial neural network, photovoltaic system and classical Delphi methods. *Sustain. Cities Soc.* **2021**, *74*, 103149. [CrossRef]
11. Serban, A.C.; Lytras, M.D. Artificial intelligence for smart renewable energy sector in europe—Smart energy infrastructures for next generation smart cities. *IEEE Access* **2020**, *8*, 77364–77377. [CrossRef]
12. Azzaoui, A.E.; Singh, S.K.; Pan, Y.; Park, J.H. Block5GIntell: Blockchain for AI-Enabled 5G Networks. *IEEE Access* **2020**, *8*, 145918–145935. [CrossRef]
13. Lee, Y.L.; Qin, D.; Wang, L.C.; Sim, G.H. 6G Massive Radio Access Networks: Key Applications, Requirements and Challenges. *IEEE Open J. Veh. Technol.* **2021**, *2*, 54–66. [CrossRef]

14. Zhang, N.; Chen, H.; Chen, X.; Chen, J. Semantic framework of internet of things for smart cities: Case studies. *Sensors* **2016**, *16*, 1501. [CrossRef] [PubMed]
15. Ramli, M.A.; Bouchekara, H.; Alghamdi, A.S. Optimal sizing of PV/wind/diesel hybrid microgrid system using multi-objective self-adaptive differential evolution algorithm. *Renew. Energy* **2018**, *121*, 400–411. [CrossRef]
16. Yang, D.; Gu, C.; Dong, Z.; Jirutitijaroen, P.; Chen, N.; Walsh, W.M. Solar irradiance forecasting using spatial-temporal covariance structures and time-forward kriging. *Renew. Energy* **2013**, *60*, 235–245. [CrossRef]
17. Oldenbroek, V.; Verhoef, L.A.; van Wijk, A.J. Fuel cell electric vehicle as a power plant: Fully renewable integrated transport and energy system design and analysis for smart city areas. *Int. J. Hydrogen Energy* **2017**, *42*, 8166–8196. [CrossRef]
18. De Luca, G.; Fabozzi, S.; Massarotti, N.; Vanoli, L. A renewable energy system for a nearly zero greenhouse city: Case study of a small city in southern Italy. *Energy* **2018**, *143*, 347–362. [CrossRef]
19. Aguilar, J.; Garces-Jimenez, A.; R-moreno, M.; García, R. A systematic literature review on the use of artificial intelligence in energy self-management in smart buildings. *Renew. Sustain. Energy Rev.* **2021**, *151*, 111530. [CrossRef]
20. Sharma, H.; Haque, A.; Blaabjerg, F. Machine learning in wireless sensor networks for smart cities: A survey. *Electronics* **2021**, *10*, 1012. [CrossRef]
21. Khan, N.; Haq, I.U.; Khan, S.U.; Rho, S.; Lee, M.Y.; Baik, S.W. DB-Net: A novel dilated CNN based multi-step forecasting model for power consumption in integrated local energy systems. *Int. J. Electr. Power Energy Syst.* **2021**, *133*, 107023. [CrossRef]
22. Le, L.T.; Nguyen, H.; Dou, J.; Zhou, J. A comparative study of PSO-ANN, GA-ANN, ICA-ANN, and ABC-ANN in estimating the heating load of buildings' energy efficiency for smart city planning. *Appl. Sci.* **2019**, *9*, 2630. [CrossRef]
23. Idowu, S.; Saguna, S.; Åhlund, C.; Schelén, O. Applied machine learning: Forecasting heat load in district heating system. *Energy Build.* **2016**, *133*, 478–488. [CrossRef]
24. Zhou, Z.; Wang, B.; Guo, Y.; Zhang, Y. Blockchain and computational intelligence inspired incentive-compatible demand response in internet of electric vehicles. *IEEE Trans. Emerg. Top. Comput. Intell.* **2019**, *3*, 205–216. [CrossRef]
25. Le, L.T.; Nguyen, H.; Zhou, J.; Dou, J.; Moayedi, H. Estimating the heating load of buildings for smart city planning using a novel artificial intelligence technique PSO-XGBoost. *Appl. Sci.* **2019**, *9*, 2714. [CrossRef]
26. Ingwersen, P.; Serrano-López, A.E. Smart city research 1990–2016. *Scientometrics* **2018**, *117*, 1205–1236. [CrossRef]
27. Rahman, A.; Srikumar, V.; Smith, A.D. Predicting electricity consumption for commercial and residential buildings using deep recurrent neural networks. *Appl. Energy* **2018**, *212*, 372–385. [CrossRef]
28. Ullah, A.; Haydarov, K.; Haq, I.U.; Muhammad, K.; Rho, S.; Lee, M.; Baik, S.W. Deep learning assisted buildings energy consumption profiling using smart meter data. *Sensors* **2020**, *20*, 873. [CrossRef]
29. Alsamhi, S.H.; Ma, O.; Ansari, M.S.; Almalki, F.A. Survey on collaborative smart drones and internet of things for improving smartness of smart cities. *IEEE Access* **2019**, *7*, 128125–128152. [CrossRef]
30. Vázquez-Canteli, J.R.; Ulyanin, S.; Kämpf, J.; Nagy, Z. Fusing TensorFlow with building energy simulation for intelligent energy management in smart cities. *Sustain. Cities Soc.* **2019**, *45*, 243–257. [CrossRef]
31. Miyasawa, A.; Akira, S.; Fujimoto, Y.; Hayashi, Y. Forecast of area-scale behaviours of behind-the-metre solar power and load based on smart-metering net demand data. *IET Smart Cities* **2023**, *5*, 19–34. [CrossRef]
32. Shafiullah, M.; Rahman, S.; Imteyaz, B.; Aroua, M.K.; Hossain, M.I.; Rahman, S.M. Review of Smart City Energy Modeling in Southeast Asia. *Smart Cities* **2023**, *6*, 72–99. [CrossRef]
33. Wu, Z.; Sun, B.; Feng, Q.; Wang, Z.; Pan, J. Physics-Informed AI Surrogates for Day-Ahead Wind Power Probabilistic Forecasting with Incomplete Data for Smart Grid in Smart Cities. *CMES Comput. Model. Eng. Sci.* **2023**, *137*, 527–554. [CrossRef]
34. Fakhar, A.; Haidar, A.M.; Abdullah, M.; Das, N. Smart grid mechanism for green energy management: A comprehensive review. *Int. J. Green Energy* **2023**, *20*, 284–308. [CrossRef]
35. Bayer, D.; Pruckner, M. A digital twin of a local energy system based on real smart meter data. *Energy Inform.* **2023**, *6*, 8. [CrossRef]
36. Alymani, M.; Mengash, H.A.; Aljebreen, M.; Alasmari, N.; Allafi, R.; Alshahrani, H.; Elfaki, M.A.; Hamza, M.A.; Abdelmageed, A.A. Sustainable residential building energy consumption forecasting for smart cities using optimal weighted voting ensemble learning. *Sustain. Energy Technol. Assess.* **2023**, *57*, 103271. [CrossRef]
37. Al-Habaibeh, A.; Yaseen, S.; Nweke, B. A comparative study of low and high resolution infrared cameras for IoT smart city applications: A comparative study of low and high resolution infrared cameras. *Ain Shams Eng. J.* **2023**, *14*, 102108. [CrossRef]
38. Selvaraj, R.; Kuthadi, V.M.; Baskar, S. Smart building energy management and monitoring system based on artificial intelligence in smart city. *Sustain. Energy Technol. Assess.* **2023**, *56*, 103271. [CrossRef]
39. Feng, Y.; Huang, Y.; Li, B.; Peng, H.; Wang, J.; Zhou, W. Connectivity Enhancement of E-VANET Based on QL-mRSU Self-Learning Energy-Saving Algorithm. *IEEE Access* **2023**, *11*, 3810–3825. [CrossRef]
40. Jiang, R.; Zeng, S.; Song, Q.; Wu, Z. Deep-Chain Echo State Network With Explainable Temporal Dependence for Complex Building Energy Prediction. *IEEE Trans. Ind. Inform.* **2023**, *19*, 426–435. [CrossRef]
41. AlHajri, I.; Ahmadian, A.; Alazmi, R. A comprehensive technical, economic, and environmental evaluation for optimal planning of renewable energy resources to supply water desalination units: Kuwait case study. *Energy* **2023**, *275*, 127416. [CrossRef]
42. Fantin Irudaya Raj, E.; Appadurai, M.; Lurthu Pushparaj, T.; Chithambara Thanu, M. Wind turbines with aramid fiber composite wind blades for smart cities like urban environments: Numerical simulation study. *MRS Energy Sustain.* **2023**, *10*, 139–156. [CrossRef]

43. Khan, N.; Khan, S.U.; Ullah, F.U.M.; Lee, M.Y.; Baik, S.W. AI-Assisted Hybrid Appr Approach for Energy Management in IoT-based Smart Microgrid. *IEEE Internet Things J.* **2023**, *10*, 18861–18875. [CrossRef]
44. Ulpiani, G.; Vetters, N.; Shtjefni, D.; Kakoulaki, G.; Taylor, N. Let's hear it from the cities: On the role of renewable energy in reaching climate neutrality in urban Europe. *Renew. Sustain. Energy Rev.* **2023**, *183*, 113444. [CrossRef]
45. Kedir, N.; Nguyen, P.H.; Pérez, C.; Ponce, P.; Fayek, A.R. Systematic Literature Review on Fuzzy Hybrid Methods in Photovoltaic Solar Energy: Opportunities, Challenges, and Guidance for Implementation. *Energies* **2023**, *16*, 3795. [CrossRef]
46. Icaza-Alvarez, D.; Jurado, F.; Tostado-Véliz, M. Long-term planning for the integration of electric mobility with 100 renewable energy generation under various degrees of decentralization: Case study Cuenca, Ecuador. *Energy Rep.* **2023**, *9*, 4816–4829. [CrossRef]
47. Moon, J.; Rho, S.; Baik, S.W. Toward explainable electrical load forecasting of buildings: A comparative study of tree-based ensemble methods with Shapley values. *Sustain. Energy Technol. Assess.* **2022**, *54*, 102888. [CrossRef]
48. Said, D. A Survey on Information Communication Technologies in Modern Demand-Side Management for Smart Grids: Challenges, Solutions, and Opportunities. *IEEE Eng. Manag. Rev.* **2023**, *51*, 76–107. [CrossRef]
49. Heidari, A.; Navimipour, N.J.; Unal, M. Applications of ML/DL in the management of smart cities and societies based on new trends in information technologies: A systematic literature review. *Sustain. Cities Soc.* **2022**, *85*, 104089. [CrossRef]
50. Chang, E.C.; Cheng, C.A.; Wu, R.C. Artificial Intelligence of Things-Based Optimal Finite-Time Terminal Attractor and Its Application to Maximum Power Point Tracking of Photovoltaic Arrays in Smart Cities. *Wirel. Commun. Mob. Comput.* **2022**, *2022*, 4213217. [CrossRef]
51. Liu, Z.; Gao, Y.; Liu, B. An artificial intelligence-based electric multiple units using a smart power grid system. *Energy Rep.* **2022**, *8*, 13376–13388. [CrossRef]
52. Khosrojerdi, F.; Akhigbe, O.; Gagnon, S.; Ramirez, A.; Richards, G. Integrating artificial intelligence and analytics in smart grids: A systematic literature review. *Int. J. Energy Sect. Manag.* **2021**, *16*, 318–338.
53. Chavhan, S.; Gupta, D.; Gochhayat, S.P.; Chandana, B.; Khanna, A.; Shankar, K.; Rodrigues, J.J.P.C. Edge Computing AI-IoT Integrated Energy-efficient Intelligent Transportation System for Smart Cities. *ACM Trans. Internet Technol.* **2022**, *22*, 4213217. [CrossRef]
54. Al-Hawawreh, M.; Elgendi, I.; Munasinghe, K. An Online Model to Minimize Energy Consumption of IoT Sensors in Smart Cities. *IEEE Sens. J.* **2022**, *22*, 19524–19532. [CrossRef]
55. Singh, S.; Nikolovski, S.; Chakrabarti, P. GWLBC: Gray Wolf Optimization Based Load Balanced Clustering for Sustainable WSNs in Smart City Environment. *Sensors* **2022**, *22*, 7113. [CrossRef] [PubMed]
56. Huang, J.; Algahtani, M.; Kaewunruen, S. Energy Forecasting in a Public Building: A Benchmarking Analysis on Long Short-Term Memory (LSTM), Support Vector Regression (SVR), and Extreme Gradient Boosting (XGBoost) Networks. *Appl. Sci.* **2022**, *12*, 9788. [CrossRef]
57. Mohamed, H.; Al-Masri, E.; Kotevska, O.; Souri, A. A Multi-Objective Approach for Optimizing Edge-Based Resource Allocation Using TOPSIS. *Electronics* **2022**, *11*, 2888. [CrossRef]
58. Ren, Y.; Xie, R.; Yu, F.R.; Huang, T.; Liu, Y. Green Intelligence Networking for Connected and Autonomous Vehicles in Smart Cities. *IEEE Trans. Green Commun. Netw.* **2022**, *6*, 1591–1603. [CrossRef]
59. Murthy Nimmagadda, S.; Harish, K.S. Review paper on technology adoption and sustainability in India towards smart cities. *Multimed. Tools Appl.* **2022**, *81*, 27217–27245. [CrossRef]
60. Islam, N.; Haseeb, K.; Ali, M.; Jeon, G. Secured Protocol with Collaborative IoT-Enabled Sustainable Communication Using Artificial Intelligence Technique. *Sustainability* **2022**, *14*, 8919. [CrossRef]
61. Zamponi, M.E.; Barbierato, E. The Dual Role of Artificial Intelligence in Developing Smart Cities. *Smart Cities* **2022**, *5*, 728–755. [CrossRef]
62. Naveed, Q.N.; Alqahtani, H.; Khan, R.U.; Almakdi, S.; Alshehri, M.; Rasheed, M.A.A. An Intelligent Traffic Surveillance System Using Integrated Wireless Sensor Network and Improved Phase Timing Optimization. *Sensors* **2022**, *22*, 3333. [CrossRef] [PubMed]
63. Akkad, M.Z.; Haidar, S.; Bányai, T. Design of Cyber-Physical Waste Management Systems Focusing on Energy Efficiency and Sustainability. *Designs* **2022**, *6*, 39. [CrossRef]
64. Garlik, B. Energy Centers in a Smart City as a Platform for the Application of Artificial Intelligence and the Internet of Things. *Appl. Sci.* **2022**, *12*, 3386. [CrossRef]
65. Saba, D.; Cheikhrouhou, O.; Alhakami, W.; Sahli, Y.; Hadidi, A.; Hamam, H. Intelligent Reasoning Rules for Home Energy Management (IRRHEM): Algeria Case Study. *Appl. Sci.* **2022**, *12*, 1861. [CrossRef]
66. Zaimen, K.; Brahmia, M.E.A.; Moalic, L.; Abouaissa, A.; Idoumghar, L. A Survey of Artificial Intelligence Based WSNs Deployment Techniques and Related Objectives Modeling. *IEEE Access* **2022**, *10*, 113294–113329. [CrossRef]
67. Serrano, W. iBuilding: Artificial intelligence in intelligent buildings. *Neural Comput. Appl.* **2022**, *34*, 875–897. [CrossRef]
68. Li, J.; Chen, J.; Yuan, Z.; Xu, L.; Zhang, Y.; Al-Bahrani, M. Multi-objective risk-constrained optimal performance of hydrogen-based multi energy systems for future sustainable societies. *Sustain. Cities Soc.* **2022**, *87*, 104176. [CrossRef]
69. Doosti, R.; Sedighizadeh, M.; Sedighizadeh, D.; Sheikhi Fini, A. Robust stochastic optimal operation of an industrial building including plug in electric vehicle, solar-powered compressed air energy storage and ice storage conditioner: A case study in the city of Kaveh, Iran. *IET Smart Cities* **2022**, *4*, 56–77. [CrossRef]
70. Vyas, M.; Chowdhury, S.; Verma, A.; Jain, V. Solar Photovoltaic Tree: Urban PV power plants to increase power to land occupancy ratio. *Renew. Energy* **2022**, *190*, 283–293. [CrossRef]

71. AlHammadi, A.; Al-Saif, N.; Al-Sumaiti, A.S.; Marzband, M.; Alsumaiti, T.; Heydarian-Forushani, E. Techno-Economic Analysis of Hybrid Renewable Energy Systems Designed for Electric Vehicle Charging: A Case Study from the United Arab Emirates. *Energies* **2022**, *15*, 6621. [CrossRef]
72. Balabel, A.; Alwetaishi, M.; Abdelhafiz, A.; Issa, U.; Sharaky, I.; Shamseldin, A.; Al-Surf, M.; Al-Harthi, M. Potential of Solatube technology as passive daylight systems for sustainable buildings in Saudi Arabia. *Alex. Eng. J.* **2022**, *61*, 339–353. [CrossRef]
73. Nuvvula, R.S.; Devaraj, E.; Madurai Elavarasan, R.; Iman Taheri, S.; Irfan, M.; Teegala, K.S. Multi-objective mutation-enabled adaptive local attractor quantum behaved particle swarm optimisation based optimal sizing of hybrid renewable energy system for smart cities in India. *Sustain. Energy Technol. Assess.* **2022**, *49*, 101689. [CrossRef]
74. Ponce, P.; Pérez, C.; Fayek, A.R.; Molina, A. Solar energy implementation in manufacturing industry using multi-criteria decision-making fuzzy TOPSIS and S4 framework. *Energies* **2022**, *15*, 8838. [CrossRef]
75. Konhäuser, W. Digitalization in Buildings and Smart Cities on the Way to 6G. *Wirel. Pers. Commun.* **2021**, *121*, 1289–1302. [CrossRef]
76. Pérez-Romero, A.; Mateo-Romero, H.F.; Gallardo-Saavedra, S.; Alonso-Gómez, V.; Alonso-García, M.D.C.; Hernández-Callejo, L. Evaluation of artificial intelligence-based models for classifying defective photovoltaic cells. *Appl. Sci.* **2021**, *11*, 4226. [CrossRef]
77. Zhou, H.; Liu, Q.; Yan, K.; Du, Y. Deep Learning Enhanced Solar Energy Forecasting with AI-Driven IoT. *Wirel. Commun. Mob. Comput.* **2021**, *2021*, 9249387. [CrossRef]
78. Saini, G.S.; Dubey, S.K.; Bharti, S.K. Tracard: A tool for smart city resource management based on novel framework algorithm. *Recent Adv. Comput. Sci. Commun.* **2021**, *14*, 134–140. [CrossRef]
79. Antonopoulos, I.; Robu, V.; Couraud, B.; Flynn, D. Data-driven modelling of energy demand response behaviour based on a large-scale residential trial. *Energy AI* **2021**, *4*, 100071. [CrossRef]
80. Calamaro, N.; Beck, Y.; Ben Melech, R.; Shmilovitz, D. An Energy-Fraud Detection-System Capable of Distinguishing Frauds from Other Energy Flow Anomalies in an Urban Environment. *Sustainability* **2021**, *13*, 10696. [CrossRef]
81. Manman, L.; Goswami, P.; Mukherjee, P.; Mukherjee, A.; Yang, L.; Ghosh, U.; Menon, V.G.; Nkenyereye, L. Distributed Artificial Intelligence Empowered Sustainable Cognitive Radio Sensor Networks: A Smart City on-demand Perspective. *Sustain. Cities Soc.* **2021**, *75*, 103265. [CrossRef]
82. Li, J.; Dai, J.; Issakhov, A.; Almojil, S.F.; Souri, A. Towards decision support systems for energy management in the smart industry and Internet of Things. *Comput. Ind. Eng.* **2021**, *161*, 107671. [CrossRef]
83. Cirella, G.T.; Russo, A.; Benassi, F.; Czermański, E.; Goncharuk, A.G.; Oniszczuk-Jastrzabek, A. Energy re-shift for an urbanizing world. *Energies* **2021**, *14*, 5516. [CrossRef]
84. Hu, Y.C.; Lin, Y.H.; Gururaj, H.L. Partitional clustering-hybridized neuro-fuzzy classification evolved through parallel evolutionary computing and applied to energy decomposition for demand-side management in a smart home. *Processes* **2021**, *9*, 1539. [CrossRef]
85. Mahmood, D.; Latif, S.; Anwar, A.; Hussain, S.J.; Jhanjhi, N.; Sama, N.U.; Humayun, M. Utilization of ICT and AI techniques in harnessing residential energy consumption for an energy-aware smart city: A review. *Int. J. Adv. Appl. Sci.* **2021**, *8*, 50–66. [CrossRef]
86. Wang, X.; Chen, Q.; Wang, J. Fuzzy rough set based sustainable methods for energy efficient smart city development. *J. Intell. Fuzzy Syst.* **2021**, *40*, 8173–8183. [CrossRef]
87. Kanase-Patil, A.B.; Kaldate, A.P.; Lokhande, S.D.; Panchal, H.; Suresh, M.; Priya, V. A review of artificial intelligence-based optimization techniques for the sizing of integrated renewable energy systems in smart cities. *Environ. Technol. Rev.* **2020**, *9*, 111–136. [CrossRef]
88. Jiang, M.; Lu, Y.; Zhu, Z.; Jia, W. Advances in smart sensing and medical electronics by self-powered sensors based on triboelectric nanogenerators. *Micromachines* **2021**, *12*, 698. [CrossRef]
89. Cheng, Y.; Xu, C.; Mashima, D.; Biswas, P.P.; Chipurupalli, G.; Zhou, B.; Wu, Y. PowerNet: A smart energy forecasting architecture based on neural networks. *IET Smart Cities* **2020**, *2*, 199–207. [CrossRef]
90. Loose, N.; Thommessen, C.; Mehlich, J.; Derksen, C.; Eicker, S. Unified energy agents for combined district heating and electrical network simulation. *Sustainability* **2020**, *12*, 9301. [CrossRef]
91. Fattahi, J.; Wright, D.; Schriemer, H. An energy internet DERMS platform using a multi-level Stackelberg game. *Sustain. Cities Soc.* **2020**, *60*, 102262. [CrossRef]
92. Sharma, S. Analysis of intelligent versus smart infrastructure. *Int. J. Adv. Res. Eng. Technol.* **2020**, *11*, 122–132. [CrossRef]
93. Marinakis, V.; Doukas, H.; Koasidis, K.; Albuflasa, H. From intelligent energy management to value economy through a digital energy currency: Bahrain city case study. *Sensors* **2020**, *20*, 1456. [CrossRef] [PubMed]
94. Shah, A.S.; Nasir, H.; Fayaz, M.; Lajis, A.; Ullah, I.; Shah, A. Dynamic user preference parameters selection and energy consumption optimization for smart homes using deep extreme learning machine and bat algorithm. *IEEE Access* **2020**, *8*, 204744–204762. [CrossRef]
95. Guo, Y.; Zhao, Z.; Zhao, R.; Lai, S.; Dan, Z.; Xia, J.; Fan, L. Intelligent Offloading Strategy Design for Relaying Mobile Edge Computing Networks. *IEEE Access* **2020**, *8*, 35127–35135. [CrossRef]
96. Zhuang, H.; Zhang, J.; Sivaparthipan, C.; Muthu, B.A. Sustainable smart city building construction methods. *Sustainability* **2020**, *12*, 4947. [CrossRef]

97. Algieri, A.; Morrone, P.; Bova, S. Techno-economic analysis of biofuel, solar and wind multi-source small-scale CHP systems. *Energies* **2020**, *13*, 3002. [CrossRef]
98. Kumar, D. Urban energy system management for enhanced energy potential for upcoming smart cities. *Energy Explor. Exploit.* **2020**, *38*, 1968–1982. [CrossRef]
99. Aghajani, D.; Abbaspour, M.; Radfar, R.; Mohammadi, A. Using Web-GIS technology as a smart tool for resiliency management to monitor wind farms performances (Ganjeh site, Iran). *Int. J. Environ. Sci. Technol.* **2019**, *16*, 5011–5022. [CrossRef]
100. Oun, A.; Benabdallah, I.; Cherif, A. Improved industrial modeling and harmonic mitigation of a grid connected steel plant in Libya. *Int. J. Adv. Comput. Sci. Appl.* **2019**, *10*, 101–109. [CrossRef]
101. Salehi, H.; Das, S.; Biswas, S.; Burgueño, R. Data mining methodology employing artificial intelligence and a probabilistic approach for energy-efficient structural health monitoring with noisy and delayed signals. *Expert Syst. Appl.* **2019**, *135*, 259–272. [CrossRef]
102. Dong, Y.; Guo, S.; Liu, J.; Yang, Y. Energy-Efficient Fair Cooperation Fog Computing in Mobile Edge Networks for Smart City. *IEEE Internet Things J.* **2019**, *6*, 7543–7554. [CrossRef]
103. Krayem, A.; Al Bitar, A.; Ahmad, A.; Faour, G.; Gastellu-Etchegorry, J.P.; Lakkis, I.; Gerard, J.; Zaraket, H.; Yeretzian, A.; Najem, S. Urban energy modeling and calibration of a coastal Mediterranean city: The case of Beirut. *Energy Build.* **2019**, *199*, 223–234. [CrossRef]
104. Marin-Perez, R.; Michailidis, I.T.; Garcia-Carrillo, D.; Korkas, C.D.; Kosmatopoulos, E.B.; Skarmeta, A. PLUG-N-HARVEST Architecture for secure and intelligent management of near-zero energy buildings. *Sensors* **2019**, *19*, 843. [CrossRef] [PubMed]
105. Aymen, F.; Mahmoudi, C. A novel energy optimization approach for electrical vehicles in a smart city. *Energies* **2019**, *12*, 929. [CrossRef]
106. Khoury, D.; Keyrouz, F. A predictive convolutional neural network model for source-load forecasting in smart grids. *WSEAS Trans. Power Syst.* **2019**, *14*, 181–189.
107. Alhussein, M.; Haider, S.I.; Aurangzeb, K. Microgrid-level energy management approach based on short-term forecasting ofwind speed and solar irradiance. *Energies* **2019**, *12*, 1487. [CrossRef]
108. Risso, C. Benefits of demands control in a smart-grid to compensate the volatility of non-conventional energies. *Rev. Fac. Ing.* **2019**, 19–31. [CrossRef]
109. Njuguna Matheri, A.; Mbohwa, C.; Ntuli, F.; Belaid, M.; Seodigeng, T.; Catherine Ngila, J.; Kinuthia Njenga, C. Waste to energy bio-digester selection and design model for the organic fraction of municipal solid waste. *Renew. Sustain. Energy Rev.* **2018**, *82*, 1113–1121. [CrossRef]
110. Chui, K.T.; Lytras, M.D.; Visvizi, A. Energy sustainability in smart cities: Artificial intelligence, smart monitoring, and optimization of energy consumption. *Energies* **2018**, *11*, 2869. [CrossRef]
111. Ashfaq, A.; Ianakiev, A. Cost-minimised design of a highly renewable heating network for fossil-free future. *Energy* **2018**, *152*, 613–626. [CrossRef]
112. Laiola, E.; Giungato, P. Wind characterization in Taranto city as a basis for innovative sustainable urban development. *J. Clean. Prod.* **2018**, *172*, 3535–3545. [CrossRef]
113. Rekik, M.; Chtourou, Z.; Mitton, N.; Atieh, A. Geographic routing protocol for the deployment of virtual power plant within the smart grid. *Sustain. Cities Soc.* **2016**, *25*, 39–48. [CrossRef]
114. De Paz, J.F.; Bajo, J.; Rodríguez, S.; Villarrubia, G.; Corchado, J.M. Intelligent system for lighting control in smart cities. *Inf. Sci.* **2016**, *372*, 241–255. [CrossRef]
115. Peña, M.; Biscarri, F.; Guerrero, J.I.; Monedero, I.; León, C. Rule-based system to detect energy efficiency anomalies in smart buildings, a data mining approach. *Expert Syst. Appl.* **2016**, *56*, 242–255. [CrossRef]
116. Huang, J.; Deng, Y.; Yang, Q.; Sun, J. An Energy-Efficient Train Control Framework for Smart Railway Transportation. *IEEE Trans. Comput.* **2016**, *65*, 1407–1417. [CrossRef]
117. Sędziwy, A.; Kotulski, L. Towards highly energy-efficient roadway lighting. *Energies* **2016**, *9*, 263. [CrossRef]
118. Lützenberger, M.; Masuch, N.; Küster, T.; Freund, D.; Voß, M.; Hrabia, C.E.; Pozo, D.; Fähndrich, J.; Trollmann, F.; Keiser, J.; et al. A common approach to intelligent energy and mobility services in a smart city environment. *J. Ambient Intell. Humaniz. Comput.* **2015**, *6*, 337–350. [CrossRef]
119. Fernández, C.; Manyà, F.; Mateu, C.; Sole-Mauri, F. Modeling energy consumption in automated vacuum waste collection systems. *Environ. Model. Softw.* **2014**, *56*, 63–73. [CrossRef]
120. Park, C.J.; Lee, J.; Kim, S.B.; Hyun, J.S.; Park, G.L. Spatio-temporal correlation analysis of wind speed and sunlight for predicting power generation in Jeju. *Int. J. Multimed. Ubiquitous Eng.* **2013**, *8*, 273–282.
121. Riaz, M.; Farid, H.M.A.; Jana, C.; Pal, M.; Sarkar, B. Efficient city supply chain management through spherical fuzzy dynamic multistage decision analysis. *Eng. Appl. Artif. Intell.* **2023**, *126*, 106712. [CrossRef]
122. Guchhait, R.; Sarkar, B. Increasing Growth of Renewable Energy: A State of Art. *Energies* **2023**, *16*, 2665. [CrossRef]

energies

MDPI

Review

Cooling Techniques for Enhanced Efficiency of Photovoltaic Panels—Comparative Analysis with Environmental and Economic Insights

Tarek Ibrahim [1], Mohamad Abou Akrouch [1], Farouk Hachem [1], Mohamad Ramadan [1,2], Haitham S. Ramadan [3,4,*] and Mahmoud Khaled [1,5]

1 Energy and Thermo-Fluid Group, Lebanese International University LIU, Bekaa P.O. Box 1801, Lebanon; tarek.toufic.ibrahim@hotmail.com (T.I.); mhammadakrouch@gmail.com (M.A.A.); farouk.hachem@liu.edu.lb (F.H.); mohamad.ramadan@liu.edu.lb (M.R.); mahmoud.khaled@liu.edu.lb (M.K.)
2 Energy and Thermo-Fluid Group, The International University of Beirut BIU, Beirut P.O. Box 146404, Lebanon
3 Electrical Power and Machines Department, Faculty of Engineering, Zagazig University, Zagazig 44519, Egypt
4 ISTHY, l'Institut International sur le Stockage de l'Hydrogène, 90400 Meroux-Moval, France
5 Center for Sustainable Energy & Economic Development (SEED), Gulf University for Science & Technology, Hawally 32093, Kuwait
* Correspondence: haitham.s.ramadan@gmail.com

Abstract: Photovoltaic panels play a pivotal role in the renewable energy sector, serving as a crucial component for generating environmentally friendly electricity from sunlight. However, a persistent challenge lies in the adverse effects of rising temperatures resulting from prolonged exposure to solar radiation. Consequently, this elevated temperature hinders the efficiency of photovoltaic panels and reduces power production, primarily due to changes in semiconductor properties within the solar cells. Given the depletion of limited fossil fuel resources and the urgent need to reduce carbon gas emissions, scientists and researchers are actively exploring innovative strategies to enhance photovoltaic panel efficiency through advanced cooling methods. This paper conducts a comprehensive review of various cooling technologies employed to enhance the performance of PV panels, encompassing water-based, air-based, and phase-change materials, alongside novel cooling approaches. This study collects and assesses data from recent studies on cooling the PV panel, considering both environmental and economic factors, illustrating the importance of cooling methods on photovoltaic panel efficiency. Among the investigated cooling methods, the thermoelectric cooling method emerges as a promising solution, demonstrating noteworthy improvements in energy efficiency and a positive environmental footprint while maintaining economic viability. As future work, studies should be made at the level of different periods of time throughout the years and for longer periods. This research contributes to the ongoing effort to identify effective cooling strategies, ultimately advancing electricity generation from photovoltaic panels and promoting the adoption of sustainable energy systems.

Keywords: photovoltaic panels; cooling techniques; environmental and economic study; future recommendations

1. Introduction

The continued population growth has resulted in the need for more energy resources to satisfy different sectors of life [1–4]. Further, the continued use of fossil fuels has led to depletion of resources and increases in price and CO_2 in the atmosphere. Therefore, current research focuses on finding alternative solutions through renewable energy resources [5,6] and heat recovery systems [7–10].

Solar energy forms an important factor in renewable energy resources, mainly through photovoltaic (PV) panels. Solar-energy-based PVs constitute a widely used technology in modern life based on the principle of converting sunlight into electricity through semiconductor materials. This technology enabled a great leap forward in the world of renewable energy resources due to its environmental impact on the reduction in CO_2 emissions, its fast payback period, and its long maintenance period (every 25–30 years). However, the need for innovative installation techniques on modern roofs, the high prices, and the low power generation on rainy days are obstacles to the installation of this technology.

The main obstacle in this technology is its low efficiency due to high temperatures. The constant contact of sun rays at the surface of the PV panel increases its temperature, thus decreasing its efficiency and output power. It was found that the efficiency of crystalline silicon solar cells falls by 0.45–0.6% for every 1 °C rise above STC (standard test conditions) in solar cell temperatures and varies according to the type of cell [11].

To increase the efficiency and the affordability of the panels, different approaches were recorded in the trial to reduce solar cell temperatures. In the literature, four cooling techniques are demonstrated with their different methods. The first technique is using passive and active cooling methods of water. The second cooling technique is the use of free and forced convection of air. The third cooling technique is the use of phase-change materials (PCM) to absorb the excess of heat produced by the PV panel. Then the last cooling technique is a sum of uncategorized and modern methods.

Table 1 portrays a collection of recent studies on different cooling techniques of photovoltaic panels using novel approaches. The studies cover research and review articles.

Table 1. Recent research conducted on cooling PV panels using different novel methods.

Objective	Methodology	Outcomes	References
Review on photovoltaic–thermal collector technology and advances in thermally driven cycles for PVT collectors.	Literature review on PVT collector types, discussion of cooling solar systems, their limitations, and future recommendations.	Electrical and thermal efficiency enhancement up to 11% and 22.02% maximum, respectively. The minimum payback period for PVT systems is 8.45–9.3 years.	Jiao et al. [12]
A comprehensive review of different cooling techniques used for concentrated PV cells.	Literature review on cooling CPV cell categories, discussion of CPV cooling systems, mentioning their advantages and disadvantages, and future recommendations.	Agreement between experimental and numerical results on enhancing the efficiency.	Ibrahim et al. [13]
Review on state-of-the-art photovoltaic thermal collectors and their abilities to increase energy production and CO_2 reduction.	Literature review on PVT systems, classification, discussion on performance enhancement, applications, and future recommendations.	The curve of emissions (Remap) could be reduced by 16% by 2030 if PV technology was used.	Herrando et al. [14]
Review on water-based PV systems and factors affecting them.	Literature review on cooling PV panels methods, classification of water-based cooling methods, discussion and analysis of these methods in a statistical manner.	Water-based cooling was shown to be effective in unused water spaces and has the potential to increase PV performance.	Ghosh [15]
A comprehensive review on cooling PV systems.	Literature discussing the different factors affecting the solar systems. Providing discussions on temperature mitigation strategies and cooling methods.	Discusses power plant performance, performance-affecting factors, and solutions to reduce the effect of those factors.	Aslam et al. [16]
Review on photovoltaic thermal systems in buildings and their application in heating, cooling, and power generation.	Literature discussing PVT systems and their integration into buildings, state-of-the-art systems designed for cooling, heating, and power production, and their limitations.	Hybrid systems showed the best performance, highlighting that PVT technology is still under development.	Herrando et al. [17]
Review on PV cooling technologies and their environmental impacts.	Literature review on PV technology, cooling techniques, advances in cooling technology, and future recommendations.	Air cooling was found to be cost-effective and simple, liquid cooling was found to be efficient but expensive, PCM cooling was found to enhance thermal efficiency but bulky, and nanomaterial was found to be efficient but expensive.	Hajjaj et al. [18]
Review on PV cooling using floating and solar tracking systems.	Literature review on PV panels, cooling methodologies, solar tracking, floating PV systems, and future recommendations.	Solar tracking and floating PV systems were found to reduce land usage and increase PV performance.	Hammoumi et al. [19]
Review of PV cooling technologies and their abilities in temperature reduction and power enhancement.	Literature review on cooling methods, discussing experimental studies and cooling systems limitations.	PCM combined with nanoparticles was found to be the most effective in cooling compared to water and air-based systems.	Sheik et al. [20]

Table 1. *Cont.*

Objective	Methodology	Outcomes	References
Review on photovoltaic thermal systems combined with PCM cooling.	A literature review was conducted about different cooling methods, traditional and advanced PV-T with PCM systems, and their potential, analyzing their performance, mentioning the challenges, and future recommendations.	Combined PV-T PCM systems are owed a 3–5% increase in electrical efficiency, 20–30% in thermal efficiency, and cost reduction by 15–20% with a payback period of less than 6 years compared to PV-T systems without PCM.	Cui et al. [21]
Review on PV passive cooling techniques.	Literature review on passive PV cooling methods, discussing the passive cooling methods while mentioning the unsolved challenges, and recommending future work.	Natural air ventilation and floatovoltaics cooling systems were found to be the most effective among the other passive cooling methods.	Mahdavi et al. [22]
Review on nano-based cooling techniques.	Literature review on nano-based PV cooling, classifying and discussing each method, and proposing designs and future recommendations.	Compared to conventional cooling methods, the hybrid nano-based cooling method could reduce PV's surface temperature by up to 16 °C and increase electrical efficiency by up to 50%.	Kandeal et al. [23]
Review on and comparison of solar tracking systems.	Literature review on PV panels, and solar tracking systems while categorizing them and focusing on dual-axis tracking, giving insights and future recommendations.	Dual-axis solar tracking systems were found to be more efficient at the level of PVs' performance compared to single-axis tracking systems and fixed systems.	Awasthi et al. [24]

In summary, this review paper aims to comprehensively explore various aspects of photovoltaic cooling methods. Most research concentrates on discussing specific cooling systems or evaluating them from a performance perspective, including photovoltaic–thermal collectors, concentrated PV cells, PVT systems in buildings, environmental impacts of cooling technologies, and various cooling methods such as air cooling, water-based systems, phase-change materials, and passive cooling techniques. The manuscript's novelty lies in its discussion of different technologies used in cooling PV panels while providing insights into the economic and environmental benefits of each cooling method.

This comprehensive review paper takes a unique and methodical approach to exploring various cooling methods for photovoltaic panels, distinguishing itself from previous research that often narrowly focused on specific systems or performance aspects. The goal is to provide a thorough and current analysis of advanced cooling technologies for solar systems, shedding light on both their economic and environmental benefits. Covering a diverse array of topics, from photovoltaic–thermal collectors to concentrated PV cells, the review showcases advancements in electrical and thermal efficiency, resulting in significant reductions in payback periods. This study emphasizes the critical role that cooling methods play in enhancing the sustainability and efficiency of PV systems. Noteworthy findings include the effectiveness of hybrid systems, thermoelectric, phase-change materials, and nano-based cooling methods in improving overall PV performance. Through this systematic categorization and assessment, coupled with insightful economic and environmental considerations, this research contributes valuable recommendations for future studies and advances in the realm of PV cooling methods, making a substantial contribution to the field.

The manuscripts mentioned in Table 1 provides valuable insights on future work and limitations that should be addressed that could be conducted in this field such as

- PVT collectors should take into consideration the available space of installation.
- Heat pipe PVT collectors are better in cooling than PVT collectors with refrigerants. However, their manufacturing and installation could be challenging.
- BIPVT collectors reduce the use of fossil fuels through offering savings at the level of electricity production and the materials that could be used.
- The PCM selection to be used for cooling could be challenging and depend on many factors. Studies should be performed at the level of the PCM to select the optimal one for this study.
- Pulsating flow for CPV cooling was found to increase the PV performance. It is suggested that this could be overcome through experimentation with the vibrations that come with pulsating flow for CPV collectors.

- It is suggested to study CPV cooling with the integration of porous media, PCM, or nanofluids.
- Building artificial intelligence devices to remove accumulated dust on PV panels as a means of cleaning and increasing efficiency.
- Despite the amount of research conducted in this field, more research needs to be performed to cover the different aspects of PV deterioration.

2. Theoretical Background

2.1. Principle

The phenomenon of photovoltaic energy was first discovered by Edmund Bequerel. The principle behind it is that when a photon reaches a semiconductor, two conductors are created: the free electron and the electron hole through rejection of the electrons by the negative transitional surface of the polarity. The released electrons flow to the upper layer. In the bottom layer, the electrons are transferred from one atom to the other in order to fill the empty spaces. Free electrons are conducted from the upper layer into the electric field, where the solar cell is located. The constant contact of sunlight on the surface of the solar panel ensures the continuity of electricity generation.

2.2. Parameters Affecting Panel Efficiency

Scientists and engineers found through experimental and numerical studies that different parameters other than panel temperature would affect its efficiency. Jathar et al. [25] reviewed the different environmental factors affecting PV panel efficiency. Environmental factors affecting panel efficiency are shown in Figure 1.

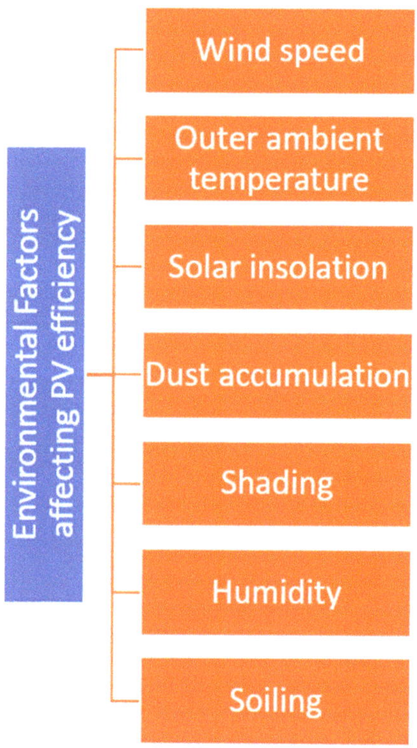

Figure 1. Environmental factors affecting the efficiency of PV panels [25].

2.3. Effect of Temperature on Panel Efficiency

Among all the mentioned parameters in Figure 1, temperature is dominant in efficiency deterioration. A PV panel absorbs approximately 80% of the incident radiation, but not all of it is converted into electricity. A definite range of wavelengths can be converted into electricity and all the others are converted into heat [26]. The remainder unconverted wavelengths can increase the solar cell temperature above the atmospheric temperature [27].

The current literature has proven the decrease in temperature coefficients (such as PV voltage and open-circuit current) with the increase in temperature [28]. Chander et al. [29] carried out an experimental study employing a solar cell simulator with varying cell temperatures, and the results showed that cell temperature has a significant effect on the PV parameters and controls the quality and performance of the solar cell.

The current literature has also shown that there are many advantages and disadvantages of using each cooling method. The advantages and disadvantages of using the standardized cooling methods of air, PCM, and water are represented in Figure 2.

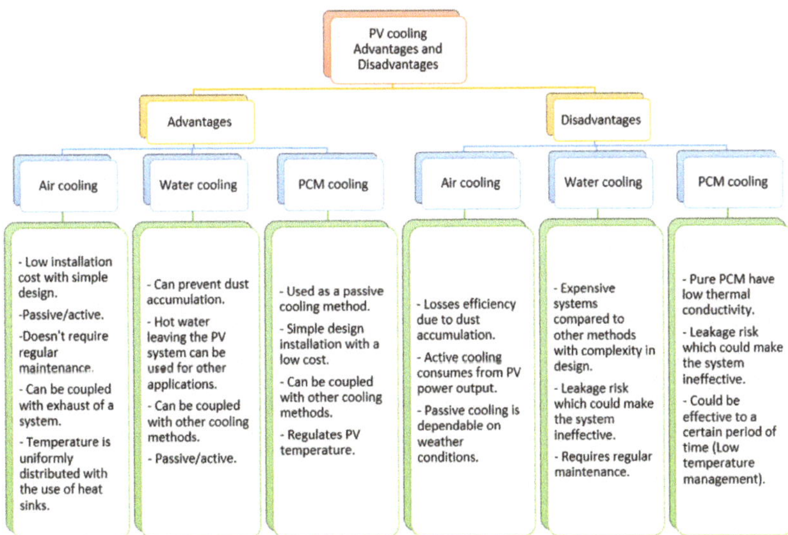

Figure 2. Advantages and disadvantages of cooling methods.

2.4. Governing Equations

Every PV panel has a length L, a width W, and a thickness t. To calculate the total area of a PV panel, then,

$$A = L \times W \tag{1}$$

where

A: Area of the P panel (m^2).

L: Length of the PV panel (m).

W: Width of the PV panel (m).

However, the effective area of the PV panel is the area which yields power. This area can be calculated as:

$$A_{eff} = A_{cell} \times nb_{cell} \tag{2}$$

where

A_{eff}: Effective area of the PV panel (m^2).

A_{cell}: Area of one cell (m^2).

nb_{cell}: Number of cells in a PV panel.

The power received from the sun is:

$$Q_{solar} = G \times A_{eff} \times \alpha \times \tau \quad (3)$$

where

Q_{solar}: Solar energy falling perpendicularly on the frontal surface of the PV panel as an input power (W).

G: Solar radiation intensity incident on the panel in (W/m^2).

α: Glass absorptivity.

τ: Glass transmissivity.

The power output of the PV panel is calculated by:

$$P_{elect} = V \times I \quad (4)$$

where

P_{elect}: Electric power output of the PV panel (W).

V: Output voltage (V)

I: Output current (A).

The output voltage and currents could be measured by mustimeters, where the voltage is measured in parallel and the current in series.

The electric efficiency of a PV panel is measured using:

$$\eta_{elect} = \frac{P_{elect}}{P_{solar}} \times 100 \quad (5)$$

where

η_{elect}: Electric efficiency $(\%)$.

P_{elect}: Output electric power (W).

P_{solar}: Input electric power (W).

The solar incident angle is the angle between the perpendicular and the incoming light from the sun. It is quantified by:

$$AOI = \cos^{-1}\left[\cos(\Theta_z)\cos(\theta_T) + \sin(\Theta_z)\sin(\theta_T)\cos(\theta_A - \theta_{A_{array}})\right] \quad (6)$$

where

Θ_z: The solar zenith angle.

θ_T: The tilt angle of the array.

θ_A: The solar azimuth angle.

$\theta_{A_{array}}$: The azimuth angle of the array.

The installation angle of a PV panel is the same as the tilt angle. It is the angle between the horizontal surface and the PV panel. It is quantified as:

For the northern hemisphere:

$$\alpha = 90^\circ - (\phi - \delta) \quad (7)$$

For the southern hemisphere:

$$\alpha = 90^\circ + (\phi - \delta) \quad (8)$$

where

ϕ: The latitude.

δ: The angle of declination.

3. PV Cooling Methods

Efficiency improvement of PV panels depends mainly on mitigating panel temperature. Figure 3 shows the three main cooling techniques in addition to other not-well-known and new techniques. The water cooling technique involves an earth water heat exchanger, solar water disinfection, a heat pipe system and an automotive radiator system. These methods are classified as either active or passive methods. The phase-change material (PCM) cooling technique is divided into organic PCM and non-organic PCM, while the air cooling method is divided into the installation of heat sinks, jet impingements, air duct or cavity air flow systems to the PV panel. These air cooling methods are classified as forced or free convection systems. Finally, non-categorized cooling methods are divided into the thermoelectric cooling method, the coating method and nanofluids. These methods are either new or not well known compared to the other cooling techniques.

Figure 3. Classification of cooling techniques.

3.1. Air Cooling Methods

The air cooling method for PV refers to the technique of dissipating heat from PV modules by circulating air around them. It can be implemented in free or forced convection, using heat sinks, fans, or blowers to increase airflow. As shown in Figure 4, natural convection occurs by the means of circulation and heat exchange between hot and cold fluids, this circulation is caused by the buoyancy effect. When the PV panel becomes hot, it warms up the layer of air surrounding it, thus the temperature of air increases, and the density increases accordingly. Consequently, hot air rises, causing a movement called a natural convection current.

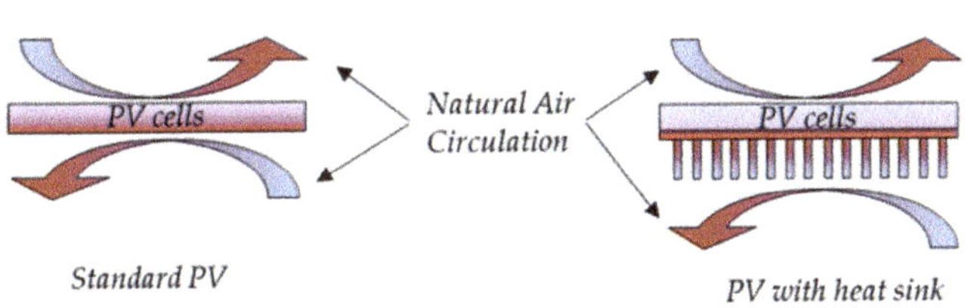

Figure 4. PV panel under free convection with or without a heat sink.

Forced convection is considered one of the most effective heat transfer mechanisms. It is characterized by using external sources such as fans, pumps, and suction devices to aid fluid transportation.

Air cooling is relatively simple and cost-effective, making it a popular choice for cooling PV systems. However, its effectiveness depends on various factors such as ambient temperature, humidity, and wind speed. Heat sinks can be used in conjunction with air cooling to further improve heat dissipation and maintain a stable operating temperature for the PV modules.

Below, we present a summary table that outlines various cooling techniques with both free and forced convection methods for photovoltaic panel cooling.

Table 2 summarizes various cooling methods applied to photovoltaic panels to enhance their efficiency under different convection conditions. The studies cover a spectrum of techniques, including forced convection with ducts and fans, free convection using multi-level fin heat sinks, and hybrid approaches combining free and forced convection with phase-change materials. Results indicate notable improvements in efficiency, ranging from 2.1% to 21.68%, with specific configurations achieving enhanced performance in different climates. Additionally, studies explore novel strategies such as curved eave and vortex generators, graphite-infused PCM, and heat spreaders with cotton wicks. Overall, the studies explore a range of cooling methods and their impacts on PV panel performance, contributing valuable insights to the field of renewable energy.

Moreover, the numerical studies in Table 2 have shown more novel approaches in the designs of the cooling methods used in cooling the PV panel. Numerical investigations shown a temperature reduction ranging between 5.89 °C and 27 °C while mainly focusing the studies on using free convection. However, experimental investigations were combining both free and forced convection and comparing their results. Air cooling was found to be effective in significant solar radiation climates, where the temperature of the air is lower than the temperature of the PV's operating temperature.

Table 2. Free and forced convection cooling methods.

Convection Method	Cooling Method	Test Methodology	Results	Climate	Author
Forced convection	PV 1: Lower duct with blower. PV 2: Duct with DC fans	Experimental and Numerical	Enhanced efficiency by 2.1% and 1.34% using fans and blower, respectively	Benha, Egypt	Hussein et al. [30]
Free Convection	Truncated multi-level fin heat sink	Numerical	Recorded a 6.13% temperature decrease and a 2.87% increase in output power	-	Ahmad et al. [31]
Free convection and forced convection	PV 1: Heat sink under free convection. PV 2: Duct under forced convection. PV 3: Fins in a duct under forced convection. PV 4: PCM (White petroleum jelly with a melting point of 37 °C)	Experimental	Improvements in efficiency by 0%, 33%, 53%, and 72% for PCM, heat sink under free convection, duct under free convection, and duct under forced convection, respectively	Kumasi, Ghana	Abdallah et al. [32]
Free convection	PV 1: Free convection using through-holes in the PV panel. PV 2: Active water spraying on the surface. PV 3: Passive and active cooling using through-holes in the PV and water spraying on surface	Numerical	Hybrid cooling resulted in an average reduction of 17.24 °C	-	Pomares-Hernández et al. [33]
Forced convection	Curved eave and vortex generators	Numerical	Achieved a 5.89 °C temperature reduction	-	Wang et al. [34]
Forced convection	PVT system under forced convection by DC fans	Experimental	Electric efficiency between 12% and 12.4% with 0.05 m channel depth and 0.018 kg/s to 0.06 kg/s air mass flow rate	Tehran, Iran	Kasaeian et al. [35]
Free convection	PV 1: 30 mm graphite-infused PCM (paraffin wax with a melting point of 40 °C). PV 2: Finned heat sink. PV 3: Finned heat sink with graphite-infused PCM	Experimental and Numerical	Finned heat sink with graphite-infused PCM demonstrated an overall efficiency increase of 12.97%	New Zealand (Laboratory)	Atkin et al. [36]

Table 2. *Cont.*

Convection Method	Cooling Method	Test Methodology	Results	Climate	Author
Free convection	Heat spreader with cotton wicks	Experimental	Recorded a 12% decrease in temperature and a 14% increase in electric output	Tamil Nadu, India	Chandrasekar et al. [37]
Free convection	Twisted baffle at the rear surface of the PV	Numerical	Efficiency increases by 1.21% and 3.36% for solar radiation of 200 W/m^2 and 1000 W/m^2, respectively	-	Benzarti et al. [38]
Free convection	PV 1: L-profile aluminum fins with parallel configuration. PV 2: L-profile aluminum fins randomly positioned	Experimental	Electric efficiency increased by 2% for L-profile aluminum fins with random distribution	Split, Croatia	Grubišić-Čabo et al. [39]
Forced convection	PV/T system with rectangular finned plate	Experimental	Recorded a maximum efficiency of 13.75% for 4 fins under solar radiation of 700 W/m^2 and a mass flow rate of 0.14 kg/s	Malaysia	Mojumder et al. [40]
Free convection	Cooling tower with PV module	Numerical	Averaged 6.83% increase in annual efficiency of the PV	-	Abdelsalam et al. [41]
Forced convection	PV 1: Air from above and water from below. PV 2: Air from above and below. PV 3: Air from above. PV 4: Air from below. PV 5: Water from below	Numerical	Water below the PV panel decreased the temperature by 21 °C	Sakaka Al-Jouf, KSA	Soliman [42]
Free convection	Effect of using the racking structure of the PV panel system as a passive heat sink for cooling	Experimental and Numerical	Achieved a 3% increase in electric efficiency with a 6.3 °C PV temperature reduction	Dammam, Saudi Arabia	El-Amri et al. [43]
Free convection	Investigated the use of different dimensions of a finned plate in cooling the PV panel	Experimental	Utilizing a 7 cm by 20 cm staggered fin array resulted in the best performance with an energy efficiency of 11.55%	Elazig, Turkey	Bayrak et al. [44]
Free convection	Studied the effect of dust accumulation density on the convective heat transfer coefficient for a large-scale PV panel array	Experimental	Increased convective heat transfer coefficient by 4.13% compared to a clean PV module	Zhongwei, Ningxia province in China	Hu et al. [45]
Free and forced convection	PV 1: PV-duct under free convection. PV 2: PV-duct under forced convection. PV 3: PV-duct under forced convection with L-shaped barrier	Experimental	The highest electric efficiency of 21.68% was recorded by the PV panel under forced convection with an L-shaped barrier in its duct	India	Kumar et al. [46]
Free convection	PV–heat sink system with different fin dimensions	Numerical	The initial heat sink model was able to cool the PV panel by 27 °C	Dubai, UAE	Mankani et al. [47]
Free convection	PV module with porous material. This study was performed on three porous fins, a porous layer, and five porous fins	Numerical	Increase of 6.73%, 8.34%, and 9.19% in efficiency for the three porous fins, porous layer, and five porous fins configurations, respectively	-	Kirwan et al. [48]
Forced convection	PV-compressed air module	Numerical	This method improved the output power of the PV panel and as a result, improved its efficiency	-	Li et al. [49]

3.2. *Water Cooling Methods*

PV water cooling methods are a set of techniques that involve the use of water or other fluids to absorb and dissipate heat from PV panels, with the goal of improving their electrical performance and prolonging their lifespan. These methods can be implemented through passive or active means and may involve the use of heat sinks, heat exchangers, direct water immersion, or other related approaches. The effectiveness of PV water cooling methods depends on various factors, such as water flow rate, temperature, and quality, as well as the design and construction of the cooling system. Table 3 represents the different cooling techniques that are either passive or active.

Table 3. Classification of passive and active water-based cooling techniques.

Passive Cooling Techniques	Active Cooling Techniques
Liquid immersion Heat pipe	Earth water heat exchanger Solar water disinfection Automotive radiator

Passive cooling techniques for cooling PV systems refer to natural methods used for reducing the temperature of PV modules without the use of mechanical or electrical devices. They rely on convection, radiation, and evaporation to dissipate heat and improve the performance and lifespan of PV modules.

Tina et al. [50] have increased the electrical efficiency by approximately 10% after experimentally submerging a PV panel inside water in a study of enhancing PV temperature. Figure 5 represents the submerged PV inside a vessel containing water.

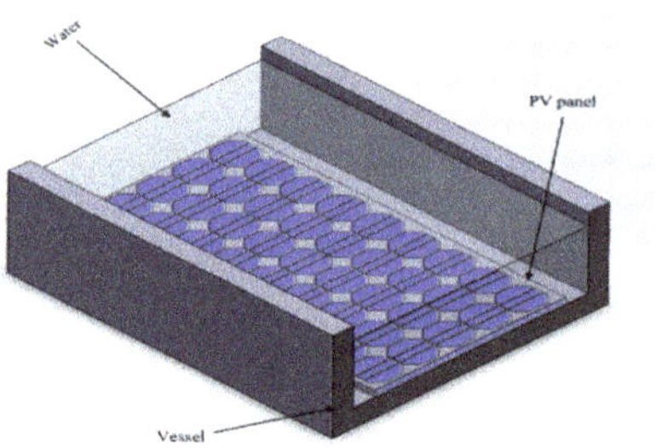

Figure 5. PV panel immersed in water [50].

On the other hand, active water cooling for PV required a mechanical or electrical devices to actively reduce the temperature of PV modules. This may include circulating water or other fluids through a heat exchanger. They are useful in hot climates or high-power output systems and provide greater cooling efficiency and control over operating temperature than passive cooling methods.

Irwan et al. [51], carried an indoor experiment in order to investigate the effect of water flowing at the surface in cooling the PV panel. Results showed that a decrease in PV temperature by 5–23 °C increases the output power of the PV panel by 9–22%.

On the other hand, Moradgholi et al. [52] experimentally investigated the effect of heat pipes in cooling PV panels, and the module used in his experimental study is represented in Figure 6. Results showed an increase of 5.67% in power when using methanol as a working fluid in spring and an increase of 7.7% in power when using acetone as a working fluid in summer.

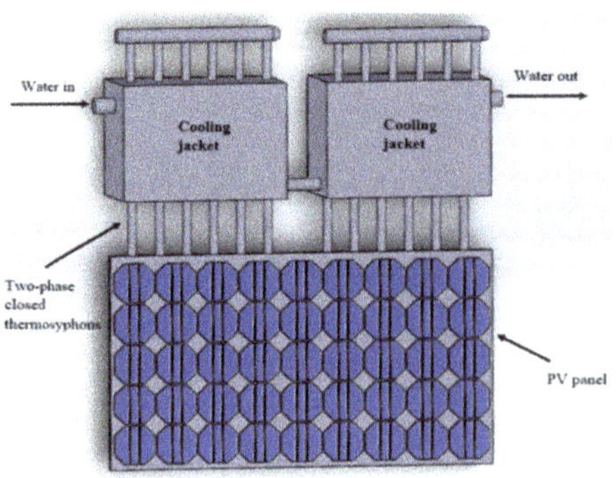

Figure 6. Heat pipes module [52].

Moreover, Sandeep Koundinya et al. [53] investigated experimentally and by simulation the effect of a finned heat pipe with water as the working fluid in cooling photovoltaic panels. Results showed a total decrease of 13.8 K in PV panel temperature and good agreement was found between experimental and computational studies.

Below, a summary table is presented for several studies about cooling PV modules with passive and active cooling techniques.

Table 4 presents a wide array of outcomes across various cooling methods for photovoltaic panels. Passive approaches, like water-saturated microencapsulated phase-change materials (MEPCM) and immersion in dielectric liquids, effectively reduce temperatures, leading to improved electric efficiency. Passive cooling techniques exhibit diverse results, with efficiency enhancements ranging from 2.7% to 12.4% and a temperature reduction of up to 13.8 K. Active cooling methods, such as spraying water and flowing water on the PV surface, consistently boost power generation and efficiency, demonstrating improvements from 8% to 9% to a significant 24 K temperature decrease. Innovative methods like floating PV on water surfaces and geothermal cooling systems show efficiency increases of 2.7% and up to 13.8%, respectively. The choice between passive and active cooling depends on factors like climate, available resources, and desired efficiency levels. These findings collectively contribute to advancing PV panel cooling, facilitating more efficient and sustainable solar power generation.

Table 4. Summary of several studies on water-based cooling techniques for PVs.

Cooling Method	Cooling Classification	Test Methodology	Key Outcomes	Climate	Author
Water-saturated microencapsulated phase-change material (MEPCM)	Passive	Numerical	A layer of PCM of 5 cm thickness with a melting temperature of 30 °C gave the best performance in enhancing the electric efficiency.	-	Ho et al. [54]
Liquid immersion of solar cells in 4 different dielectric liquids.	Passive	Numerical	Immersing the solar cells in the dielectric liquids maintained a low temperature in the solar cells.	-	Liu et al. [55]
Spraying water on frontal and rear surfaces.	Passive	Experimental	Increase in power and efficiency by 16.3% and 14.1%, respectively.	Croatia	Nizetic et al. [56]
Finned heat pipe system with water as a working fluid.	Passive	Experimental and Numerical	A total decrease of 13.8 K in PV panel temperature and good agreement was found between experimental and computational studies.	India	Koundinya et al. [53]
PV/T system with water and ethylene glycol as working fluids.	Passive	Experimental and Numerical	Water was found to be a better coolant than ethylene glycol with an overall efficiency enhancement by 25%.		Joy et al. [57]
Spraying water on surface.	Active	Experimental and Numerical	Cooling system have good performance in hot and dusty regions.	Egypt	Moharram et al. [58]
Flat-plate PV/T system with and without glass cover.	Active	Numerical	Empirical correlations were performed and conclusions were conducted.	-	Bajestan et al. [59]
Flowing water on PV surface.	Active	Experimental	Increase in power by 8–9%	Laboratory	Krauter [60]
Heat pipe.	Active	Numerical	As the convective heat, transfer coefficient increases the solar cells temperatures decreases when operating at low flow rates and at high optical concentration ratios.	-	Sabry [61]
Spraying water on the PV surface.	Active	Experimental	Increase of 2.7% in electrical efficiency and 21 W in power.	Alexandria, Egypt	Elnozahy et al. [62]
Flowing water on the surface.	Active	Experimental	Increase in the power generated and in total efficiency.	Iran	Kordzadeh et al. [63]
Water system with air blowing to the back of the PV.	Active	Numerical	Yearly improvement of 5% in efficiency.	-	Arcuri et al. [64]
Earth water heat exchanger.	Active	Numerical	Increasing the length of the feed pipe to 60 m would decrease PV temperature by 23 °C.	Pilani, Rajhasthan, India	Jakhar et al. [65]

Table 4. *Cont.*

Cooling Method	Cooling Classification	Test Methodology	Key Outcomes	Climate	Author
Concentrated PV/T system.	Active	Numerical	Empirical correlations were performed and conclusions were conducted.	-	Mittelman et al. [66]
Automotive radiator.	Active	Experimental and Numerical	Theoretical heat rejection by 91% and experimental efficiency increased by 4.46%.	Kuala Lumpur, Malaysia	Chong et al. [67]
Solar desalination combined with an intermittent solar-operated cooling unit.	Active	Experimental	A 13.75% energy efficiency for the system.	Cairo, Egypt	Ibrahim et al. [68]
PV/T system laminated with polymer matrix composite with water as a coolant.	Active	Experimental and Numerical	The maximum efficiency recorded was 20.8% with a 53.5% thermal efficiency.	-	Korkut et al. [69]
PV 1: single-pass ducts. PV 2: multi-pass ducts. PV 3: tube-type heat absorber. Water is used as a fluid.	Active	Numerical	Cell temperature achieved a maximum of 38.310 °C.	Islamabad, Pakistan	Sattar et al. [70]
Flowing water on the PV surface.	Active	Experimental (laboratory and real-life conditions)	The system showed a temperature decrease of 24 K with a power generation increase of 10% with a return on investment of less than 10 years.	Krakow, Poland	Sornek et al. [71]
Floating PV on the water surface.	Passive	Experimental	An efficiency increase of 2.7% was recorded with a temperature decrease of 2.7 °C	Cagliari, Italy	Majumder et al. [72]
A new innovative cooling box acting as a thermal collector.	Active	Numerical	Electric efficiency of 17.79% and a thermal efficiency of 76.13% when the system was studied with a mass flow rate of 0.014 kg/s and an inlet water temperature of 15 °C.	-	Yildirim et al. [73]
Water flows on the surface of the PV panel.	Passive	Experimental	An increase in exergy efficiency from 2.91% to 12.76%.	Sisattanark district, Vientiane Capital, Laos	Chanphavong et al. [74]
Comparison between water flowing on the surface of the PV panel and wet grass cooling.	Passive	Experimental	Running water on the upper surface of the PV helps in cooling it and increasing its efficiency.	Gwalior, India	Panda et al. [75]
Comparison between conventional PV panels, concentrated PV systems, and water-cooled concentrated PV systems.	Active	Experimental and Numerical	Significant increase in the efficiency and power output of the water-cooled CPV system to 17% and 23%, respectively. The overall output power of the water-cooled CPV was 24.4%.	Duhok, North of Iraq	Zubeer et al. [76]
A geothermal cooling system containing a mixture of water and ethylene glycol.	Active	Experimental	An increase in electric efficiency up to 13.8% using a constant coolant flow rate of 1.8 L/min.	Alcalá de Henares, Madrid, Spain	Lopez-Pascual et al. [77]
Radiative cooling module.	Active	Experimental	Increase in efficiency by 1.21% and 0.96% in summer and autumn, respectively, for the system without cold storage. For the system with a cold storage, the efficiency increased by 1.69% and 1.51% in summer and autumn, respectively.	China	Li et al. [78]
Porous media with water as a cooling fluid.	Active	Experimental and Numerical	Decrease by 35.7% of PV's surface temperature and increase by 9.4% in the output power under a volume flow rate of 3 L/m with a porosity of 0.35.	Jordan	Masalha et al. [79]
Geothermal heat exchanger with water and ethylene glycol as cooling fluids.	Active	Experimental and Numerical	Increase in PV's electric power generation by 9.8%.	Turkey	Jafari et al. [80]
Water cooling system and phase-change material (PCM) module with OM35 as a PCM with a melting point of 35 °C.	Passive	Experimental	Increase in the electric efficiency by 12.4% compared to the other configurations.	Chennai, India	Sudhakar et al. [81]

The use of evaporative cooling could be more beneficial than vapor compression at the level of the cost. However, the system is not reliable or needs more design work [82].

Moreover, it was noticed in the water-cooled methods that the experimental studies mentioned in Table 4 were greater than the numerical studies and the climates the water cooling methods were studied in are hot such as India, Egypt, and Saudi Arabia.

3.3. PCM Cooling Methods

Phase-change materials (PCMs) are substances used in cooling systems for photovoltaic modules to absorb and store heat from the panels during peak sunlight hours. PCMs have a high latent heat of fusion, which means they can absorb large amounts of heat without a significant increase in temperature. PCMs can be integrated into PV panels, or used in a separate thermal management system to enhance the overall efficiency and lifetime of the PV system.

In a typical PV–PCM hybrid system, illustrated in Figure 7, the PCM functions as a heat sink that absorbs excess heat from the PV panel, thereby reducing its temperature. During the peak sun hours, the temperature of the PV panel exceeds the melting temperature of the PCM. As a result, the PCM absorbs excess heat from the PV panel and maintain a stable operating temperature for the PV system until it completely melts, transitioning from solid to liquid phase. During the low sunlight period, as the ambient temperature decreases and the temperature of the PV panel drops below the melting point of the PCM, the PCM releases excess heat and solidifies again.

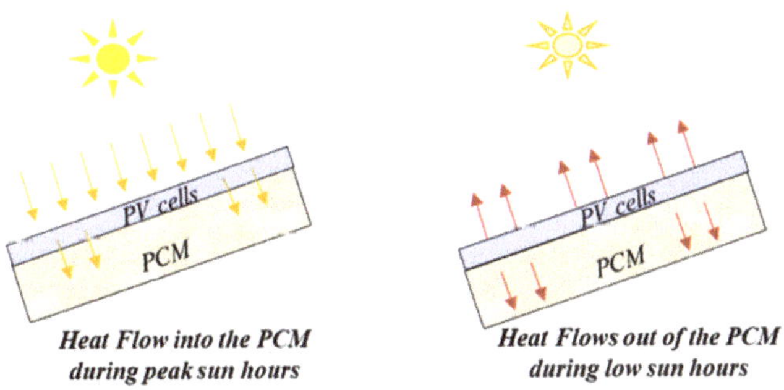

Figure 7. Typical PV–PCM system [83].

Table 5 summarizes the various cooling techniques using PCM with different combinations and materials.

Table 5. Summary of PV cooling techniques based on PCMs.

PCM Used	PCM Melting Point	Cooling Method	Test Methodology	Key Outcomes	Climate	Author
Pure PCM: white petroleum jelly. Combined PCM: white petroleum jelly + graphite + copper	36–60	Pure and combined PCM	Exp.	Efficiency increased by an average of 3% when using pure PCM and by an average of 5.8% when using combined PCM.	Bekaa Valley, Lebanon.	Hachem et al. [84]
-	0–50	PV panel containing an integrated layer of PCM	Num.	Efficiency exceeds 6% in some regions.	-	Smith et al. [85]
RT25	25	Impure PCM layer integrated into the PV panel	Num.	Maintain panel operating temperature under 40 °C for 80 min under solar radiation of 1000 W/m^2.	-	Biwole et al. [86]
Salt hydrate, $CaCl_2{\cdot}6H_2O$ and eutectic of capric acid–palmitic acid	$CaCl_2{\cdot}6H_2O$: 29.8 and eutectic of capric acid-palmitic acid: 22.5	PCM layer with aluminum alloy fins integrated into the PV panel	Exp.	$CaCl_2{\cdot}6H_2O$ showed an increased power output of 3% compared to capric-palmitic acid in Pakistan. The two PCMs showed better results in Vehari, Pakistan than in Dublin, Ireland with a total of 13% in power saving.	Dublin, Ireland and Vehari, Pakistan	Hasan et al. [87]
Paraffin wax	37.5–42.5	PV–PCM system	Exp.	Average maximum efficiency and power were increased by 1.63% and 1.35 W, respectively.	Laboratory	Xu et al. [88]
Rubitherm 28 HC and Rubitherm 35 HC	Rubitherm 28 HC: 27–29 and Rubitherm 35 HC: 34–36	PV–PCM system	Num.	Increase by 10% in peak power and 3.5% in energy produced throughout the whole year round.	-	Aneli et al. [89]

Table 5. *Cont.*

PCM Used	PCM Melting Point	Cooling Method	Test Methodology	Key Outcomes	Climate	Author
Docosane paraffin wax	42	PV–PCM system	Num. and Exp.	An increase of 1.05% in efficiency and a 34% increase in life span.	Doha, Qatar.	Amalu et al. [90]
RT35HC	36	PV–PCM system	Num.	Temperature reduction by 24.9 °C and an increase of 11.02% in electric output.	-	Zhao et al. [91]
RT35	35	PV–PCM system	Num.	Total increase of 5% in productivity.	-	Kant et al. [92]
-	24.85	PV–PCM system	Num. and Exp.	Increase of 1–1.5% in electric efficiency.	Song-do, Incheon, South Korea	Park et al. [93]
RT28HC	28	PV–PCM system	Exp. and Num.	Increase in power by 9.2% experimentally and 4.3–8.7% numerically.	Ljubljana, Slovenia	Stropnik et al. [94]
-	23	BIPV–PCM	Exp. and Num.	Maximum electric and thermal efficiencies recorded were 10% and 12%, respectively.	Lisbon, Portugal	Aelenei et al. [95]
Paraffin wax	34.9–42	PV–PCM system and PV–PCM thermal system	Exp.	An electric output increase of 5.18% in the PV–PCM system and 30.4% electric sum was recorded in the PV–PCM-T system.	Shanghai, China	Li et al. [96]
RT42	38–43	PV–PCM system	Exp.	Annual enhancement of 5.9% in electric yield in hot climate.	Al Ain, United Arab Emirates	Hasan et al. [97]
RT27 & RT31	RT27: 25–28 and RT31: 27–33	PV–PCM systems	Exp.	Enhancement in energy by 4.19% and 4.24% when using RT27 and RT31, respectively.	Chania, Greece	Savvakis et al. [98]
Eutectic of capricpalmitic acid and calcium chloride hexahydrate and RT20 and RT25 and RT35	Eutectic of capricpalmitic acid: 22.5 and calcium chloride hexahydrate: 29.8 and RT20: 25.73 and RT25 26.6 and RT35: 29–36	PV-T-nano-PCM system	Num.	Increase in electric efficiency by 6.9% and 22% in winter and summer weather, respectively.	Dhahran, Saudi Arabia	Abdelrazik et al. [99]
Paraffin A44	44	PVT–PCM system	Exp. and Num.	Electric performance increased by 7.2% numerically and 7.6% experimentally.	Kuala Lumpur, Malaysia	Fayaz et al. [100]
Lauric acid	44–46	PVT–PCM system	Exp.	PVT–PCM system increased the electric efficiency of the PV by 1.2% under a volume flow rate of 4 LPM.	Kuala Lumpur, Malaysia	Hossain et al. [101]
Paraffin wax	46–48	PVT–PCM system with pure water and ethylene glycol as working fluids	Exp.	Energy loss percentage was decreased by 9.28%, 23.33%, and 48.58% for the PVT/water, PVT/ethylene glycol (50%), and PVT/ethylene glycol (100%).	Mashhad, Iran	Kazemian et al. [102]
RT25	26	PV–PCM and PV–PCM–fins systems	Num.	Increase of 2.5% and 3.5% in electric efficiency in fair and sunny weather, respectively.	Different weather conditions	Metwally et al. [103]
RT58, RT42, and C22-C40	RT58: 58 and RT42: 42	PV–PCM–heat sink system	Num.	Temperature drop of 18.3 °K, 21.2 °K, and 26.1 °K when using C22-C40, RT58, and RT42, respectively.	Oujda, Morocco	Bria et al. [104]
-	273.15 K	PV–PCM matrix absorber system	Num.	Analytical and numerical results were in agreement.	-	Hassabou et al. [105]
-	-	PV/T system with nano-enhanced MXene-PCM and R407C working fluid	Num.	Power output increased by 535 KWh/year and electric efficiency increased by 3.01%.	Derby, United Kingdom	Cui et al. [106]
-	-	PV–PCM system with a heat sink with convex/concave dimples	Num.	PV cell temperature decreased by 7.14%, 4.65%, and 2.22% when studying the PV cells at inclinations of 90°, 60°, and 30°, respectively.	-	Soliman et al. [107]
Paraffin wax	38–43	PV–PCM system	Exp.	Efficiency was improved by 14.4% when using a PCM thickness of 3 cm and tilting the PV at an angle of 30°.	Qena, Egypt	Maghrabie et al. [108]
Paraffin wax and vaseline	Paraffin wax: 45 Vaseline: 25	Water-cooled PVT system with PCM	Exp.	Increase in electric and thermal efficiencies up to 13.7% and 39%, respectively.	Indoor (Simulating Iraq's weather)	Chaichan et al. [109]
RT28HC	25–29	PV–PCM system	Exp.	Enhancement by 2.5% in the power output.	Mediterranean climate	Nizetic et al. [110]

Table 5 presents a comprehensive overview of different phase-change materials utilized in conjunction with photovoltaic (PV) panels. Each entry includes details on the specific PCM used, its melting point, cooling method, test methodology, key outcomes, climate conditions, and the contributing authors. Noteworthy PCMs like white petroleum jelly, paraffin wax, and specialized formulations such as Rubitherm 28 HC and Rubitherm 35 HC are explored across various cooling systems. Passive cooling methods incorporating PCMs exhibit efficiency gains ranging from 1.05% to 12.4%. On the other hand, active systems like PV–PCM configurations and PV/T systems consistently showcase improvements in electric efficiency and power output, reaching up to 24%. These findings underscore the impact of factors such as climate, location, and PCM composition on the effectiveness of these cooling techniques. Overall, Researchers have identified optimal PCM parameters, thicknesses, and integration methods, contributing to the advancement of efficient photovoltaic cooling strategies.

Moreover, cooling by PCM is shown to be used in hot climates where solar irradiation is considered to be high such as in Egypt, Iran, Saudi Arabia, and United Arab Emirates.

3.4. Other Cooling Methods

There are various cooling methods for photovoltaic systems other than air, water, and phase-change materials. One of these methods is using encapsulated PCM, which improves the PCM's expansion property during melting. Another method is thermoelectric cooling, which uses the Peltier effect to create a temperature difference and transfer heat away from the PV module. Additionally, researchers have explored the use of nanofluids, which are liquids containing nanoparticles that can improve the thermal properties of the cooling fluid. Other methods include using refrigeration systems or hybrid systems that combine multiple cooling methods. Each of these cooling methods has its own advantages and disadvantages and can be suitable for different types of PV systems and operating conditions.

Saleh et al. [111] numerically studied the effect of nanofluid and water in cooling photovoltaic–thermal (PVT) collectors. Results showed that the use of 1% volumetric fraction of nanofluids increases the thermal efficiency up to 19.5% and the electric efficiency up to 55.45%.

Ghadiri et al. [112], experimentally studied the effect of cooling a PVT system by ferrofluids shown in Figure 8. Different fluids were used at a constant and an alternating magnetic field in order to discuss the effect of magnetic field on ferrofluids. Results showed an increase of 45% in the overall efficiency when using ferrofluid and a total increase of 50% in the overall efficiency when using an alternating magnetic field of 50 Hz frequency. Also, a total of 48 W of exergy was increased after using ferrofluid.

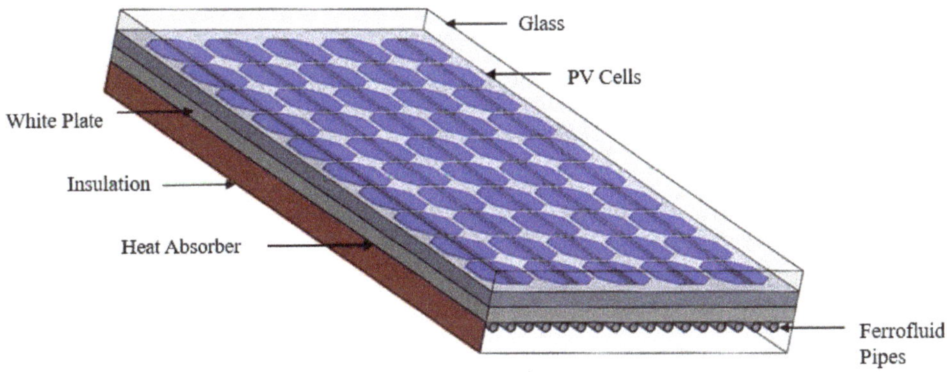

Figure 8. Ferrofluids cooling system [112].

Figure 9 illustrates the use of thermoelectric water-nanofluid cooling and thermoelectric finned heat sink cooling in PV/T and PV systems respectively [113].

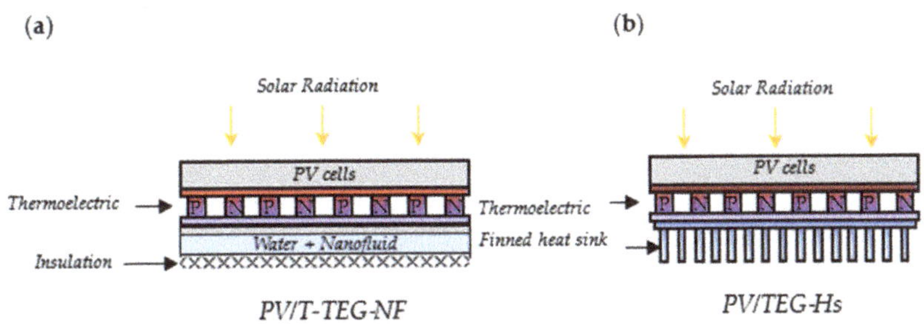

Figure 9. PV with thermoelectric plate and: (**a**) photovoltaic/thermal-thermoelectric water-nanofluid (PV/T-TEG-NF) cooling system and (**b**) photovoltaic-thermoelectric finned heat sink (PV/TEG-Hs) cooling system [113].

Table 6 shows the recent studies performed on cooling the PV panel using different methods.

Table 6. Different cooling techniques.

Cooling Method	Test Methodology	Key Outcomes	Climate	Author
Microencapsulated PCM heat sink with a thermoelectric generator	Experimental	Increase in efficiency by 2% in the intermediate season and by 2.5% in the summer.	Republic of Korea	Kang et al. [114]
PCM-integrated PV system with fins and nanofluid (CPV/T/NF/FPCM)	Experimental	Electric efficiency was improved up to 17.02% while thermal efficiency was improved up to 61.25%.	Tehran, Iran	Kouravand et al. [115]
Photovoltaic thermal collector with a nano-PCM and micro-fin tube nanofluid system	Experimental	The micro-fins, nanofluids, and nano-PCM PV had a thermal efficiency of 77.5% with an increase in electric power of 4.01 W.	Indoor (Solar Simulator)	Bassam et al. [116]
Micro-fin tube counterclockwise twisted tape nanofluid and nano-PCM	Experimental	Increase of 44.5% in electric power.	Indoor (Solar Simulator)	Bassam et al. [117]
PV/nano-enhanced PCM heat sink system	Experimental	The GNP-CuO 3% mixture has enhanced the thermal conductivity by 91.81%, reduced temperature by 6.6 °C, and enhanced the electricity output by 3%.	Iran	Moein-Jahromi et al. [118]
PCM, thermoelectric cooling, and installing fins made of aluminum in cooling the PV panel	Experimental	The PV panel with aluminum fins had the highest power generation enhancement of 47.88 watts.	Elazig, Turkey	Bayrak et al. [119]
PV/T with spectrum-splitting module	Numerical	Conversion efficiency exceeded 43%.	-	Xu et al. [120]

The outcomes presented in Table 6 highlight the diverse and innovative cooling methods for photovoltaic panels. The utilization of a microencapsulated phase-change material combined with a heat sink, and a thermoelectric generator, demonstrated a 2% efficiency increase in the intermediate season and 2.5% in summer. The integration of PCM with fins and nanofluid (CPV/T/NF/FPCM) showed significant improvements, achieving an electric efficiency of 17.02% and a thermal efficiency of 61.25%. Indoor experiments involving a photovoltaic thermal collector with Nano-PCM and micro-fin tube nanofluid revealed a remarkable thermal efficiency of 77.5% and a 4.01 W increase in electric power. Another noteworthy system, incorporating a micro-fin tube counter clockwise twisted tape nanofluid and nano-PCM, demonstrated a substantial 44.5% increase in electric power. The PV/nano-enhanced PCM heat sink system displayed enhancements, including a 91.81% increase in thermal conductivity, a 6.6 °C temperature reduction, and a 3% improvement in electricity output. Further experiments, incorporating PCM, thermoelectric cooling, and aluminum fins, yielded the highest power generation enhancement of 47.88 Watts. Additionally, a numerical simulation of a PV/T system with a spectrum-splitting module revealed an impressive conversion efficiency exceeding 43%. These advancements hold promise for improving the energy efficiency and sustainability of photovoltaic panels and increasing the adoption of renewable energy sources.

4. Discussion and Analysis

The literature has provided numerous methods for cooling the PV panel and increasing its efficiency, resulting in methods with more effectiveness over others. With different methods of cooling, different ranges of efficiency arise that were obtained and illustrated in Figure 10.

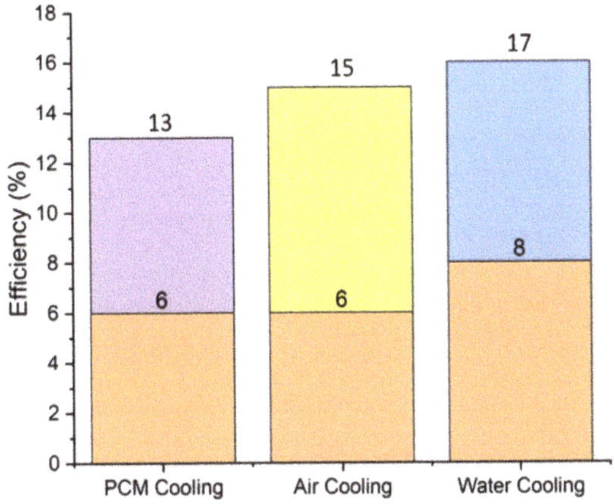

Figure 10. Efficiencies of cooling methods.

According to the literature, efficiency ranged between 6% and 13% when PCM was used as a cooling technique. This method had both advantages and disadvantages. Through all the previous studies, the main problem that faced the researchers is the low thermal conductivity of PCM, and the change in volume when PCM melts, which in turn leads to poor temperature management. Researchers tried to solve these problems through several ways such as mixing PCM with graphite and developing a shape-stabilized PCM. In contrast, the advantages of this method were the simplicity of the cooling system, the low cost, and the long lifetime. With the need for only a tank filled with PCM attached to the back of the PV panel, the price of this system was low, and with the absence of electrical instruments, there will be no need for maintenance.

The air cooling techniques literature revealed a range of efficiency between 6% and 15%, and several methods were tested experimentally and numerically based on natural convection and forced convection. Several systems were tested by scientists such as finned plates, fans and air ducts, finned plates combined with fans and air ducts, and jet impingements. Others tried to combine the effects of the latent heat storage of PCM along with finned plates under natural and forced convection. The simplest method and most effective was using finned plates under forced or natural convection. Under forced convection, a better efficiency was recorded but a higher cost compared with the use of finned plates under natural convection. Therefore, there is no method better than another in general; but in specific conditions, optimization between efficiency and cost can be achieved. In a windy location, a finned free convective system will give great efficiency with low cost, while in a non-windy location, a finned forced convective system would cost a little bit more but will give a higher efficiency.

The highest efficiency was recorded when water cooling systems were tested. Different techniques were taken into consideration, spraying water over the surface of the panel, immersion of the panel in water, using water as a circulation fluid in heat pipes attached to the back of the PV, etc. Efficiency with water systems ranged in the literature between 8% and 17%, but designing systems to deal with water had a high cost because of the need for pumps, pipes, fittings, etc. In addition, when taking the location of the project into consideration, a water cooling system was the best technique in dusty or sandy places, where high efficiency could be maintained by removing dust from the front surface of the panel which would otherwise reduce the amount of irradiation received.

5. Economic Study

After cooling the PV panels, cooling techniques showed an increase in power for each PV panel with different increased values. This increase in power showed a remarkable increase financially when compared to the standard PV. Economic and environmental analyses were conducted on a PV panel with an area of 0.218 m^2.

The governing equations used in the economic study are presented as follows.

$$E = I \times A_{effective} \times \eta_{electrical} \times 30 \quad (9)$$

where

E: The energy produced by the PV panel in kWh.

I: The average solar insolation per day in $\frac{kWh}{m^2 \times day}$.

$A_{effective}$: The effective area of the PV panel in m^2.

$\eta_{electrical}$: The electrical efficiency of the PV panel.

The relative efficiency is the relative difference between the cooled PV efficiency and the standard PV. This relative efficiency is used to calculate the absolute electrical efficiency of each cooled PV case and is quantified as:

$$\eta_{relative} = \frac{\eta_2 - \eta_1}{\eta_1} \times 100 \quad (10)$$

where

η_2: The efficiency of cooled PV (%).

η_1: The efficiency of standard PV (%).

Savings were quantified as:

$$Savings = E \times S \quad (11)$$

where

E: The energy produced by the PV panel in kWh.

S: The price of each kWh.

5.1. Water Cooling

The sun hours vary according to the months of the year. Figure 11 shows the variation in the sun hours with respect to the months in Lebanon. As shown in the following figure, July month reached the maximum of 438.2 h.

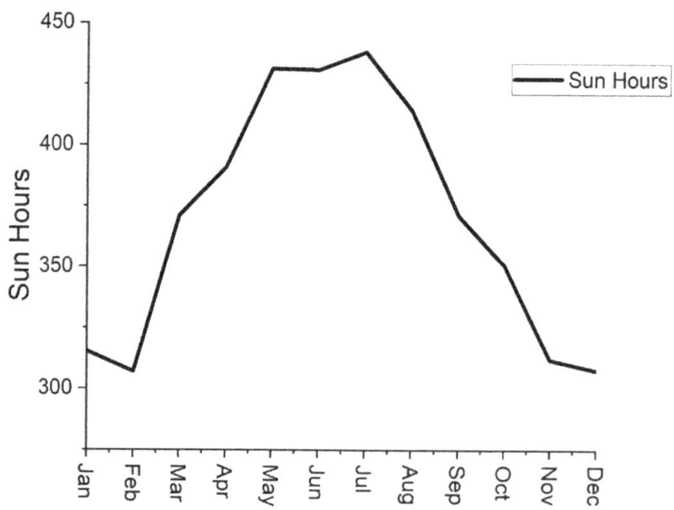

Figure 11. Number of sun hours versus months.

The solar insolation in Beirut, Lebanon is shown in Figure 12 [121].

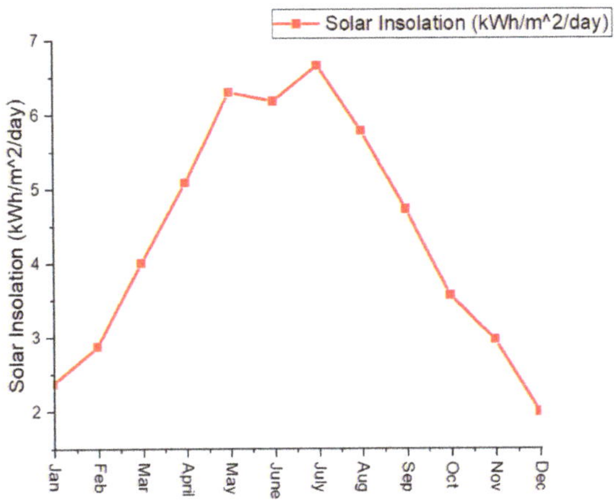

Figure 12. Solar insolation in Beirut, Lebanon [121].

As shown in Figure 12, the minimum solar insolation was recorded in December, with a value of 2 $\frac{kWh}{m^2}$ per day. The highest solar insolation was recorded in July, with a value of 6.67 $\frac{kWh}{m^2}$ per day.

Water cooling methods were found to be effective in cooling the PV panels. As shown in Figure 13, flowing water on the surface of the PV panel was found to produce the maximum energy, with an average of 32.29 kWh compared to the other cooling methods. Following this method, the liquid immersion method, with an average of 32.17 kWh, proved to be the next best. Also, the heat pipe cooling system recorded an average of 31 kWh, while the automotive radiator system recorded the least energy between the cooling methods, with an average of 30.55 kWh. The standard PV panel recorded an average of 29.24 kWh.

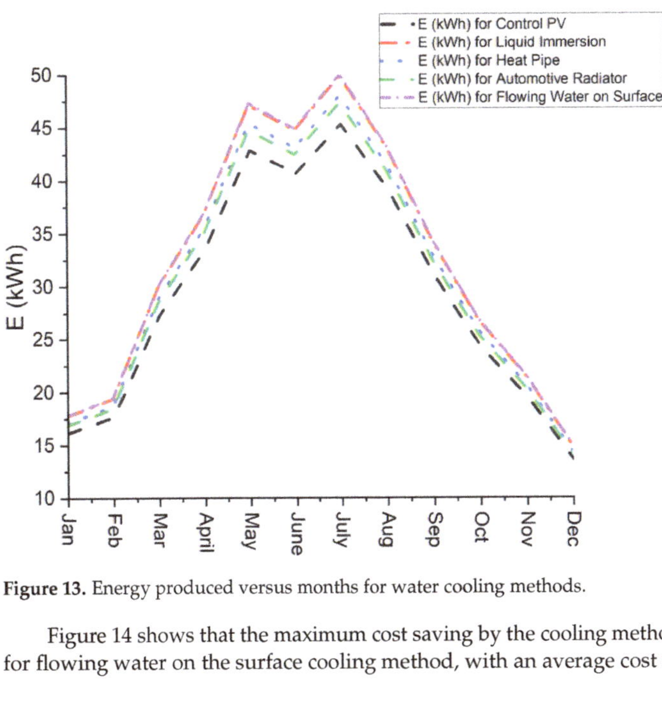

Figure 13. Energy produced versus months for water cooling methods.

Figure 14 shows that the maximum cost saving by the cooling methods was recorded for flowing water on the surface cooling method, with an average cost saving of United

States Dollar (USD) 0.273. The liquid immersion method follows, with an average cost saving of USD 0.263, and the heat pipe cooling method showed an average cost saving of USD 0.157. The automotive radiator cooling method showed the lowest average cost saving, as shown in the following figure with an average compared to a standard PV panel of USD 0.117.

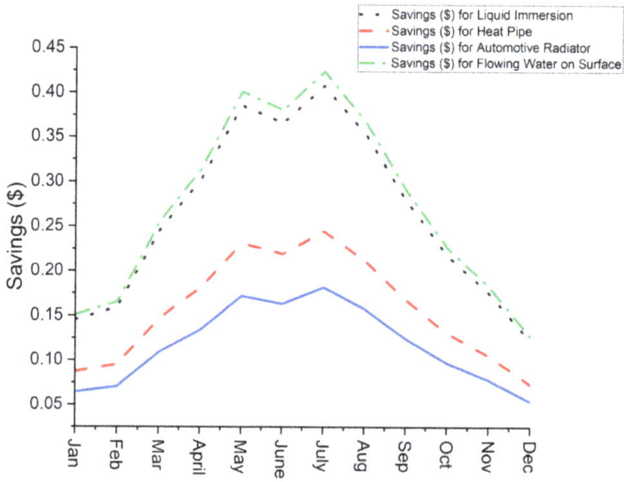

Figure 14. Cost savings versus months for water cooling methods.

5.2. Air Cooling

In air cooling methods, the exhaust air cooling method was found to produce the maximum energy output, with an average of 32.201 kWh. Figure 15 shows the variation in energy produced for air cooling methods with respect to the months of the year.

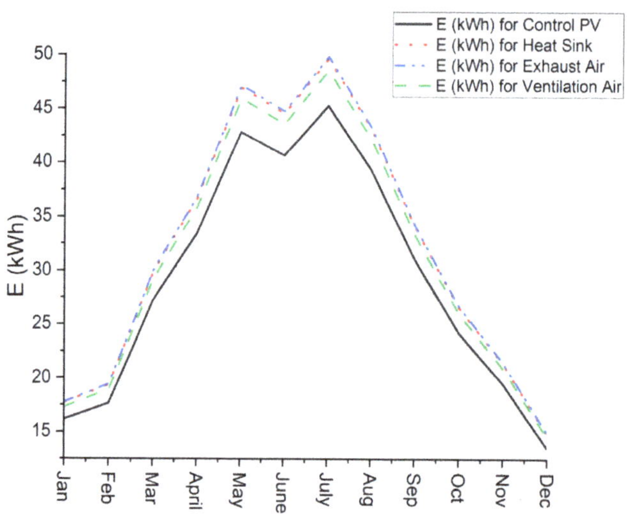

Figure 15. Energy produced versus months for air cooling methods.

As shown in Figure 16, the exhaust air cooling method showed the highest cost savings, with an average of USD 0.265.

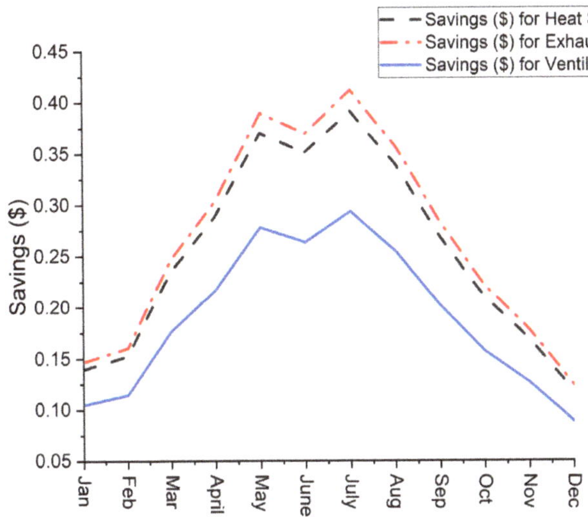

Figure 16. Cost savings versus months for air cooling methods.

5.3. PCM Cooling

Figure 17 shows that cooling by PCM increased the total energy produced compared to the standard PV panel. An average of 31.733 kWh was recorded for cooling by PCM compared to an average of 37 kWh recorded for the standard PV panel.

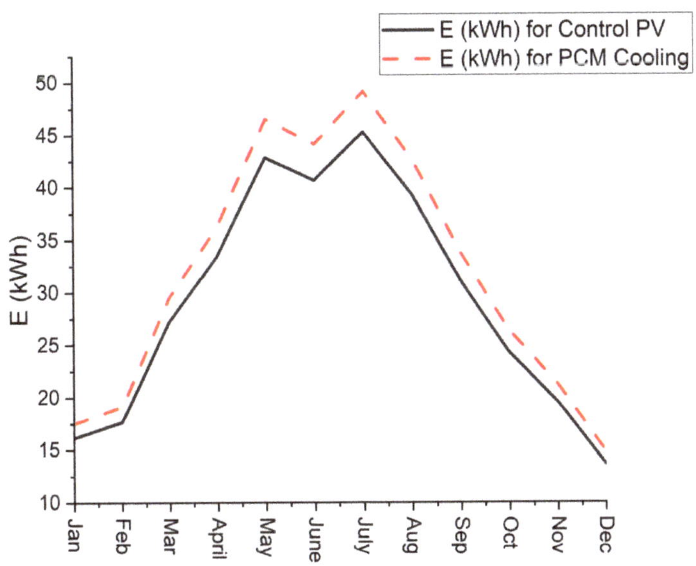

Figure 17. Energy produced versus months for PCM cooling method.

Cooling by PCM increased cost savings compared to the standard PV panel. Figure 18 shows that the maximum amount of money saved by PCM cooling was USD 0.223.

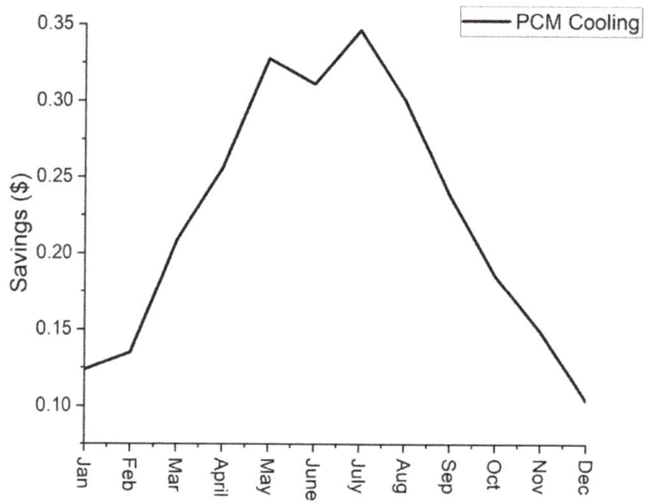

Figure 18. Cost savings versus months for PCM cooling method.

5.4. Other Cooling Methods

As shown in Figure 19, the thermoelectric cooling method was found to produce the maximum energy, with an average of 34.512 kWh.

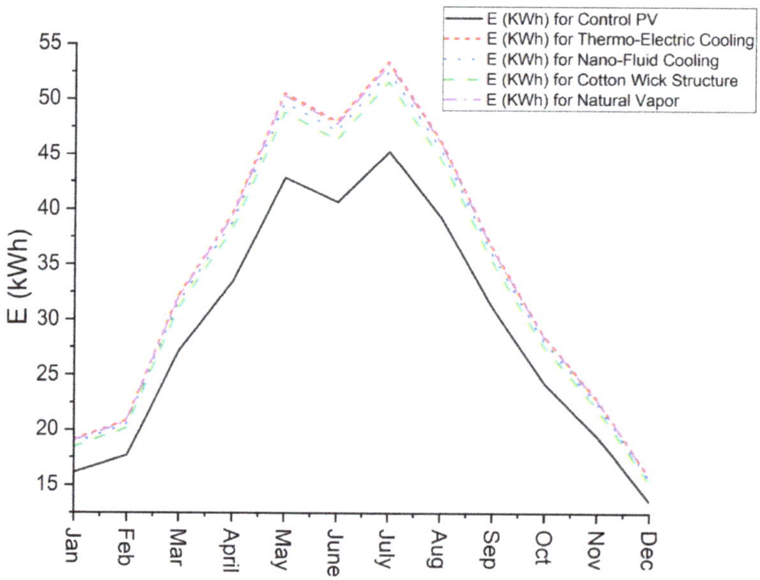

Figure 19. Energy produced versus months for other cooling methods.

The maximum cost saving was recorded by the thermoelectric cooling method, with an average of USD 0.473 recorded in July, as shown in Figure 20.

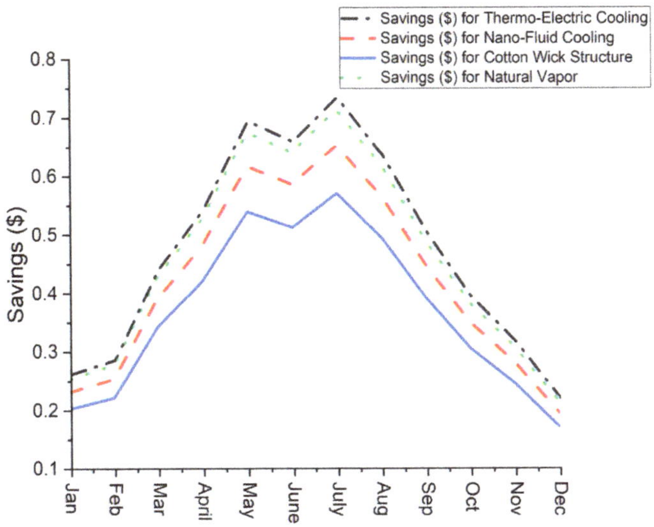

Figure 20. Cost savings versus months for other cooling methods.

In the realm of photovoltaic panel cooling methods, the economic evaluation highlighted the significant benefits of these technologies, both in terms of increased energy production and cost savings compared to standard PV systems. Water-based cooling methods, exemplified by flowing water on the PV panel, have exhibited the highest energy production, yielding an average of 32.29 kWh. This translated into significant financial gains, with cost savings averaging USD 0.273. Liquid immersion and heat pipe cooling systems also demonstrated promising results, while automotive radiator-based cooling methods exhibited slightly lower energy gains and cost savings. In the air cooling category, exhaust air cooling proved to be the most effective, generating an average of 32.201 kWh and yielding the highest cost savings of USD 0.265. Additionally, phase-change material cooling strategies contributed to increased energy production, with an average of 31.73 kWh, resulting in notable cost savings of up to USD 0.223. Among various cooling methods, thermoelectric cooling emerged as the leader in energy production, delivering an average of 34.512 kWh and recording the highest cost savings—particularly in July, with an average of USD 0.473. These results confirm the economic feasibility and financial advantages of applying advanced cooling technologies in PV panel systems, enhancing their ability to drive sustainable and cost-effective energy solutions.

6. Environmental Study

The increased use of fossil fuels has increased CO_2 emissions, which pollutes the air and leads to many serious problems, mainly global warming. Photovoltaic panels were found to reduce CO_2 emissions to the atmosphere as a renewable energy resource.

The governing equations used in the environmental study are as follows.

The CO_2 reduction value is quantified as:

$$CO_{2\ reduced} = E \times P \tag{12}$$

where

$CO_{2\ reduced}$: The amount of CO_2 reduced in kg.

E: Energy produced by a PV panel in kWh.

P: The amount of CO_2 produced per 1 kWh of electricity $\frac{kg}{kWh}$.

6.1. Water Cooling

Water cooling methods had a major impact on cooling techniques in reducing CO_2 emissions as a renewable energy resource. A system of nozzles flowing water on the surface of the PV panel was found to result in the maximum CO_2 reduction of 26.509 kg with respect to the other water cooling methods, as shown in Figure 21.

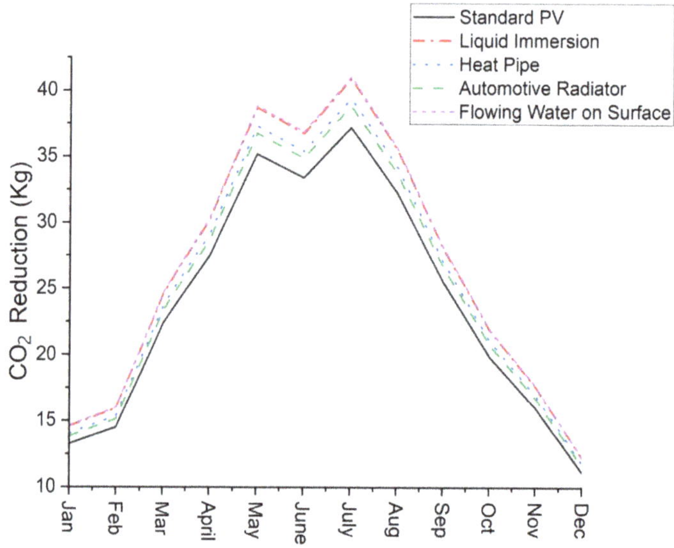

Figure 21. CO_2 emission reduction versus month for water cooling methods.

6.2. Air Cooling

The air cooling technique CO_2 emission reduction varies between the methods. However, as shown in Figure 22, the exhaust air system had the maximum CO_2 emission reduction, with an average of 26.437 kg, compared the other methods.

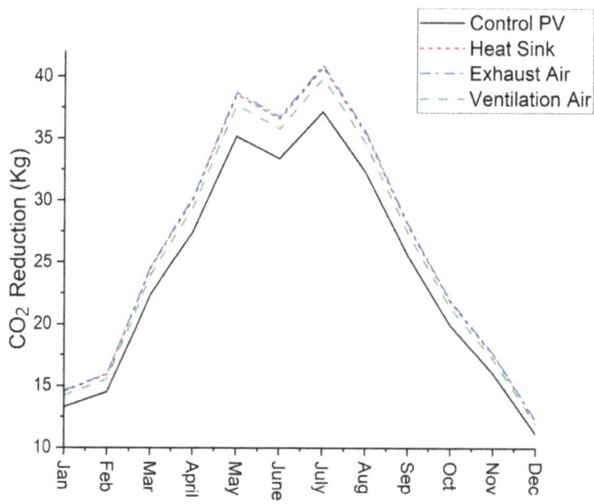

Figure 22. CO_2 emission reduction versus months for air cooling methods.

6.3. PCM Cooling

The PCM cooling method was found to reduce CO_2 emissions by an average of 26.053 kg, as shown in Figure 23.

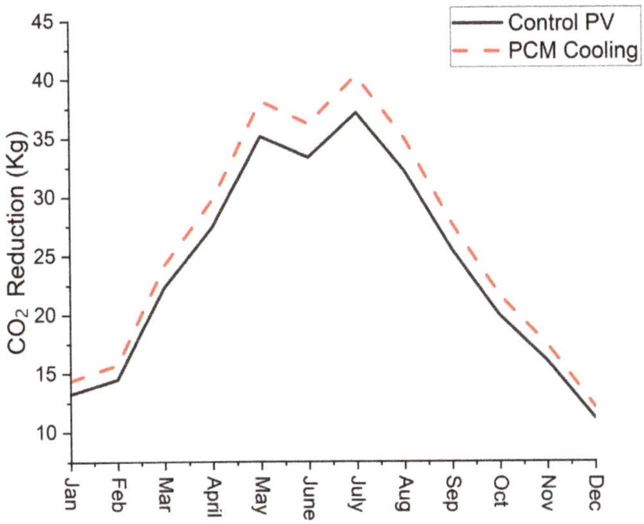

Figure 23. CO_2 emission reduction versus months for PCM cooling method.

6.4. Other Cooling Methods

Other uncategorized cooling techniques had a good impact on the reduction in CO_2 emissions. The thermoelectric cooling system had a maximum reduction in CO_2 emissions, with an average of 28.334 kg, as shown in Figure 24.

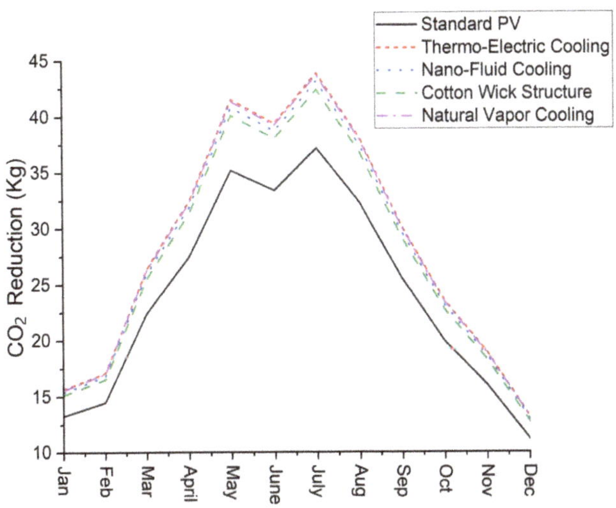

Figure 24. CO_2 emission reduction versus months for other cooling methods.

In short, the escalating use of fossil fuels has led to an alarming rise in carbon dioxide emissions, which has greatly contributed to worsening environmental issues such as global warming. Photovoltaic panels have emerged as a renewable energy resource with the potential to mitigate these emissions. This study investigated different cooling technologies and their effectiveness in reducing carbon dioxide emissions. Of these, water cooling methods, particularly the nozzle-based system, showed the greatest impact, reducing emissions by 26.509 kg. Air cooling technologies, especially the exhaust air system, also played a decisive role, achieving an average reduction of 26.437 kg. PCM cooling methods contributed to an average weight reduction of 26.053 kg. Even other unclassified cooling technologies, such as the thermoelectric cooling system, succeeded in reducing CO_2 emissions, with an average reduction of 28.334 kg. These results underscore the pivotal role of photovoltaic panels not only in generating renewable energy but also in combating carbon dioxide emissions. As the world grapples with the necessity of tackling climate change, innovative cooling strategies offer promising ways to reduce carbon dioxide emissions, thus promoting a more sustainable and environmentally conscious future.

7. Payback Period

The payback period of each system was studied as investments in order to reveal how much each system approximately costs and how much time it would need to pay the initial investment.

The *payback period* is quantified by the following equation.

$$Payback\ period = \frac{Income}{Cost} \tag{13}$$

where

Income: Profit produced by the system.

Cost: Initial cost of the system.

Figure 25 shows the payback period for the systems consisting of one PV panel each. As shown in the following figure, the automotive radiator system needs approximately 7.576 years in order to pay the initial investment while the standard PV needs approximately 1.9 years.

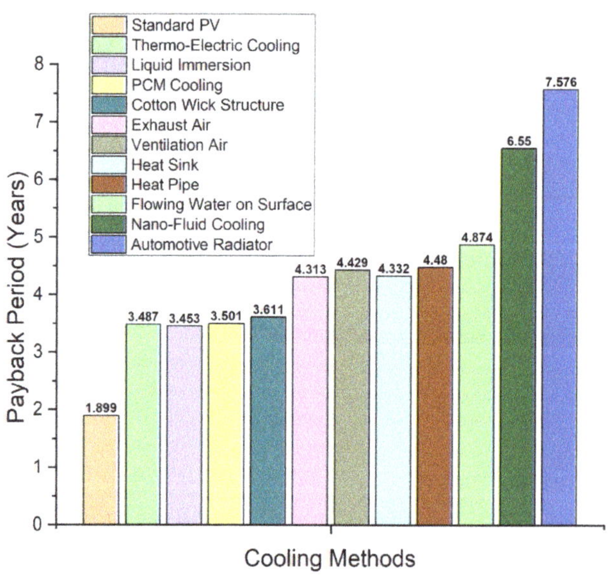

Figure 25. Payback period of each system.

As shown in Figure 25, the return on invest period differs drastically based on the initial investment paid for each system. The standard PV needs approximately 1.9 years to return the initial investment, while the automotive radiator and nanofluid cooling systems need approximately 7.576 years to return the initial investment paid from their enhanced electric output.

It is true that the payback period has increased when constructing a cooling technique for the PV panel; however, the benefits of the cooling technique on the PV are far more beneficial. The PV panel lifespan increases whenever a cooling system is used because a cooling system decreases its temperature with time. The increase in green energy produced by the PV panel with a cooling system could benefit the environment and be a smart investment on bigger systems, where in the case of cooling, the system needs fewer PV panels to operate and produce higher power outputs, while contributing with a decrease in CO_2 emissions.

8. Conclusions and Recommendations

This review paper addresses the importance of effective cooling strategies to enhance the efficiency of photovoltaic panels. It highlights the negative impact of high temperatures on the performance of photovoltaic panels and emphasizes the necessity of efficient cooling technologies. This review thoroughly explores and discusses a variety of cooling methods, including traditional methods such as water and air cooling, along with innovative solutions such as incorporating phase-change materials, thermoelectric cooling, heat pipes, evaporative cooling, and nanofluids. Furthermore, this review takes into account environmental and economic factors to comprehensively assess the impact of cooling on the performance of photovoltaic panels.

Additionally, the findings of this review emphasize that all evaluated cooling methods have the potential to improve the electrical efficiency of PV panels. However, specific techniques stand out for their superior performance. Notably, among these approaches, the automatic water spraying system, exhaust ventilated air, phase-change materials, and thermoelectric cooling methods exhibited the highest energy production levels. In terms of cost-effectiveness, thermoelectric cooling outperformed evaporative cooling, water-nanofluid cooling, and the automatic spraying system. Furthermore, thermoelectric cooling, evaporative cooling, exhaust-ventilated air, and automatic water spraying demonstrated the greatest reductions in CO_2 emissions.

Moreover, the evaporative cooling technique, along with thermoelectric and PCM cooling methods, showed the shortest payback period. Consequently, evaporative and thermoelectric cooling emerge as particularly promising choices, offering substantial energy improvements, positive environmental effects, and favorable returns on investment. These results emphasize the importance of integrating cooling strategies to improve the efficiency of photovoltaic panels and to maximize the generation of eco-friendly electricity. Given the essential role of renewable energy in addressing climate change and the transition to sustainable energy systems, the integration of efficient cooling technologies can contribute significantly to the advancement of the renewable energy sector.

The development of a highly conductive phase-change material would increase the effect of PCM cooling, enhancing the efficiency and performance of the PV panel. Studies should be targeted on testing different combinations of PCM with other materials, and on the PCM itself to reach a formula where the thermal conductivity is as high as possible and the melting point of the PCM is as close as possible to the standard test conditions of the PV panel. Also, as a future recommendation, the period of analysis at the level of the cooling techniques and methods could be for longer periods, meaning that each cooling method should be studied over different periods of time and for longer hours. Moreover, not many review studies combine all of the cooling methods in one paper. This study mentions nearly all of the cooling methods and a parametric investigation was conducted at the level of environmental and economic analysis, a state-of-the art analysis.

Funding: This research received no external funding.

Conflicts of Interest: The authors declare no conflict of interest.

References

1. Khaled, M.; Harambat, F.; Hage, H.E.; Peerhossaini, H. Spatial Optimization of an Underhood Cooling Module—Towards an Innovative Control Approach. *Appl. Energy* **2011**, *88*, 3841–3849. [CrossRef]
2. Khaled, M.; Harambat, F.; Peerhossaini, H. Towards the Control of Car Underhood Thermal Conditions. *Appl. Therm. Eng.* **2011**, *31*, 902–910. [CrossRef]
3. Khaled, M.; Harambat, F.; Peerhossaini, H. Temperature and Heat Flux Behavior of Complex Flows in Car Underhood Compartment. *Heat Transf. Eng.* **2010**, *31*, 1057–1067. [CrossRef]
4. Taher, R.; Ahmed, M.M.; Haddad, Z.; Abid, C. Poiseuille-Rayleigh-Bénard Mixed Convection Flow in a Channel: Heat Transfer and Fluid Flow Patterns. *Int. J. Heat Mass Transf.* **2021**, *180*, 121745. [CrossRef]
5. Razi, F.; Dincer, I. Renewable Energy Development and Hydrogen Economy in MENA Region: A Review. *Renew. Sustain. Energy Rev.* **2022**, *168*, 112763. [CrossRef]
6. Haddad, A.; Ramadan, M.; Khaled, M.; Ramadan, H.S.M.; Becherif, M. Triple Hybrid System Coupling Fuel Cell with Wind Turbine and Thermal Solar System. *Int. J. Hydrogen Energy* **2020**, *45*, 11484–11491. [CrossRef]
7. Khaled, M.; Ramadan, M.; Hage, H.E. Parametric Analysis of Heat Recovery from Exhaust Gases of Generators. *Energy Procedia* **2015**, *75*, 3295–3300. [CrossRef]
8. Du, W.; Yin, Q.; Cheng, L. Experiments on Novel Heat Recovery Systems on Rotary Kilns. *Appl. Therm. Eng.* **2018**, *139*, 535–541. [CrossRef]
9. Hage, H.E.; Ramadan, M.; Jaber, H.; Khaled, M.; Olabi, A.G. A Short Review on the Techniques of Waste Heat Recovery from Domestic Applications. *Energy Sources Part A Recovery Util. Environ. Eff.* **2019**, *42*, 3019–3034. [CrossRef]
10. Akbari, A.; Kouravand, S.; Chegini, G. Experimental Analysis of a Rotary Heat Exchanger for Waste Heat Recovery from the Exhaust Gas of Dryer. *Appl. Therm. Eng.* **2018**, *138*, 668–674. [CrossRef]
11. Gomaa, M.R.; Ahmed, M.; Rezk, H. Temperature Distribution Modeling of PV and Cooling Water PV/T Collectors through Thin and Thick Cooling Cross-Fined Channel Box. *Energy Rep.* **2022**, *8*, 1144–1153. [CrossRef]
12. Jiao, C.; Li, Z. An updated review of solar cooling systems driven by Photovoltaic–Thermal collectors. *Energies* **2023**, *16*, 5331. [CrossRef]
13. Ibrahim, K.A.; Luk, P.; Luo, Z. Cooling of Concentrated Photovoltaic Cells—A review and the perspective of Pulsating flow Cooling. *Energies* **2023**, *16*, 2842. [CrossRef]
14. Herrando, M.; Wang, K.; Huang, G.; Otanicar, T.; Mousa, O.B.; Agathokleous, R.A.; Ding, Y.; Kalogirou, S.A.; Ekins-Daukes, N.J.; Taylor, R.A.; et al. A review of solar hybrid photovoltaic-thermal (PV-T) collectors and systems. *Prog. Energy Combust. Sci.* **2023**, *97*, 101072. [CrossRef]
15. Ghosh, A. A Comprehensive Review of Water Based PV: Flotavoltaics, under Water, Offshore & Canal Top. *Ocean Eng.* **2023**, *281*, 115044. [CrossRef]
16. Aslam, A.; Ahmed, N.; Qureshi, S.A.; Assadi, M.; Ahmed, N. Advances in Solar PV Systems; A Comprehensive Review of PV Performance, Influencing Factors, and Mitigation Techniques. *Energies* **2022**, *15*, 7595. [CrossRef]
17. Herrando, M.; Ramos, A.C. Photovoltaic-Thermal (PV-T) systems for combined cooling, heating and power in buildings: A review. *Energies* **2022**, *15*, 3021. [CrossRef]
18. Hajjaj, S.S.H.; Aqeel, A.A.K.A.; Sultan, M.T.H.; Shahar, F.S.; Shah, A.U.M. Review of recent efforts in cooling photovoltaic panels (PVs) for enhanced performance and better impact on the environment. *Nanomaterials* **2022**, *12*, 1664. [CrossRef]
19. Hammoumi, A.E.; Chtita, S.; Motahhir, S.; Ghzizal, A.E. Solar PV energy: From material to use, and the most commonly used techniques to maximize the power output of PV systems: A focus on solar trackers and floating solar panels. *Energy Rep.* **2022**, *8*, 11992–12010. [CrossRef]
20. Sheik, M.S.; Kakati, P.; Dandotiya, D.; Udaya Ravi, M.; Ramesh, C.S. A Comprehensive Review on Various Cooling Techniques to Decrease an Operating Temperature of Solar Photovoltaic Panels. *Energy Nexus* **2022**, *8*, 100161. [CrossRef]
21. Cui, Y.; Zhu, J.; Zhang, F.; Shao, Y.; Xue, Y. Current status and future development of hybrid PV/T system with PCM module: 4E (energy, exergy, economic and environmental) assessments. *Renew. Sustain. Energy Rev.* **2022**, *158*, 112147. [CrossRef]
22. Mahdavi, A.; Farhadi, M.; Gorji-Bandpy, M.; Mahmoudi, A.H. A review of passive cooling of photovoltaic devices. *Clean. Eng. Technol.* **2022**, *11*, 100579. [CrossRef]
23. Kandeal, A.; Algazzar, A.M.; Elkadeem, M.R.; Thakur, A.K.; Abdelaziz, G.B.; El-Said, E.M.; Elsaid, A.M.; An, M.; Kandel, R.; Fawzy, H.E.; et al. Nano-enhanced cooling techniques for photovoltaic panels: A systematic review and prospect recommendations. *Sol. Energy* **2021**, *227*, 259–272. [CrossRef]
24. Awasthi, A.; Shukla, A.K.; Manohar, S.R.M.; Dondariya, C.; Shukla, K.K.; Porwal, D.; Richhariya, G. Review on sun tracking technology in solar PV system. *Energy Rep.* **2020**, *6*, 392–405. [CrossRef]
25. Jathar, L.D.; Ganesan, S.; Awasarmol, U.; Nikam, K.C.; Shahapurkar, K.; Soudagar, M.E.M.; Fayaz, H.; El-Shafay, A.; Kalam, M.A.; Boudila, S.; et al. Comprehensive review of environmental factors influencing the performance of photovoltaic panels: Concern over emissions at various phases throughout the lifecycle. *Environ. Pollut.* **2023**, *326*, 121474. [CrossRef]
26. Şahin, G. Effect of Wavelength on the Electrical Parameters of a Vertical Parallel Junction Silicon Solar Cell Illuminated by Its Rear Side in Frequency Domain. *Results Phys.* **2016**, *6*, 107–111. [CrossRef]
27. Maka, A.O.; O'Donovan, T.S. Effect of thermal load on performance parameters of solar concentrating photovoltaic: High-efficiency solar cells. *Energy Built Environ.* **2022**, *3*, 201–209. [CrossRef]

28. Paudyal, B.R.; Imenes, A.G. Investigation of temperature coefficients of PV modules through field measured data. *Sol. Energy* **2021**, *224*, 425–439. [CrossRef]
29. Chander, S.; Purohit, A.; Sharma, A.; Arvind, A.; Nehra, S.; Dhaka, M.S. A study on photovoltaic parameters of mono-crystalline silicon solar cell with cell temperature. *Energy Rep.* **2015**, *1*, 104–109. [CrossRef]
30. Hussien, A.A.; Eltayesh, A.; El-Batsh, H.M. Experimental and numerical investigation for PV cooling by forced convection. *Alex. Eng. J.* **2022**, *64*, 427–440. [CrossRef]
31. Ahmad, E.Z.; Fazlizan, A.; Jarimi, H.; Sopian, K.; Ibrahim, A. Enhanced heat dissipation of truncated multi-level fin heat sink (MLFHS) in case of natural convection for photovoltaic cooling. *Case Stud. Therm. Eng.* **2021**, *28*, 101578. [CrossRef]
32. Abdallah, A.; Opoku, R.; Sekyere, C.; Boahen, S.; Amoabeng, K.O.; Uba, F.; Obeng, G.Y.; Forson, F. Experimental investigation of thermal management techniques for improving the efficiencies and levelized cost of energy of solar PV modules. *Case Stud. Therm. Eng.* **2022**, *35*, 102133. [CrossRef]
33. Pomares-Hernández, C.; Zuluaga-García, E.A.; Escorcia Salas, G.E.; Robles-Algarín, C.; Sierra Ortega, J. Computational Modeling of Passive and Active Cooling Methods to Improve PV Panels Efficiency. *Appl. Sci.* **2021**, *11*, 11370. [CrossRef]
34. Wang, Y.; Zhao, T.; Cao, Z.; Zhai, C.; Zhou, Y.; Lv, W.; Xu, T.; Wu, S. Numerical study on the forced convection enhancement of flat-roof integrated photovoltaic by passive components. *Energy Build.* **2023**, *289*, 113063. [CrossRef]
35. Kasaeian, A.; Khanjari, Y.; Golzari, S.; Mahian, O.; Wongwises, S. Effects of forced convection on the performance of a photovoltaic thermal system: An experimental study. *Exp. Therm. Fluid Sci.* **2017**, *85*, 13–21. [CrossRef]
36. Atkin, P.; Farid, M. Improving the efficiency of photovoltaic cells using PCM infused graphite and aluminium fins. *Sol. Energy* **2015**, *114*, 217–228. [CrossRef]
37. Chandrasekar, M.; Senthilkumar, T. Experimental demonstration of enhanced solar energy utilization in flat PV (photovoltaic) modules cooled by heat spreaders in conjunction with cotton wick structures. *Energy* **2015**, *90*, 1401–1410. [CrossRef]
38. Benzarti, S.; Chaabane, M.; Mhiri, H.; Bournot, P. Performance improvement of a naturally ventilated building integrated photovoltaic system using twisted baffle inserts. *J. Build. Eng.* **2022**, *53*, 104553. [CrossRef]
39. Grubišić-Čabo, F.; Nižetić, S.; Čoko, D.; Kragić, I.; Papadopoulos, A.M. Experimental investigation of the passive cooled free-standing photovoltaic panel with fixed aluminum fins on the backside surface. *J. Clean. Prod.* **2018**, *176*, 119–129. [CrossRef]
40. Mojumder, J.C.; Chong, W.T.; Show, P.L.; Leong, K.W.; Abdullah-Al-Mamoon. An experimental investigation on performance analysis of air type photovoltaic thermal collector system integrated with cooling fins design. *Energy Build.* **2016**, *130*, 272–285. [CrossRef]
41. Abdelsalam, E.; Alnawafah, H.; Almomani, F.; Mousa, A.; Jamjoum, M.; Alkasrawi, M. Efficiency Improvement of Photovoltaic Panels: A Novel Integration Approach with Cooling Tower. *Energies* **2023**, *16*, 1070. [CrossRef]
42. Soliman, A.M. A Numerical Investigation of PVT System Performance with Various Cooling Configurations. *Energies* **2023**, *16*, 3052. [CrossRef]
43. Al-Amri, F.; Saeed, F.; Mujeebu, M.A. Novel dual-function racking structure for passive cooling of solar PV panels–thermal performance analysis. *Renew. Energy* **2022**, *198*, 100–113. [CrossRef]
44. Bayrak, F.; Oztop, H.F.; Selimefendigil, F. Effects of different fin parameters on temperature and efficiency for cooling of photovoltaic panels under natural convection. *Sol. Energy* **2019**, *188*, 484–494. [CrossRef]
45. Hu, W.; Li, X.; Wang, J.; Tian, Z.; Zhou, B.; Wu, J.; Li, R.; Li, W.; Ma, N.; Kang, J.; et al. Experimental research on the convective heat transfer coefficient of photovoltaic panel. *Renew. Energy* **2021**, *185*, 820–826. [CrossRef]
46. Kumar, P.S.; NaveenKumar, R.; Sharifpur, M.; Issakhov, A.; Ravichandran, M.; Maridurai, T.; Al-Sulaiman, F.A.; Banapurmath, N.R. Experimental investigations to improve the electrical efficiency of photovoltaic modules using different convection mode. *Sustain. Energy Technol. Assess.* **2021**, *48*, 101582. [CrossRef]
47. Mankani, K.L.; Chaudhry, H.N.; Calautit, J.K. Optimization of an air-cooled heat sink for cooling of a solar photovoltaic panel: A computational study. *Energy Build.* **2022**, *270*, 112274. [CrossRef]
48. Kiwan, S.; Khlefat, A.M. Thermal cooling of photovoltaic panels using porous material. *Case Stud. Therm. Eng.* **2021**, *24*, 100837. [CrossRef]
49. Li, D.; King, M.; Dooner, M.S.; Guo, S.; Wang, J. Study on the cleaning and cooling of solar photovoltaic panels using compressed airflow. *Sol. Energy* **2021**, *221*, 433–444. [CrossRef]
50. Tina, G.M.; Rosa-Clot, M.; Rosa-Clot, P.; Scandura, P.F. Optical and thermal behavior of submerged photovoltaic solar panel: SP2. *Energy* **2012**, *39*, 17–26. [CrossRef]
51. Irwan, Y.; Leow, W.; Irwanto, M.; Fareq, M.; Amelia, A.; Gomesh, N.; Safwati, I. Indoor Test Performance of PV Panel through Water Cooling Method. *Energy Procedia* **2015**, *79*, 604–611. [CrossRef]
52. Moradgholi, M.; Nowee, S.M.; Abrishamchi, I. Application of heat pipe in an experimental investigation on a novel photovoltaic/thermal (PV/T) system. *Sol. Energy* **2014**, *107*, 82–88. [CrossRef]
53. Koundinya, S.; Vigneshkumar, N.; Krishnan, A.S. Experimental Study and Comparison with the Computational Study on Cooling of PV Solar Panel Using Finned Heat Pipe Technology. *Mater. Today Proc.* **2017**, *4*, 2693–2700. [CrossRef]
54. Ho, C.; Chou, W.Y.; Lai, C.M. Thermal and electrical performance of a water-surface floating PV integrated with a water-saturated MEPCM layer. *Energy Convers. Manag.* **2015**, *89*, 862–872. [CrossRef]
55. Liu, L.; Zhu, L.; Wang, Y.; Huang, Q.; Sun, Y.; Yin, Z. Heat dissipation performance of silicon solar cells by direct dielectric liquid immersion under intensified illuminations. *Sol. Energy* **2011**, *85*, 922–930. [CrossRef]

56. Nižetić, S.; Čoko, D.; Yadav, A.; Grubišić-Čabo, F. Water spray cooling technique applied on a photovoltaic panel: The performance response. *Energy Convers. Manag.* **2016**, *108*, 287–296. [CrossRef]
57. Joy, B.; Philip, J.; Zachariah, R. Investigations on serpentine tube type solar photovoltaic/thermal collector with different heat transfer fluids: Experiment and numerical analysis. *Sol. Energy* **2016**, *140*, 12–20. [CrossRef]
58. Moharram, K.A.; Kandil, H.A.; El-Sherif, H. Enhancing the performance of photovoltaic panels by water cooling. *Ain Shams Eng. J.* **2013**, *4*, 869–877. [CrossRef]
59. Yazdanifard, F.; Ebrahimnia-Bajestan, E.; Ameri, M. Investigating the performance of a water-based photovoltaic/thermal (PV/T) collector in laminar and turbulent flow regime. *Renew. Energy* **2016**, *99*, 295–306. [CrossRef]
60. Krauter, S. Increased electrical yield via water flow over the front of photovoltaic panels. *Sol. Energy Mater. Sol. Cells* **2004**, *82*, 131–137. [CrossRef]
61. Sabry, M. Temperature optimization of high concentrated active cooled solar cells. *NRIAG J. Astron. Geophys.* **2016**, *5*, 23–29. [CrossRef]
62. Elnozahy, A.; Rahman, A.K.A.; Ali, A.H.H.; Abdel-Salam, M.; Ookawara, S. Performance of a PV module integrated with standalone building in hot arid areas as enhanced by surface cooling and cleaning. *Energy Build.* **2015**, *88*, 100–109. [CrossRef]
63. Kordzadeh, A. The effects of nominal power of array and system head on the operation of photovoltaic water pumping set with array surface covered by a film of water. *Renew. Energy* **2010**, *35*, 1098–1102. [CrossRef]
64. Arcuri, N.; Reda, F.; De Simone, M. Energy and thermo-fluid-dynamics evaluations of photovoltaic panels cooled by water and air. *Sol. Energy* **2014**, *105*, 147–156. [CrossRef]
65. Jakhar, S.; Soni, M.; Gakkhar, N. Performance Analysis of Earth Water Heat Exchanger for Concentrating Photovoltaic Cooling. *Energy Procedia* **2016**, *90*, 145–153. [CrossRef]
66. Mittelman, G.; Kribus, A.; Mouchtar, O.; Dayan, A. Water desalination with concentrating photovoltaic/thermal (CPVT) systems. *Sol. Energy* **2009**, *83*, 1322–1334. [CrossRef]
67. Chong, K.; Tan, W. Study of automotive radiator cooling system for dense-array concentration photovoltaic system. *Sol. Energy* **2012**, *86*, 2632–2643. [CrossRef]
68. Ibrahim, A.M.; Dincer, I. Experimental performance evaluation of a combined solar system to produce cooling and potable water. *Sol. Energy* **2015**, *122*, 1066–1079. [CrossRef]
69. Korkut, T.B.; Gören, A.; Rachid, A. Numerical and Experimental Study of a PVT Water System under Daily Weather Conditions. *Energies* **2022**, *15*, 6538. [CrossRef]
70. Sattar, M.; Rehman, A.; Ahmad, N.; Mohammad, A.; Alahmadi, A.; Ullah, N. Performance Analysis and Optimization of a Cooling System for Hybrid Solar Panels Based on Climatic Conditions of Islamabad, Pakistan. *Energies* **2022**, *15*, 6278. [CrossRef]
71. Sornek, K.; Goryl, W.; Figaj, R.D.; Dąbrowska, G.B.; Brezdeń, J. Development and Tests of the Water Cooling System Dedicated to Photovoltaic Panels. *Energies* **2022**, *15*, 5884. [CrossRef]
72. Majumder, A.; Innamorati, R.; Ghiani, E.; Kumar, A.; Gatto, G. Performance Analysis of a Floating Photovoltaic System and Estimation of the Evaporation Losses Reduction. *Energies* **2021**, *14*, 8336. [CrossRef]
73. Yildirim, M.; Cebula, A.; Sułowicz, M. A cooling design for photovoltaic panels—Water-based PV/T system. *Energy* **2022**, *256*, 124654. [CrossRef]
74. Chanphavong, L.; Chanthaboune, V.; Phommachanh, S.; Vilaida, X.; Bounyanite, P. Enhancement of performance and exergy analysis of a water-cooling solar photovoltaic panel. *Total Environ. Res. Themes* **2022**, *3–4*, 100018. [CrossRef]
75. Panda, S.; Panda, B.; Jena, C.; Nanda, L.; Pradhan, A. Investigating the similarities and differences between front and back surface cooling for PV panels. *Mater. Today Proc.* **2022**, *74*, 358–363. [CrossRef]
76. Zubeer, S.A.; Ali, O. Experimental and numerical study of low concentration and water-cooling effect on PV module performance. *Case Stud. Therm. Eng.* **2022**, *34*, 102007. [CrossRef]
77. Lopez-Pascual, D.; Valiente-Blanco, I.; Manzano-Narro, O.; Fernandez-Munoz, M.; Diez-Jimenez, E. Experimental characterization of a geothermal cooling system for enhancement of the efficiency of solar photovoltaic panels. *Energy Rep.* **2022**, *8*, 756–763. [CrossRef]
78. Li, S.; Zhou, Z.; Liu, J.; Zhang, J.; Tang, H.; Zhuofen, Z.; Na, Y.; Jiang, C. Research on indirect cooling for photovoltaic panels based on radiative cooling. *Renew. Energy* **2022**, *198*, 947–959. [CrossRef]
79. Masalha, I.; Masuri, S.; Badran, O.; Ariffin, M.; Talib, A.A.; Alfaqs, F. Outdoor experimental and numerical simulation of photovoltaic cooling using porous media. *Case Stud. Therm. Eng.* **2023**, *42*, 102748. [CrossRef]
80. Jafari, R. Optimization and energy analysis of a novel geothermal heat exchanger for photovoltaic panel cooling. *Sol. Energy* **2021**, *226*, 122–133. [CrossRef]
81. Sudhakar, P.; Santosh, R.; Asthalakshmi, B.; Kumaresan, G.; Velraj, R. Performance augmentation of solar photovoltaic panel through PCM integrated natural water circulation cooling technique. *Renew. Energy* **2021**, *172*, 1433–1448. [CrossRef]
82. Pezzutto, S.; Quaglini, G.; Rivière, P.; Kranzl, L.; Novelli, A.; Zambito, A.; Wilczynski, E. Screening of cooling technologies in Europe: Alternatives to vapour compression and possible market developments. *Sustainability* **2022**, *14*, 2971. [CrossRef]
83. Waqas, A.; Ji, J.; Xu, L.; Ali, M.; Zeashan; Alvi, J.Z. Thermal and Electrical Management of Photovoltaic Panels Using Phase Change Materials—A Review. *Renew. Sustain. Energy Rev.* **2018**, *92*, 254–271. [CrossRef]

84. Hachem, F.; Abdulhay, B.; Ramadan, M.; Hage, H.E.; Rab, M.G.E.; Khaled, M. Improving the performance of photovoltaic cells using pure and combined phase change materials—Experiments and transient energy balance. *Renew. Energy* **2017**, *107*, 567–575. [CrossRef]
85. Smith, C.; Forster, P.M.; Crook, R. Global analysis of photovoltaic energy output enhanced by phase change material cooling. *Appl. Energy* **2014**, *126*, 21–28. [CrossRef]
86. Biwole, P.H.; Eclache, P.; Kuznik, F. Phase-change materials to improve solar panel's performance. *Energy Build.* **2013**, *62*, 59–67. [CrossRef]
87. Hasan, A.; McCormack, S.; Huang, M.; Sarwar, J.; Norton, B. Increased photovoltaic performance through temperature regulation by phase change materials: Materials comparison in different climates. *Sol. Energy* **2015**, *115*, 264–276. [CrossRef]
88. Xu, Z.; Kong, Q.; Qu, H.; Wang, C. Cooling characteristics of solar photovoltaic panels based on phase change materials. *Case Stud. Therm. Eng.* **2023**, *41*, 102667. [CrossRef]
89. Aneli, S.; Arena, R.; Gagliano, A. Numerical Simulations of a PV Module with Phase Change Material (PV-PCM) under Variable Weather Conditions. *Int. J. Heat Technol.* **2021**, *39*, 643–652. [CrossRef]
90. Amalu, E.H.; Fabunmi, O.A. Thermal control of crystalline silicon photovoltaic (c-Si PV) module using Docosane phase change material (PCM) for improved performance. *Sol. Energy* **2022**, *234*, 203–221. [CrossRef]
91. Zhao, J.; Li, Z.; Ma, T. Performance analysis of a photovoltaic panel integrated with phase change material. *Energy Procedia* **2019**, *158*, 1093–1098. [CrossRef]
92. Kant, K.; Shukla, A.; Sharma, A.; Biwole, P.H. Heat transfer studies of photovoltaic panel coupled with phase change material. *Sol. Energy* **2016**, *140*, 151–161. [CrossRef]
93. Park, J.; Kim, T.; Leigh, S.B. Application of a phase-change material to improve the electrical performance of vertical-building-added photovoltaics considering the annual weather conditions. *Sol. Energy* **2014**, *105*, 561–574. [CrossRef]
94. Stropnik, R.; Stritih, U. Increasing the efficiency of PV panel with the use of PCM. *Renew. Energy* **2016**, *97*, 671–679. [CrossRef]
95. Aelenei, L.; Pereira, R.N.C.; Gonçalves, H.; Athienitis, A.K. Thermal Performance of a Hybrid BIPV-PCM: Modeling, Design and Experimental Investigation. *Energy Procedia* **2014**, *48*, 474–483. [CrossRef]
96. Li, Z.; Ma, T.; Zhao, J.; Song, A.; Cheng, Y. Experimental study and performance analysis on solar photovoltaic panel integrated with phase change material. *Energy* **2019**, *178*, 471–486. [CrossRef]
97. Hasan, A.; Sarwar, J.; Alnoman, H.; Abdelbaqi, S. Yearly energy performance of a photovoltaic-phase change material (PV-PCM) system in hot climate. *Sol. Energy* **2017**, *146*, 417–429. [CrossRef]
98. Savvakis, N.; Dialyna, E.; Tsoutsos, T. Investigation of the operational performance and efficiency of an alternative PV + PCM concept. *Sol. Energy* **2020**, *211*, 1283–1300. [CrossRef]
99. Abdelrazik, A.; Al-Sulaiman, F.A.; Saidur, R. Numerical investigation of the effects of the nano-enhanced phase change materials on the thermal and electrical performance of hybrid PV/thermal systems. *Energy Convers. Manag.* **2020**, *205*, 112449. [CrossRef]
100. Fayaz, H.; Rahim, N.A.; Hasanuzzaman, M.; Rivai, A.K.; Nasrin, R. Numerical and outdoor real time experimental investigation of performance of PCM based PVT system. *Sol. Energy* **2019**, *179*, 135–150. [CrossRef]
101. Hossain, M.S.; Pandey, A.K.; Selvaraj, J.; Rahim, N.A.; Islam, M.S.; Tyagi, V. Two side serpentine flow based photovoltaic-thermal-phase change materials (PVT-PCM) system: Energy, exergy and economic analysis. *Renew. Energy* **2019**, *136*, 1320–1336. [CrossRef]
102. Kazemian, A.; Taheri, A.; Sardarabadi, A.; Ma, T.; Passandideh-Fard, M.; Peng, J. Energy, exergy and environmental analysis of glazed and unglazed PVT system integrated with phase change material: An experimental approach. *Sol. Energy* **2020**, *201*, 178–189. [CrossRef]
103. Metwally, H.; Mahmoud, N.A.; Aboelsoud, W.; Ezzat, M. Comprehensive analysis of PCM container construction effects PV panels thermal management. *Adv. Environ. Waste Manag. Recycl.* **2022**, *5*, 326–338.
104. Bria, A.; Raillani, B.; Chaatouf, D.; Salhi, M.; Amraqui, S.; Mezrhab, A. Numerical investigation of the PCM effect on the performance of photovoltaic panels; comparison between different types of PCM. In Proceedings of the 2022 2nd International Conference on Innovative Research in Applied Science, Engineering and Technology (IRASET), Meknes, Morocco, 3–4 March 2022. [CrossRef]
105. Hassabou, A.; Isaifan, R.J. Simulation of Phase Change Material Absorbers for Passive Cooling of Solar Systems. *Energies* **2022**, *15*, 9288. [CrossRef]
106. Cui, Y.; Zhu, J.; Zoras, S.; Hassan, K.; Tong, H. Photovoltaic/Thermal Module Integrated with Nano-Enhanced Phase Change Material: A Numerical Analysis. *Energies* **2022**, *15*, 4988. [CrossRef]
107. Soliman, A.S.; Xu, L.; Dong, J.; Cheng, P. A novel heat sink for cooling photovoltaic systems using convex/concave dimples and multiple PCMs. *Appl. Therm. Eng.* **2022**, *215*, 119001. [CrossRef]
108. Maghrabie, H.M.; Mohamed, A.; Fahmy, A.M.; Samee, A.a.A. Performance enhancement of PV panels using phase change material (PCM): An experimental implementation. *Case Stud. Therm. Eng.* **2023**, *42*, 102741. [CrossRef]
109. Chaichan, M.T.; Kazem, H.A.; Al-Waeli, A.H.; Sopian, K. Controlling the melting and solidification points temperature of PCMs on the performance and economic return of the water-cooled photovoltaic thermal system. *Sol. Energy* **2021**, *224*, 1344–1357. [CrossRef]
110. Nižetić, S.; Jurčević, M.; Čoko, D.; Arıcı, M. A novel and effective passive cooling strategy for photovoltaic panel. *Renew. Sustain. Energy Rev.* **2021**, *145*, 111164. [CrossRef]

111. Saleh, A.H.; Hussein, A.M.; Danook, S.H. Efficiency Enhancement of Solar Cell Collector Using Fe_3O_4/Water Nanofluid. *IOP Conf. Ser.* **2021**, *1105*, 012059. [CrossRef]
112. Ghadiri, M.; Sardarabadi, M.; Pasandideh-Fard, M.; Moghadam, A.J. Experimental Investigation of a PVT System Performance Using Nano Ferrofluids. *Energy Convers. Manag.* **2015**, *103*, 468–476. [CrossRef]
113. Lekbir, A.; Hassani, S.; Ghani, M.R.A.; Gan, C.K.; Mekhilef, S.; Saidur, R. Improved Energy Conversion Performance of a Novel Design of Concentrated Photovoltaic System Combined with Thermoelectric Generator with Advance Cooling System. *Energy Convers. Manag.* **2018**, *177*, 19–29. [CrossRef]
114. Kang, Y.; Joung, J.; Kim, M.; Jeong, J. Energy impact of heat pipe-assisted microencapsulated phase change material heat sink for photovoltaic and thermoelectric generator hybrid panel. *Renew. Energy* **2023**, *207*, 298–308. [CrossRef]
115. Kouravand, A.; Kasaeian, A.; Pourfayaz, F.; Rad, M.A.V. Evaluation of a nanofluid-based concentrating photovoltaic thermal system integrated with finned PCM heatsink: An experimental study. *Renew. Energy* **2022**, *201*, 1010–1025. [CrossRef]
116. Bassam, A.M.; Sopian, K.; Ibrahim, A.; Fauzan, M.F.; Al-Aasam, A.B.; Abusaibaa, G.Y. Experimental analysis for the photovoltaic thermal collector (PVT) with nano PCM and micro-fins tube nanofluid. *Case Stud. Therm. Eng.* **2022**, *41*, 102579. [CrossRef]
117. Bassam, A.M.; Sopian, K.; Ibrahim, A.; Al-Aasam, A.B.; Dayer, M. Experimental analysis of photovoltaic thermal collector (PVT) with nano PCM and micro-fins tube counterclockwise twisted tape nanofluid. *Case Stud. Therm. Eng.* **2023**, *45*, 102883. [CrossRef]
118. Moein-Jahromi, M.; Rahmanian-Koushkaki, H.; Rahmanian, S.; Jahromi, S.P. Evaluation of nanostructured GNP and CuO compositions in PCM-based heat sinks for photovoltaic systems. *J. Energy Storage* **2022**, *53*, 105240. [CrossRef]
119. Bayrak, F.; Oztop, H.F.; Selimefendigil, F. Experimental study for the application of different cooling techniques in photovoltaic (PV) panels. *Energy Convers. Manag.* **2020**, *212*, 112789. [CrossRef]
120. Xu, Q.; Ji, Y.; Riggs, B.C.; Ollanik, A.; Farrar-Foley, N.; Ermer, J.; Romanin, V.; Lynn, P.; Codd, D.S.; Escarra, M.D. A transmissive, spectrum-splitting concentrating photovoltaic module for hybrid photovoltaic-solar thermal energy conversion. *Sol. Energy* **2016**, *137*, 585–593. [CrossRef]
121. Bayssary, A.; Hajjar, C. Solar Irradiation Data for Lebanon. The Lebanese Center for Energy Conservation (LCEC). 2020. Available online: https://lcec.org.lb/sites/default/files/2021-02/Solar%20Irradiation%20Data%20for%20Lebanon%20August%202020.pdf (accessed on 8 November 2023).

energies

MDPI

Article

Novel Multi-Criteria Decision Analysis Based on Performance Indicators for Urban Energy System Planning

Benjamin Kwaku Nimako [1,2], Silvia Carpitella [3] and Andrea Menapace [4,5,*]

1 Sustainable Development and Climate Change, University School for Advanced Studies IUSS, Piazza della Vittoria 15, 27100 Pavia, Italy; bnimako@unibz.it
2 Faculty of Engineering, Free University of Bozen-Bolzano, Piazza Domenicani 3, 39100 Bolzano, Italy
3 Department of Manufacturing Systems Engineering and Management (MSEM), California State University, 18111 Nordhoff St., Los Angeles, CA 91330, USA; silvia.carpitella@csun.edu
4 Eurac Research, Institute for Renewable Energy, 39100 Bolzano, Italy
5 Faculty of Agricultural, Environmental and Food Sciences, Free University of Bozen-Bolzano, Piazza Università 5, 39100 Bolzano, Italy
* Correspondence: andrea.menapace@eurac.edu

Abstract: Urban energy systems planning presents significant challenges, requiring the integration of multiple objectives such as economic feasibility, technical reliability, and environmental sustainability. Although previous studies have focused on optimizing renewable energy systems, many lack comprehensive decision frameworks that address the complex trade-offs between these objectives in urban settings. Addressing these challenges, this study introduces a novel Multi-Criteria Decision Analysis (MCDA) framework tailored for the evaluation and prioritization of energy scenarios in urban contexts, with a specific application to the city of Bozen-Bolzano. The proposed framework integrates various performance indicators to provide a comprehensive assessment tool, enabling urban planners to make informed decisions that balance different strategic priorities. At the core of this framework is the Technique for Order of Preference by Similarity to Ideal Solution (TOPSIS), which is employed to systematically rank energy scenarios based on their proximity to an ideal solution. This method allows for a clear, quantifiable comparison of diverse energy strategies, facilitating the identification of scenarios that best align with the city's overall objectives. The flexibility of the MCDA framework, particularly through the adjustable criteria weights in TOPSIS, allows it to accommodate the shifting priorities of urban planners, whether they emphasize economic, environmental, or technical outcomes. The study's findings underscore the importance of a holistic approach to energy planning, where trade-offs are inevitable but can be managed effectively through a structured decision-making process. Finally, the study addresses key gaps in the literature by providing a flexible and adaptable tool that can be replicated in different urban contexts to support the transition toward 100% renewable energy systems.

Keywords: renewable energy; urban energy systems; multi-criteria decision analysis; EnergyPLAN; performance indicator

1. Introduction

Recent events have underscored the urgent need to transition to a sustainable energy system, including extreme weather phenomena [1,2], geopolitical tensions affecting fossil fuel supplies [3–5], energy market volatility [6,7], the increasing frequency of natural disasters [1,8], and rising pollution levels [9,10]. These events highlight the vulnerabilities of current energy systems from both technical and economic perspectives [11,12]. As a consequence, the ongoing energy transition needs an acceleration to buffer these pressing current issues and, thus, to shift towards renewable and resilient energy solutions [7,13]. However, achieving such an ambitious goal as a 100% renewable energy system requires appropriate planning, taking into account the continuously changing socio-economic-environmental conditions and the evolution of various technologies [14,15]. In this respect, many aspects of the energy planning process need to be adapted to the new requirements, starting from energy modelling [16] to the optimization methodology [17] for the various objectives and finally to the decision-making process [18]. Thus, effective change toward a renewable-based future requires concerted efforts and actions from countries, cities, and individuals alike [19–21].

The 100% renewable energy system at the global and regional scale represents the final goal and, as a consequence, exhibits various challenges and opportunities. Studies emphasize the importance of transitioning towards cleaner energy sources to reduce CO_2 emissions and combat climate change [22–24]. The integration of renewable energy sources like solar and wind power is crucial, with islands serving as ideal environments for showcasing technical solutions and transition pathways [25]. Furthermore, the role of energy storage technologies, such as batteries and hydrogen storage, is essential in managing fluctuations in renewable energy sources and electricity demand, contributing to the feasibility of a fully renewable energy system [26]. While there is a growing trend towards 100% renewable energy systems, it is acknowledged that achieving this goal poses technical and economic challenges that require innovative solutions and comprehensive planning [27].

On a city scale, there have been several works to understand the technical and economic feasibility of transitioning to 100% renewable energy in the urban environment, which highlights the importance of factors like energy job sector growth, land requirements, and investment recovery [28,29]. The potential benefits of this transition have been further stressed in these works, which include the reductions in primary energy consumption, cost and greenhouse gas emission while showing the inconsistencies in investment recovery and emission reductions across different renewable energy systems in urban settings [29]. There is also a need for robust policies, infrastructure development, and multi-governance approaches to accelerate the energy transition and achieve climate neutrality in cities by leveraging renewable energy sources [28,30]. Additionally, there are other efforts to achieve a 100% renewable energy system in places like the residential community of Huanglong Township Island in Zhejiang province, China [31], the campus of Cornell University in the United States of America (USA) [32] and the local municipalities in Fukuoka, Japan [33].

This paper bridges the gab that exist in the energy system modelling phase and the decision making phase. It deals with the investigation of 100% solutions at an urban scale for optimal planning through decision analysis; although it is the final step, it is a crucial stage in the urban energy design process supporting the sensitive process of the decision-making [34]. Decision analysis aims to analyze a set of solutions identified through rigorous studies, which typically include energy system modelling and the search for near-optimal configurations of scenarios [35] or pathways [36]. Multi-Criteria Decision Analysis (MCDA) evaluates a pool of near-optimal solutions, enabling a comprehensive analysis of trade-offs between various energy strategies. This analysis considers multiple evaluation objectives (e.g., technical, economic, and environmental) and different sources of uncertainty (e.g., climate scenarios and demographic trends) [37,38].

In energy systems planning, decision analysis plays a crucial role in managing the complexities of shifting to sustainable energy sources [39]. Among the various approaches, MCDA stands out for its ability to evaluate and balance multiple conflicting criteria [40], such as technical feasibility [41], economic viability [42], and environmental impact [43]. This approach is particularly beneficial when planning for renewable energy systems, where diverse factors have to be considered to develop effective and sustainable strategies [44]. MCDA allows planners to systematically assess different energy scenarios, weighing the pros and cons of each option. It provides a structured framework that helps decision-makers identify the best solutions from a pool of alternatives [45], considering not only immediate costs and benefits but also long-term implications [46]. This comprehensive evaluation is essential for ensuring that selected energy strategies are both practical and aligned with broader sustainability goals [47].

The Technique for Order Preference by Similarity to Ideal Solution (TOPSIS) is especially useful within the MCDA framework. TOPSIS ranks alternative solutions based on their closeness to an ideal solution that maximizes desirable attributes and minimizes undesirable ones [48]. This method is advantageous because it offers clear, interpretable rankings of different alternatives, making it easier for decision-makers to make informed decisions [49]. By applying TOPSIS, we aim to provide energy planners with a tool capable of evaluating various energy scenarios in a systematic and transparent manner. This method helps to identify the most balanced and effective options, considering all relevant criteria [50]. As a result, TOPSIS supports robust, evidence-based decision-making, ensuring that the chosen solutions are technically sound, economically feasible, and environmentally sustainable [51]. This approach can be strategic for achieving the goal of effectively transitioning to renewable energy systems.

In this paper, the authors aim to seek the challenge of analyzing and interpreting the near-optimal solutions for urban energy systems planning in the case of multi-objective technical-economical-environmental optimization incorporating climate change. The main novelty lies in the proposed MCDA-based methodology, which uses performance indicators to flexibly analyze various conflicting aspects of energy planning (e.g., the cost of investing in and implementing state-of-the-art technologies, the use of renewable resources such as biomass and ensuring their sustainable use, and the environmental impact of the renewable energy system configurations) while also providing robust solutions to support decision-making processes. The proposed methodology is developed based on an urban case study of the alpine city of Bozen-Bolzano, where a pool of near-optimal configurations of the energy system according to technical, economic and environmental has been studied in [52,53]. Based on these solutions, the authors implemented the MCDA using a series of performance indicators covering different technical-economic-environmental aspects of the solutions considered. The indicators involved covered critical aspects of the design of urban renewable-based energy systems, such as Mismatch Compensation Factor, Emissions Reduction Effectiveness, Biomass System Efficiency, and Curtailment Fraction, among others. Instead, the MCDA proposes a comprehensive and robust investigation of the best solutions with respect to different types of targets.

The remainder of the paper details a case study of the energy system in Bozen-Bolzano, describing the materials and methodology, which include performance indicators and multi-criteria decision analysis in Section 2. The study's results are presented in Section 3, followed by the final remarks in Section 4.

2. Case Study, Materials and Methods

2.1. Case Study

The Municipality of Bozen-Bolzano is the case study employed for developing and testing the proposed methodology in Section 2.3. Bozen-Bolzano is a city found in the North-Eastern part of Italy, in the centre of the Alps, with a population of 106,000. The climate is characterized by cold winters and hot summers, with 2328 °C Heating Degree Days (HDD) and 222 °C Cooling Degree Days (CDD)for the typical year [52,54,55]. This city offers distinct opportunities and challenges for the energy transition due to unique geographical and extreme climatic conditions [56].

The actual energy system of Bozen-Bolzano city is characterized by a transition towards smart energy city (SEC) concepts spearheaded by the integration of renewable energy sources, which includes photovoltaic and thermal solar panels, a growing district heating system fed mainly by a waste-to-energy plant, hydropower storage and run of the river plants, which aims to accelerate decarbonization efforts [57]. The city's energy system emphasizes the use of renewable energy sources to improve overall energy efficiency and reduce its environmental impacts [58]. By combining various renewable energy sources, energy storage devices and advanced technologies, Bozen-Bolzano is striving towards sustainable economic development and energy independence, setting a modern example for other urban cities to follow [57,59]. More details are reported in [52].

2.2. Background Information

The background information forming the basis of the proposed decision analysis is derived from two previous studies. The final outcomes of these studies, which serve as inputs for the current decision analysis methodology, consist of a set of near-optimal solutions representing different configurations of 100% renewable-based energy systems. Menapace et al. (2020) [52] address the first steps in designing a 100% renewable energy system in Bozen-Bolzano by 2050. They first proposed a path by focusing on the operation between the sustainable use of biomass, replacement of traditional flexibility residing in fossil fuels with modern ones based on smart energy systems, balancing import and export of electricity and management of exchange of energy between the local energy system and it surrounding systems. Their approach integrated the energy system modelling using EnergyPLAN (a tool designed by the Sustainable Energy Planning Research Group at the Aalborg University to support the analysis of complex energy system factoring in advance technologies [60–62]) with multi simulations to accurately investigate the best technical alternative to achieve the main aim of a 100% renewable energy system.

Building on this work, Battini et al. (2024) [53] further integrated energy system design with the impacts of climate change on energy demand and renewable production. Then, the optimal scenarios are searched by varying the installed capacity of Photo Voltaics (PV), Combined Heat and Power (CHP) and Heat Pumps (HP) through a multi-objective optimization approach considering economic, environmental and technical targets. Two methods were adopted for the best scenarios identification: Grid Search and Non-dominated Sorting Genetic Algorithm-II (NSGA2). For the purposes of this work, we consider the scenarios resulting from the grid search as near-optimal due to their performance comparable to NSGA2 but in a regular grid.

2.3. Decision Analysis Methodology

The overall energy planning strategy is described in Figure 1 from the beginning with the data collection and the modelling of the actual energy system to the final best scenarios identification. Specifically, steps 1–3 regard the energy modelling of the system, including the actual year and climate projections, the optimization and selection of the near-optimal scenarios [52,53]. The next steps 4–6 belong to the proposed work and include the definition of the performance indicator for a wide-ranging evaluation of the different scenarios and a comparison analysis of the best scenario through a new MCDA methodology.

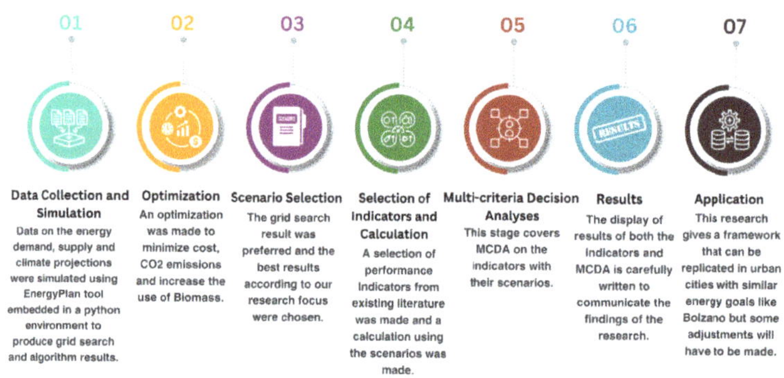

Figure 1. Methodology scheme for urban energy planning.

Thus, this decision analysis includes these two main phases that merge the robustness of the performance indicator and the flexibility of MCDA. Thus, specifically, the former phase consists of the evaluation of the selected performance indicators that reflect the key targets of the 100% renewable energy system: environmental impact, economic feasibility and technical reliability.

The performance indicators were carefully selected from a pool of advanced performance indicators [60] well known for their efficiency and ability to access, analyze and revolutionize the renewable energy space. For the benefit of this case study, we try to regenerate and adapt them as well as group them into our targets, i.e., technical, economic and environmental.

2.3.1. Technical Targets

To ensure a sustainable and reliable energy system, the technical targets primarily aim to reduce the reliance on biomass and minimize electricity imports. For this purpose, we have selected three key indicators, namely Relocation Coefficient (RC), Flexibility Factor or System Flexibility (FF) and finally, Biomass System Efficiency (BSE).

1. Relocation Coefficient (RC) is defined as the measure of comparison between the ability of different technologies in the supply system flexibility. It is the ratio between the net electricity exchange between the plant and system and the electricity demand minus intermittent electricity production. This indicator essentially helps evaluate how well the energy system can adjust to changes in energy demand and production, particularly when integrating renewable energy sources that may have fluctuating output. Its formula is reported as follows:

$$\text{RC} = \frac{\text{Net electricity exchange between plant and system}}{\text{Electricity demand - intermittent electricity production}} \tag{1}$$

2. Flexibility Factor or System Flexibility (FF) is an indicator first described by Paul Denholm and Robert M. Margolis to be the lowest hourly value over the year divided by the maximum hourly value with regard to the output of a simulation [63]. Thus, this indicator was used to assess the flexibility of the system over the year used in the simulation. We gave it a range between 0 and 1 with a value close to 0, which means the system is not flexible, and a value close to 1 means the system is flexible. In general terms, this metric helps determine how well the energy system can maintain consistent performance despite fluctuations in energy production and demand throughout the year.

$$\text{FF} = \frac{\text{Lowest hourly value of the year}}{\text{Maximum hourly value of the year}} \tag{2}$$

3. Biomass System Efficiency (BSE) is used to assess the importance of biomass in the energy system without the transportation system [64]. This indicator was helpful in this work since it could help in the quantification and reduction of biomass in the system. To attain this, the output from synthetic fuel is subtracted from the production of all the fuel by biomass, which is then divided by the biomass used for transportation subtracted from the input amount of biomass. Essentially, this efficiency metric shows how effectively the system uses biomass resources, helping to minimize waste and maximize energy output from the available biomass.

$$\text{BSE} = \frac{\text{Output of all fuel by Biomass} - \text{Output from synthetic fuel}}{\text{Input Biomass amount} - \text{Biomass used for transportation}} \tag{3}$$

2.3.2. Economic Targets

Economic targets are characterized by the minimization of the annual cost of the system and consist of Mismatch Compensation Factor (MCF) and Marginal Economic Efficiency (MEE).

1. Mismatch Compensation Factor (MCF) was developed by Lund et al. [65] with respect to zero-energy buildings. It relates cost balance (i.e., the installed capacity of

renewable energy sources where the import costs and export incomes are balanced) to energy balance (i.e., the installed capacity of renewable energy source (RES) balancing aggregated annual imports to exports from the energy system). This indicator helps measure how well the energy system can balance its energy production with its costs, ensuring that it produces enough renewable energy to meet its own needs while minimizing external energy purchases.

$$\text{MCF} = \frac{\text{Cost Balance}}{\text{Energy Balance}} = \frac{\text{Total RES Production} - \text{RES Used}}{\text{Total Energy Demand} - \text{RES Used}} \quad (4)$$

2. Marginal Economic Efficiency (MEE) shows how the added cost of RES contributes to the total cost of the system. It is expressed by dividing the change in the total system cost by the change in the cost of RES [60]. In simpler terms, this indicator helps assess how cost-effective the system is when adding renewable energy sources, showing whether the investment in renewable technologies leads to efficient use of resources and overall cost savings.

$$\text{MEE} = \frac{\Delta\text{Total System Cost}}{\Delta\text{Renewable Energy Sources Costs}} \quad (5)$$

2.3.3. Environmental Targets

The environmental target goal is to minimize the CO_2 emissions of the system. It is also related to the system's renewable energy sources. It has three indicators, namely, Curtailment Fraction, Marginal Primary Energy Supply and Marginal Export.

1. When a system is not able to hold excess production of RES within a given period, the percentage of the RES production lost by the technology is called Curtailment Fraction (CF). When the percentage is equal to 100%, we say the system has the capacity to integrate the excess RES produced and vice versa. It is calculated by subtracting the realized RES production from the potential RES production, and the results are divided by the potential RES production. This indicator practically measures how much renewable energy is wasted because the system cannot fully utilize or store it, with higher curtailment indicating greater energy loss.

$$\text{CF} = \frac{\text{Potential RES production} - \text{Realized RES production}}{\text{Potential RES production}} \quad (6)$$

2. Marginal Primary Energy Supply (MPES) compares the different RES where the factors may be determined by marginal effects. Specifically, the MPES indicates how the marginal Primary Energy Supply (PES) of the system is affected by a marginal change in the PES from RES. If it is less than 1, the system cannot fully integrate marginal RES production [60]. In other therms, this indicator shows how efficiently the system can incorporate small increases in renewable energy supply, helping to assess the system's ability to handle additional renewable energy without performance losses. This is represented by the formula below.

$$\text{MPES} = \frac{\Delta\text{Primary Energy Sources}}{(\Delta\text{Primary Energy Source} \times \text{Renewable Energy Sources})} \quad (7)$$

3. Marginal Export (ME) is used to determine the relationship between marginal export and marginal changes in PES, which are biomass-based [64].

$$\text{ME} = \frac{\Delta\text{Export}}{\Delta\text{Primary Energy Sources}} \quad (8)$$

These indicators were placed in a range between 0 and 1, where 0 indicates a poor performance and 1 optimal performance. Furthermore, after obtaining the results for each indicator across their respective scenarios, the data were normalized between 0 and 1 to reflect the aforementioned range. Therefore, values close to 1 represent favourable indicators, while values close to 0 represent unfavourable indicators. Table 1 represents the first 10 indicators before normalization and after normalization to throw more light on the above statement.

Table 1. Values of the performance indicators in different scenarios, the first 10 selected as examples.

Before Normalization								
Scenarios	RC	FF	CF	MCF	MPES	MEE	ME	BSE
Scenario 1	0.981722	0.025282	0.104913	0.792006	2.189719	17.02612	0.963444	0.196561
Scenario 2	0.914448	0.010581	0.049434	0.871856	2.395090	1.176235	1.064378	1.098251
Scenario 3	0.957826	0.020870	0.092879	0.917171	2.333604	1.151019	0.915653	1.099818
Scenario 4	0.967399	0.018738	0.086294	0.871856	2.406525	1.152334	0.934797	1.049301
Scenario 5	0.894655	0.009454	0.045639	0830840	2.470161	1.210174	1.081707	1.048504
Scenario 6	0.977985	0.015756	0.075988	0.830840	2.506582	1.408031	0.955969	1.043592
Scenario 7	0.960285	0.013243	0.065025	0.793491	2.537576	1.840971	1.027957	0.947787
Scenario 8	0.983609	0.017858	0.086074	0.793491	2.518343	2.206353	0.967219	0.883702
Scenario 9	0.964461	0.021068	0.100287	0.793491	2.505681	2.551852	0.928922	0.840727
Scenario 10	0.952290	0.023275	0.109848	0.793491	2.496894	2.835778	0.904580	0.812299
After Normalization								
Scenarios	RC	FF	CF	MCF	MPES	MEE	ME	BSE
Scenario 1	0.850946	0.277390	0.361767	0.451797	0.111556	1.000000	0.309250	0.000000
Scenario 2	0.286335	0.151160	0.218751	0.567825	0.202887	0.056067	0.839516	0.998265
Scenario 3	0.650396	0.239502	0.330747	0.633672	0.175544	0.054565	0.058171	1.000000
Scenario 4	0.730733	0.221197	0.313773	0.567825	0.207973	0.054643	0.158749	0.944072
Scenario 5	0.120217	0.141479	0.208970	0.508226	0.236273	0.058088	0.930552	0.943190
Scenario 6	0.819577	0.195596	0.287204	0.508226	0.252470	0.069871	0.269977	0.937751
Scenario 7	0.671030	0.174015	0.258943	0.453955	0.266253	0.095655	0.648173	0.831685
Scenario 8	0.866785	0.213642	0.313203	0.453955	0.257700	0.117415	0.329079	0.760737
Scenario 9	0.706079	0.241208	0.349842	0.453955	0.252069	0.137991	0.127884	0.713158
Scenario 10	0.603932	0.260161	0.374491	0.453955	0.248161	0.154900	0.000000	0.681686

2.4. TOPSIS-Based MCDA approach

The proposed approach consists in employing the TOPSIS to evaluate and rank different scenarios based on relevant criteria. This method is herein suggested because it considers both the best and worst possible scenarios, providing a clear and balanced comparison of alternatives. The step-by-step procedure to implement the TOPSIS method is outlined as follows.

- Construct the assessment matrix: first, compile the quantitative evaluations g_{ij} for each alternative i across each criterion j. This matrix provides a comprehensive overview of how each alternative performs under each criterion.
- Compute the normalized matrix, with the generic element z_{ij} representing the normalized evaluation of alternative i under criterion j as:
  Normalize the matrix: next, standardize the values in the assessment matrix to make them comparable across criteria. The normalized value z_{ij} for each alternative i and criterion j is calculated as:
  $$z_{ij} = \frac{g_{ij}}{\sqrt{\sum_{i=1}^{n} g_{ij}^2}}. \tag{9}$$
  This step removes the units of measurement and scales the data, ensuring that each criterion contributes equally to the analysis.
- Calculate the weighted normalized matrix: apply the assigned weights to the normalized values to reflect the importance of each criterion. The weighted normalized value u_{ij} is given by:
  $$u_{ij} = w_j \times z_{ij}, \forall i, \forall j; \tag{10}$$
  where w_j is the weight assigned to criterion j. This step adjusts the normalized values according to the significance of each criterion.
- Determine the ideal solutions: identify the best possible (positive ideal) and worst possible (negative ideal) values for each criterion. The positive ideal solution A^* and the negative ideal solution A^- are defined as:

$$A^* = (u_1^*, \ldots, u_k^*) = \{(\max_i u_{ij} | j \in I'), (\min_i u_{ij} | j \in I'')\}; \quad (11)$$

$$A^- = (u_1^-, \ldots, u_k^-) = \{(\min_i u_{ij} | j \in I'), (\max_i u_{ij} | j \in I'')\}; \quad (12)$$

where I' includes criteria to be maximized and I'' includes criteria to be minimized. These ideal solutions serve as reference points for comparison.

- Calculate the distances to the ideal solutions: measure the distances of each alternative from the positive and negative ideal solutions. The distances S^* and S^- for each alternative i are computed as:

$$S^* = \sqrt{\sum_{j=1}^{k} (u_{ij} - u_{ij}^*)^2}, i = 1, \ldots, n; \quad (13)$$

$$S^- = \sqrt{\sum_{j=1}^{k} (u_{ij} - u_{ij}^-)^2}, i = 1, \ldots, n. \quad (14)$$

These distances quantify how far each alternative is from the ideal solutions.

- Calculate the closeness coefficient: determine the closeness coefficient C_i^* for each alternative i, which indicates its relative proximity to the ideal solutions. The closeness coefficient is calculated as:

$$C_i^* = \frac{S^-}{S^- + S^*}, 0 < C_i^* < 1, \forall i. \quad (15)$$

This coefficient shows how closely an alternative aligns with the best possible scenario while avoiding the worst.

- Rank the alternatives: finally, rank the alternatives based on their closeness coefficients in descending order. For example, in comparison between two generic alternatives i and z, if $C_i^* \geq C_z^*$, then alternative i is preferred over alternative z. This ranking helps in making informed decisions by highlighting the most favorable options.

By following these steps, the TOPSIS method provides a systematic and objective way to evaluate and rank multiple alternatives based on a set of criteria, ensuring balanced and well-informed decision-making.

After obtaining the final ranking by initially assigning equal weights to all criteria, we will conduct a sensitivity analysis by varying the weights assigned to each criterion. This analysis will help us understand the robustness of the rankings and the impact of different criteria on the overall assessment. The criteria will be grouped based on their nature, such as environmental, technical, and economic targets, allowing us to systematically explore how changes within these groups affect the final rankings.

3. Results and Discussion

This section presents the findings of the proposed MCDA across various scenarios. Firstly, the performance indicators are analyzed and grouped according to different targets, i.e., environmental, technical, and economic. The relationship between each indicator and the capacity of the key technologies is then illustrated. Subsequently, the MCDA of this section dives deep into the performance of the scenarios across all targets and the sensitivity of each scenario is analyzed. The analysed scenarios are the 52 near-optimal energy system configurations resulting from the grid search in [53].

3.1. Performance Indicators of the Energy Scenarios

This particular session of the paper is dedicated to the performance indicator results, where each scenario performance with each indicator is accessed. The indicator analysis presented in Figure 2 focuses on three primary targets: Economic, Technical, and Environmental. Each indicator within these targets offers insight into various aspects of the scenarios' performance, allowing for a comprehensive evaluation of its efficiency, reliability, and adaptability across all the indicators.

Economic targets are depicted through the Mismatch Compensation Factor (MCF) and the Marginal Economic Efficiency (MEE) in Figure 2a. The MCF values exhibit a variation ranging from 0.8 to 1.4, indicating shifts in the cost and energy balance within the system. MCF values suggest effective compensation for mismatches in the various scenarios, indicating the best solution that is closest to 1. On the other hand, the MEE shows high variability with significant peaks over 4 in a few cases. This indicates that the share of renewable costs on overall cost varies among the different scenarios.

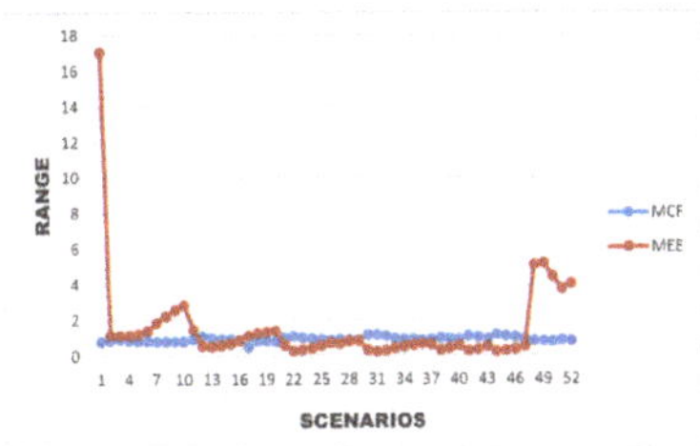

(**a**) Economic Target

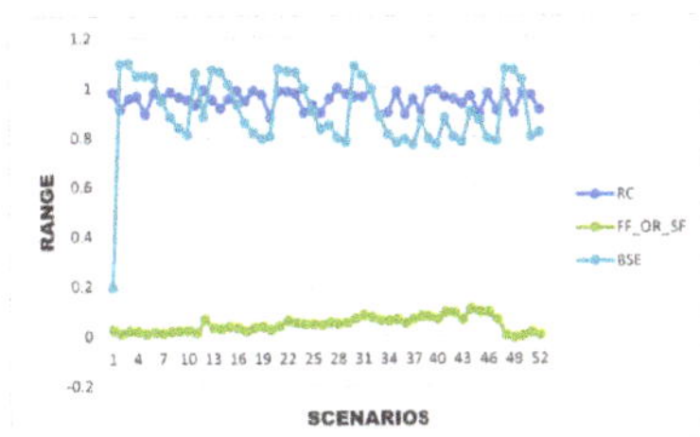

(**b**) Technical Target

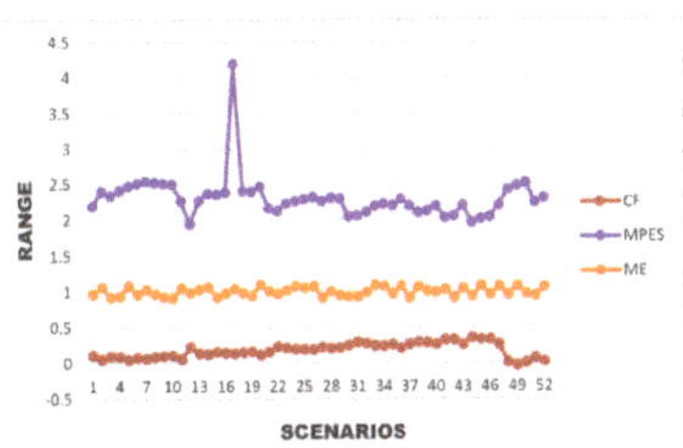

(**c**) Environmental Target

Figure 2. Performance indicator divided in three targets: Economic Target - Mismatch Compensation Factor (MCF) and Marginal Economic Efficiency (MEE), Technical Target—Reliability Coefficient (RC), Flexibility Factor or System Flexibility (FF) and Biomass System Efficiency (BSE), and Environmental Target—Curtailment Fraction (CF), Marginal Primary Energy Supply (MPES), Marginal Export (ME).

Technical targets are evaluated in Figure 2a using the Reliability Coefficient (RC), Flexibility Factor (FF), and Biomass System Efficiency (BSE). The RC values fluctuate moderately between 0.95 and 1.02, indicating a high-reliability level of all the scenarios with minor variability among them. This consistency underscores the system's dependable performance across different scenarios. The FF shows values between 0 and 0.2, indicating the overall scarce flexibility of all the scenarios due to the specific characteristics of the analysed urban case study. The BSE values range from 0.8 to 1.2, suggesting that the biomass system efficiency consistently varies across scenarios. Despite this variability, values close to 1 indicate overall effectiveness in biomass utilization.

Environmental targets are then examined by means of through the Curtailment Fraction (CF), Marginal Primary Energy Supply (MPES), and Marginal Export (ME) reported in Figure 2c. The CF is characterized by values ranging between 0 and 0.5, with an important variation among the different scenarios. The highest values indicate scenarios with a moderate capacity to integrate RES. The MPES values range around 2–2.5 with a spike of 4. This consistency reflects the system's stability in integrating renewable energy sources, with the spike indicating a potential anomaly or specific scenario causing a significant reduction. The ME values are quite stable, fluctuating slightly around 0.9 to 1.1. This stability suggests that the system maintains efficient performance in terms of energy export across different scenarios, with only minor expected fluctuations. In summary, the analyses of these indicators across economic, technical, and environmental targets reveal a complex interplay of factors influencing its performance for each scenario. The varying trends and fluctuations observed in each indicator highlight the complex dynamic of energy system configurations, emphasizing the nature of the analyzed case study.

The dependency between the eight indicators and the capacities of three key technologies (i.e., Combined Heat and Power (CHP), Heat Pump (HP), and Photovoltaic (PV)) is investigated. The relationships are illustrated through scatter plots, providing a comprehensive representation of how each indicator is affected by the installed capacities.

The analysis of the Mismatch Compensation Factor (MCF) in relation to capacities (Figure 3a) shows that the data points for CHP and HP are clustered without any pattern in relation to MCF. Instead, PV capacities, while more broadly distributed, exhibit a negative correlation with MCF. Thus, the MCF indicator, which focuses on cost balance and energy balance, appears to be relatively independent of the capacities of CHP and HP, but not from PV, suggesting that changes in PV capacity can significantly affect the marginal capacity factor.

The Marginal Economic Efficiency (MEE) indicator (Figure 3b) shows different behaviour with the different technologies. Specifically, PV and HP present a slight positive correlation with MEE, while CHP seems to not have any influence. This means that both HP and PV capacity play a positive effects on MEE.

The Reliability Coefficient (RC) indicator (Figure 3c) is again analyzed against the capacities of CHP, HP, and PV. The scatter plot indicates that all three technologies do not have a clear pattern on RC. This means that the contained variation of reliability does not depend on a single technology but on the combination of all of them together with other minor system configurations.

The Flexibility Factor or System Flexibility (FF) is also analyzed (Figure 3d). In this case, the scatter plot demonstrates a pattern for all the technologies. Specifically, CHP and HP are positively correlated with FF, with CHP showing a stronger relationship. Instead, PV is negatively correlated with FF, highlighting a clearly different impact compared with CHP and HP.

The Biomass System Efficiency (BSE) is analyzed through the scatter plot in Figure 3e). This reveals that the data points for PV have no trend but are randomly spread among various capacities over BSE values. A slight negative correlation is presented between CHP and BSE. Instead, HP shows a more evident negative relationship with biomass system efficiency. This suggests that the BSE indicator is relatively independent of PV but not from CHP and HP, which the latter has a more important impact on the biomass system's efficiency.

Next, the Curtailment Fraction (CF) is examined. The corresponding scatter plot (Figure 3f) shows that both CHP and HP capacities exhibit a positive correlation with CF capacity. On the contrary, PV presents a negative correlation, highlighting an antithetical behaviour between PV and the other two technologies.

In examining the Marginal Primary Energy Supply (MPES) (Figure 3g), the scatter plot reveals that data points for CHP and HP are tightly clustered with no apparent variation in MPES. PV capacities show a broader range, with a slight positive relationship with MPES. This indicates that the primary energy sources and renewable energy integration are not directly affected by the installed capacities of CHP, HP, and PV, suggesting that marginal production efficiency is quite independent of the single technology capacities.

Finally, for the Marginal Export (ME) indicator (Figure 3h), the scatter plot shows that data points for CHP, HP and HP capacities have no pattern with ME.

Briefly, the dependency analysis between these performance indicators and the capacities of CHP, HP, and PV systems reveals the complexity of energy system behaviour across different technical, economic, and environmental dimensions. This finding suggests that the performance captured by these indicators is essential for accurately describing the complexity of various scenario configurations. Moreover, it highlights the importance of thoroughly analysing these solutions using an MCDA methodology.

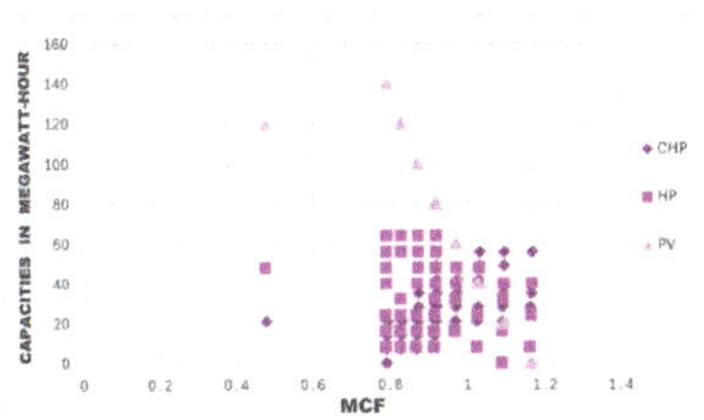

(**a**) Mismatch Compensation Factor (MCF)

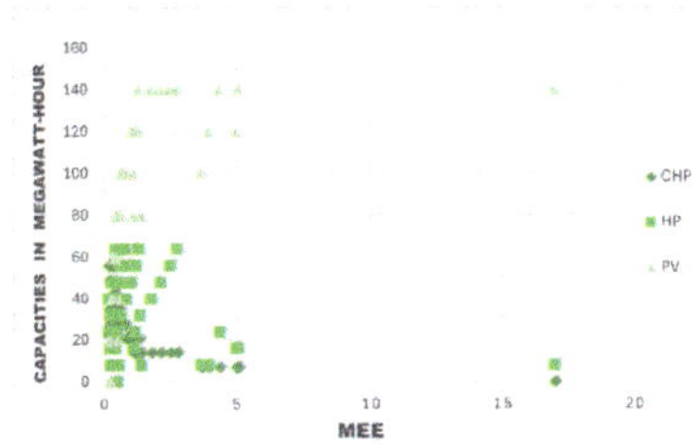

(**b**) Marginal Economic Efficiency (MEE)

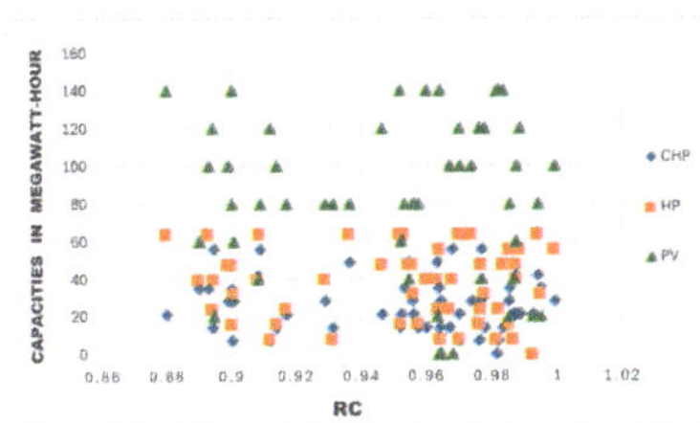

(**c**) Reliability Coefficient (RC)

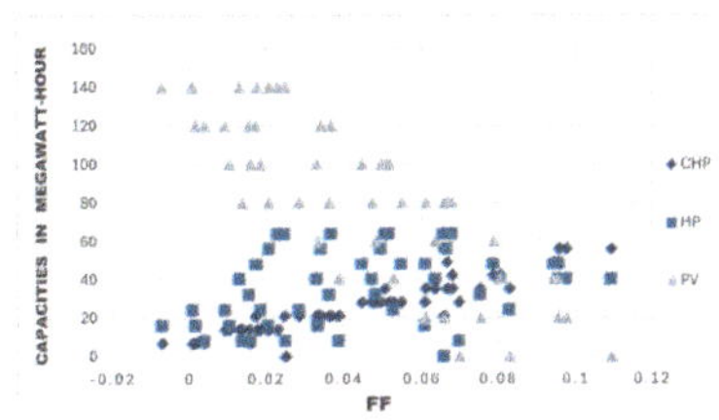

(**d**) Flexibility Factor or System Flexibility (FF)

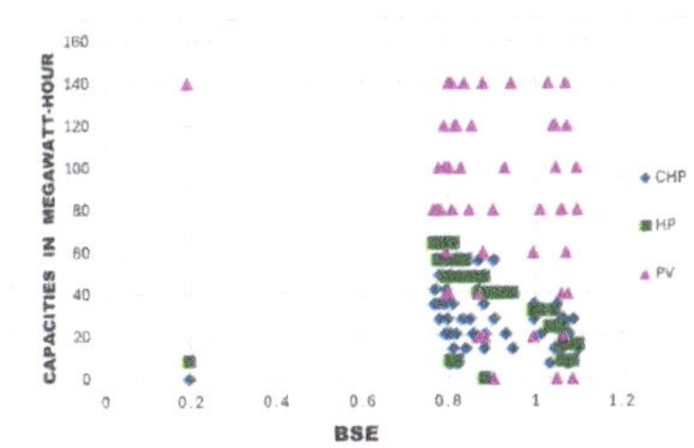

(**e**) Biomass System Efficiency (BSE)

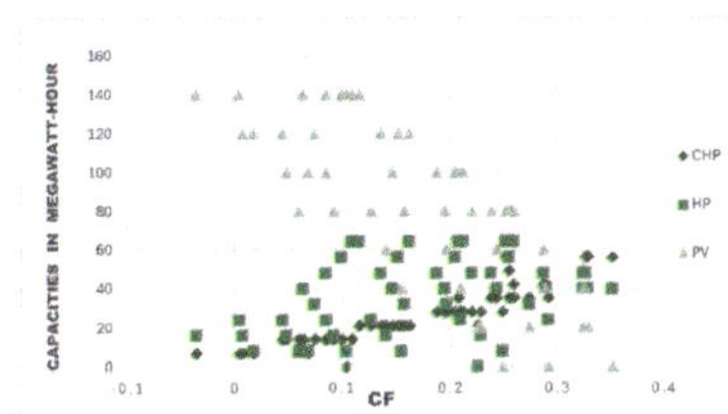

(**f**) Curtailment Fraction (CF)

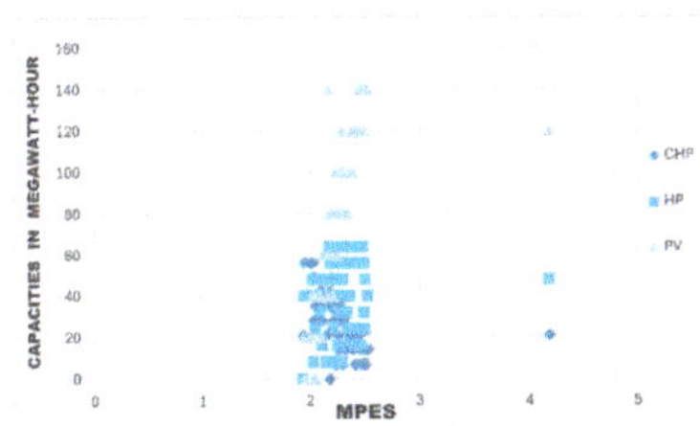

(**g**) Marginal Primary Energy Supply (MPES)

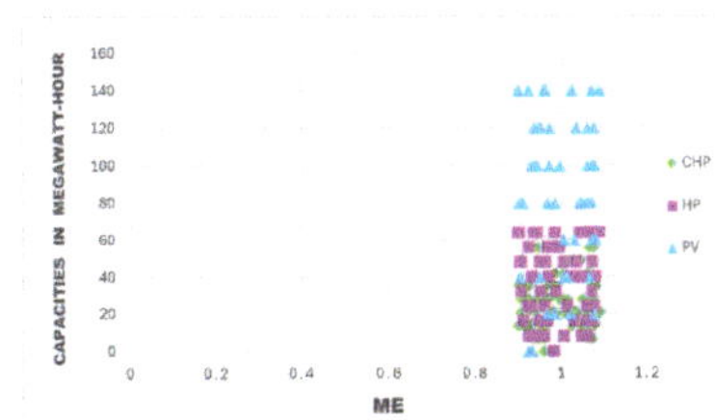

(**h**) Marginal Export (ME)

Figure 3. Indicator analyses based on targets: Economic Target (Mismatch Compensation Factor (MCF) and Marginal Economic Efficiency (MEE)), Technical Target (Reliability Coefficient (RC), Flexibility Factor or System Flexibility (FF or SF) and Biomass System Efficiency (BSE)) Environmental Target (Curtailment Fraction (CF), Marginal Primary Energy Supply (MPES), Marginal Export (ME))

3.2. Multi-Criteria Decision Analysis (MCDA) and Sensitivity Analysis

In this study, a MCDA was conducted to evaluate the performance of various energy scenarios for urban energy system planning. The MCDA approach employed in this analysis integrates economic, technical, and environmental targets, enabling a comprehensive assessment of each scenario's performance under different prioritization schemes. Results are illustrated in the series of bar graphs and line charts provided, which depict the top 10 performing scenarios and the overall performance across all scenarios. In these figures, *CC* values have been rescaled to a 0–100 range for improved interpretability.

The baseline scenario assigns equal weight (1/3) to each of the three macro groups of criteria, which are economic, technical, and environmental. Various indicators were distributed among the groups. The performance of this scenario is illustrated in the top 10 CC values bar graph in Figure 4a. Scenario 41 emerged as the highest-ranking scenario, consistently demonstrating superior performance across multiple criteria. This scenario's robust performance can be attributed to its balanced approach, which ensures that none of the criteria groups are disproportionately prioritized, leading to a well-rounded energy system design.

In the environmental target scenario, 60% of the weight was allocated to environmental indicators, with the remaining 40% equally divided between technical and economic indicators. The results (Figure 4b) indicate that this scenario favours configurations that maximize environmental benefits, such as reduced emissions and efficient resource utilization. Scenario 17, which stands out, is characterized by a substantial emphasis on curtailment reduction and optimal biomass system efficiency, which are critical for minimizing the environmental footprint of the energy system.

For the technical target scenario, 60% of the weight was assigned to technical indicators, with the remaining weight equally split between economic and environmental indicators. The top-performing scenarios (Figure 4c) in this case highlight the importance of system reliability and flexibility. Scenario 41 in this case appears as the 2nd best while 32, which ranks highest in this target, reflects its superior adaptability and technical robustness, which are crucial for maintaining system stability and efficiency under varying operational conditions.

In the economic target scenario, 60% of the weight was allocated to economic indicators, with the technical and environmental indicators each receiving 20%. The top 10 scenarios in this case (Figure 4d) underline the significance of cost-effectiveness and economic efficiency in the system design. Scenario 45, which ranks highest in this scenario, showcases a highly cost-efficient configuration, effectively balancing the trade-offs between investment costs and system performance. In this case, we also see Scenario 41 performing well by placing 5th.

The comparative analysis of the different scenarios (Figure 5) shows significant variability in Closeness Coefficient (CC) values across all scenarios, indicating that no single scenario consistently outperforms the others under all target conditions. The sensitivity analysis further reveals how changes in the weighting of criteria affect the rankings of the scenarios, providing insights into the robustness of the decision-making process. Notably, Scenario 41 frequently appears in the top ranks across multiple targets, suggesting that it strikes an effective balance across all three target; technical, economic, and environmental. Unlike other scenarios that may perform well in one area but fall short in others, Scenario 41 consistently delivers high performance across multiple indicators. Its configuration offers optimal flexibility and reliability while maintaining economic efficiency and minimizing environmental impact, making it a well-rounded choice for urban energy planning. This scenario's balanced approach aligns well with the overarching goals of transitioning to a zero-carbon future without compromising on technical stability or affordability.

This MCDA result has highlighted the importance of considering multiple perspectives when planning urban energy systems. By adjusting the weights assigned to different criteria, decision-makers can tailor the energy system design to prioritize specific objectives, whether environmental sustainability, technical reliability, or economic feasibility. The flexibility of the MCDA approach ensures that the chosen scenario aligns with broader strategic goals while accommodating the unique needs of the urban energy system under study.

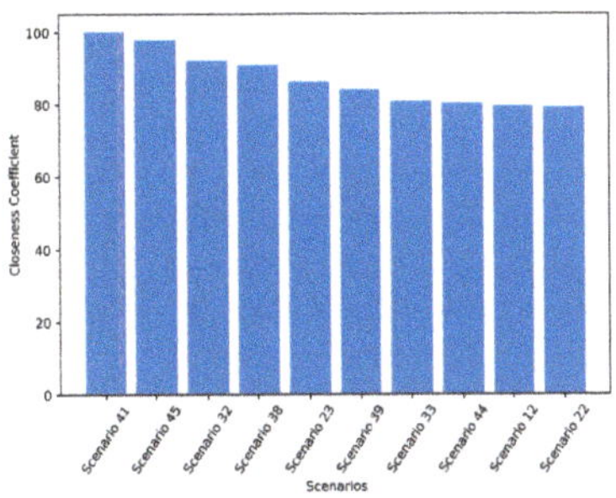

(**a**) Top 10 CC Values - Baseline Scenario

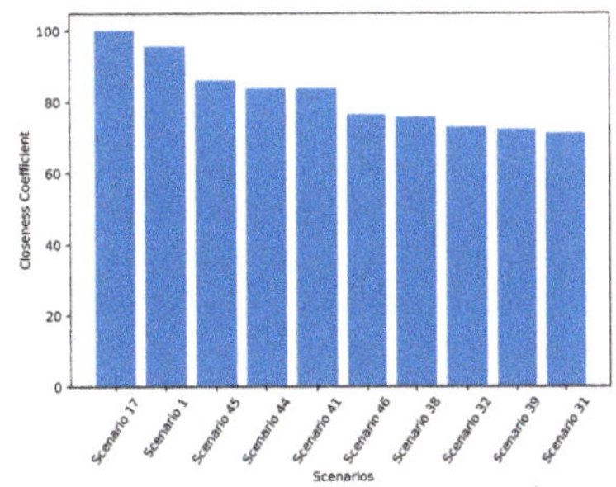

(**b**) Top 10 CC Values - Environmental Target

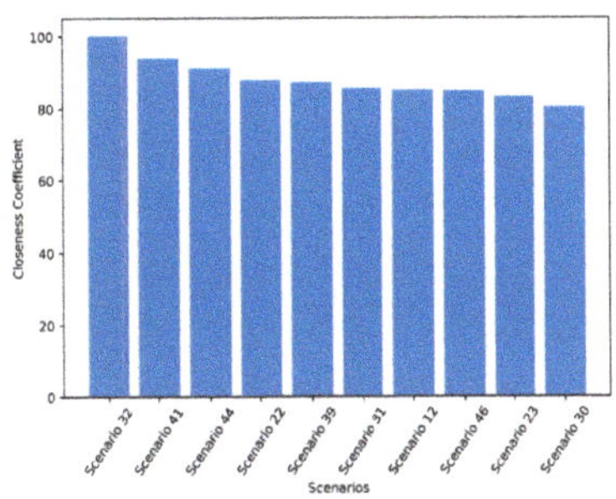

(**c**) Top 10 CC Values - Technical Target

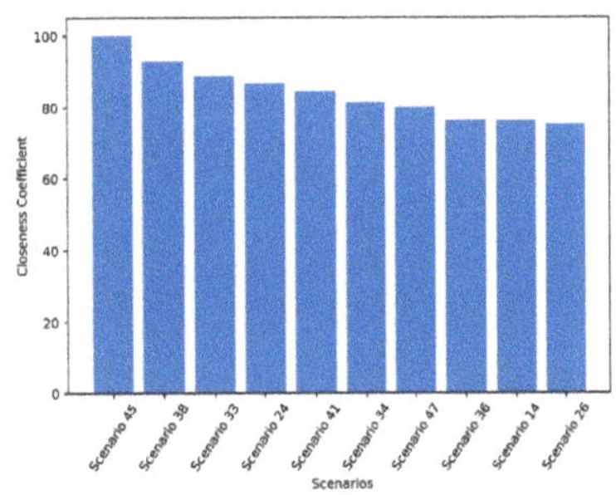

(**d**) Top 10 CC Values - Economic Target

Figure 4. Closeness Coefficient (CC) values for the best 10 scenarios of the MCDA for the baseline scenario, environmental target, technical target and economic target.

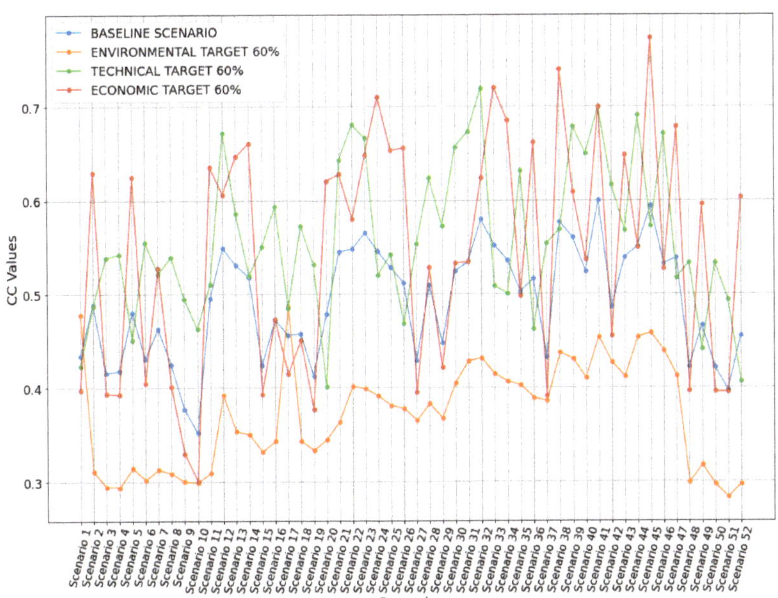

Figure 5. Overall Performance of Different Scenarios Across All Targets.

Upon analyzing the sensitivity of each indicator to changes in criteria weights, we further synthesized results in Table 2, including the best 10 scenarios across all targets, and Table 3, related to the worst 10 scenarios across all targets. It is worth noting that other scenarios may be preferable if greater importance is placed on specific targets, such as Scenario 17, which performs particularly well under environmental considerations. Additionally, certain scenarios may be more suitable depending on specific constraints, such as the maximum installable capacity of a technology or budget limitations, highlighting the adaptability of the proposed methodology to different planning contexts.

Table 2. Performance of Best Scenarios Across Different Targets.

BS	CC	ENV 60%	CC	TEC 60%	CC	ECO 60%	CC
Scenario 41	0.5994	Scenario 17	0.4868	Scenario 32	0.7190	Scenario 45	0.7729
Scenario 45	0.5936	Scenario 1	0.4777	Scenario 41	0.6992	Scenario 38	0.7389
Scenario 32	0.5790	Scenario 45	0.4580	Scenario 44	0.6903	Scenario 33	0.7197
Scenario 38	0.5762	Scenario 44	0.4537	Scenario 22	0.6802	Scenario 24	0.7098
Scenario 23	0.5648	Scenario 41	0.4536	Scenario 39	0.6783	Scenario 41	0.6993
Scenario 39	0.5596	Scenario 46	0.4385	Scenario 31	0.6730	Scenario 34	0.6847
Scenario 33	0.5517	Scenario 38	0.4369	Scenario 12	0.6715	Scenario 47	0.6781
Scenario 44	0.5502	Scenario 32	0.4313	Scenario 46	0.6708	Scenario 36	0.6613
Scenario 12	0.5485	Scenario 39	0.4300	Scenario 23	0.6657	Scenario 14	0.6601
Scenario 22	0.5476	Scenario 31	0.4277	Scenario 30	0.6564	Scenario 26	0.6552

Table 3. Performance of worst Scenarios Across Different Targets.

BS	CC	ENV 60%	CC	TEC 60%	CC	ECO 60%	CC
Scenario 8	0.4238	Scenario 8	0.3081	Scenario 2	0.4868	Scenario 50	0.3952
Scenario 15	0.4233	Scenario 6	0.3016	Scenario 17	0.4847	Scenario 51	0.3948
Scenario 48	0.4223	Scenario 9	0.2995	Scenario 26	0.4683	Scenario 27	0.3945
Scenario 50	0.4215	Scenario 10	0.2984	Scenario 36	0.4627	Scenario 3	0.3932
Scenario 4	0.4172	Scenario 48	0.2981	Scenario 10	0.4624	Scenario 15	0.3929
Scenario 3	0.4154	Scenario 52	0.2962	Scenario 5	0.4506	Scenario 4	0.3922
Scenario 19	0.4121	Scenario 50	0.2959	Scenario 49	0.4407	Scenario 37	0.3914
Scenario 51	0.3972	Scenario 3	0.2941	Scenario 1	0.4226	Scenario 19	0.3768
Scenario 9	0.3766	Scenario 4	0.2935	Scenario 52	0.4064	Scenario 9	0.3295
Scenario 10	0.3516	Scenario 51	0.2823	Scenario 20	0.4010	Scenario 10	0.3003

4. Conclusions

In the face of rapidly advancing urbanization and the escalating demand for sustainable energy solutions, this study has significantly contributed to urban energy system planning. By introducing a novel Multi-Criteria Decision Analysis (MCDA) framework tailored to Bozen-Bolzano, we have established an integrated approach to evaluating and selecting energy scenarios.

The proposed framework balances economic, environmental, and technical targets in decision-making. The economic aspect minimizes costs while maximizing energy output, the environmental side reduces emissions and protects natural resources, and technical indicators ensure system flexibility and reliability amid fluctuating renewable energy sources. The MCDA framework allows for a careful comparison of energy options, considering the efficiency, cost, and environmental impact of different technologies, ensuring that the chosen solutions are practical and sustainable.Moreover, the tool's flexibility makes it highly useful for decision-makers and energy policy-makers, allowing them to adjust the weighting of different criteria based on their specific priorities and constraints. However, it is important to note that this work does not delve into the integration of decision-making processes or explore the implications of energy policy, as these topics are beyond the scope of the current study.

The integration of the TOPSIS method into the MCDA framework was crucial, providing a quantifiable way to rank energy scenarios. TOPSIS simplified the decision-making process by identifying scenarios closest to the ideal solution, offering flexibility to adjust criteria weights based on evolving priorities and conditions. Scenario 41 stood out, performing well across multiple criteria, making it a compelling choice for urban planners aiming to balance economic viability, environmental sustainability, and technical reliability.

Our findings have far-reaching implications for urban areas striving to transition to 100% renewable energy. The study emphasizes the necessity of a holistic approach, ensuring that selected energy systems are sustainable, economically viable, and technically sound. This is critical to achieving broader goals of climate neutrality and energy independence. Scenario analysis also played a key role in our study, allowing urban planners to anticipate uncertainties and ensure that chosen strategies are resilient and adaptable.

Beyond Bozen-Bolzano, the MCDA framework we developed can be adapted to other urban contexts, making it a versatile tool for cities facing similar challenges in energy system planning. This study contributes methodologically by demonstrating the effectiveness of integrating TOPSIS in MCDA to evaluate energy scenarios. This combination offers structured decision-making guidance while allowing for adjustments based on specific city needs, enhancing its practical application.

While our research offers valuable insights, it opens avenues for future exploration. Future research could integrate additional factors like social acceptance and technological innovation to provide a more comprehensive evaluation. Applying the MCDA framework to other cities with different socio-economic and climatic conditions could yield further insights, enhancing its generalizability.

Additionally, the dynamic nature of urban energy systems presents an opportunity to develop adaptive frameworks using real-time data and advanced computational methods like machine learning. This would allow planners to make informed, timely decisions, enhancing resilience and sustainability. Although this framework provides a comprehensive tool for evaluating energy systems, its reliance on predefined performance indicators may not fully capture future uncertainties. The study's focus on Bozen-Bolzano, a relatively small city, also means that implementing the proposed methodology on larger and complex cities can bring challenging issues, such as data collection and energy systems modelling.

Future updates should address these limitations by incorporating real-time data and diverse socio-political factors. In conclusion, this study represents a significant advancement in urban energy planning. The framework offers a novel, adaptable tool to support the transition to 100% renewable energy systems, with insights that are relevant not only to Bozen-Bolzano but to urban energy planning worldwide.

Author Contributions: Conceptualization, A.M. and S.C.; methodology, A.M., B.K.N. and S.C.; software, A.M., B.K.N. and S.C.; validation, A.M., B.K.N. and S.C.; formal analysis, A.M., B.K.N. and S.C.; investigation, A.M., B.K.N. and S.C.; resources, A.M.; data curation, A.M., B.K.N. and S.C.; writing—original draft preparation, A.M., B.K.N. and S.C.; writing—review and editing, A.M., B.K.N. and S.C.; visualization, B.K.N. and S.C.; supervision, A.M. and S.C.; project administration, A.M.; funding acquisition, A.M. All authors have read and agreed to the published version of the manuscript.

Funding: This research received no external funding.

Informed Consent Statement: Not applicable.

Data Availability Statement: Data available on request due to restrictions.

Conflicts of Interest: The authors declare no conflicts of interest.

References

1. Gonçalves, A.C.; Costoya, X.; Nieto, R.; Liberato, M.L. Extreme weather events on energy systems: A comprehensive review on impacts, mitigation, and adaptation measures. *Sustain. Energy Res.* **2024**, *11*, 4. [CrossRef]
2. Xu, L.; Feng, K.; Lin, N.; Perera, A.; Poor, H.V.; Xie, L.; Ji, C.; Sun, X.A.; Guo, Q.; O'Malley, M. Resilience of renewable power systems under climate risks. *Nat. Rev. Electr. Eng.* **2024**, *1*, 53–66. [CrossRef]
3. Cheikh, N.B.; Zaied, Y.B. Renewable energy deployment and geopolitical conflicts. *J. Environ. Manag.* **2023**, *344*, 118561. [CrossRef] [PubMed]
4. Yasmeen, R.; Shah, W.U.H. Energy Uncertainty, Geopolitical Conflict, and Militarization Matters for Renewable and Non-renewable Energy Development: Perspectives from G7 Economies. *Energy* **2024**, *306*, 132480. [CrossRef]
5. Ibekwe, K.I.; Etukudoh, E.A.; Nwokediegwu, Z.Q.S.; Umoh, A.A.; Adefemi, A.; Ilojianya, V.I. Energy security in the global context: A comprehensive review of geopolitical dynamics and policies. *Eng. Sci. Technol. J.* **2024**, *5*, 152–168. [CrossRef]
6. Frilingou, N.; Xexakis, G.; Koasidis, K.; Nikas, A.; Campagnolo, L.; Delpiazzo, E.; Chiodi, A.; Gargiulo, M.; McWilliams, B.; Koutsellis, T.; et al. Navigating through an energy crisis: Challenges and progress towards electricity decarbonisation, reliability, and affordability in Italy. *Energy Res. Soc. Sci.* **2023**, *96*, 102934. [CrossRef]
7. Adelekan, O.A.; Ilugbusi, B.S.; Adisa, O.; Obi, O.C.; Awonuga, K.F.; Asuzu, O.F.; Ndubuisi, N.L. Energy transition policies: A global review of shifts towards renewable sources. *Eng. Sci. Technol. J.* **2024**, *5*, 272–287. [CrossRef]
8. Ahmed, N.; Ore Areche, F.; Saenz Arenas, E.R.; Cosio Borda, R.F.; Javier-Vidalón, J.L.; Silvera-Arcos, S.; Ober, J.; Kochmańska, A. Natural disasters and energy innovation: Unveiling the linkage from an environmental sustainability perspective. *Front. Energy Res.* **2023**, *11*, 1256219. [CrossRef]
9. Wu, D.; Xie, Y.; Liu, D. Rethinking the complex effects of the clean energy transition on air pollution abatement: Evidence from China's coal-to-gas policy. *Energy* **2023**, *283*, 128413. [CrossRef]
10. Dincer, I.; Aydin, M.I. New paradigms in sustainable energy systems with hydrogen. *Energy Convers. Manag.* **2023**, *283*, 116950. [CrossRef]
11. Johansson, B.; Jonsson, D.K.; Veibäck, E.; Sonnsjö, H. Assessing the capabilites to manage risks in energy systems–analytical perspectives and frameworks with a starting point in Swedish experiences. *Energy* **2016**, *116*, 429–435. [CrossRef]
12. Cuisinier, E.; Bourasseau, C.; Ruby, A.; Lemaire, P.; Penz, B. Techno-economic planning of local energy systems through optimization models: A survey of current methods. *Int. J. Energy Res.* **2021**, *45*, 4888–4931. [CrossRef]
13. Hasselqvist, H.; Renström, S.; Strömberg, H.; Håkansson, M. Household energy resilience: Shifting perspectives to reveal opportunities for renewable energy futures in affluent contexts. *Energy Res. Soc. Sci.* **2022**, *88*, 102498. [CrossRef]
14. Lund, H. Renewable energy strategies for sustainable development. *Energy* **2007**, *32*, 912–919. [CrossRef]
15. Kalair, A.; Abas, N.; Saleem, M.S.; Kalair, A.R.; Khan, N. Role of energy storage systems in energy transition from fossil fuels to renewables. *Energy Storage* **2021**, *3*, e135. [CrossRef]
16. Lopion, P.; Markewitz, P.; Robinius, M.; Stolten, D. A review of current challenges and trends in energy systems modeling. *Renew. Sustain. Energy Rev.* **2018**, *96*, 156–166. [CrossRef]
17. Kotzur, L.; Nolting, L.; Hoffmann, M.; Groß, T.; Smolenko, A.; Priesmann, J.; Büsing, H.; Beer, R.; Kullmann, F.; Singh, B.; et al. A modeler's guide to handle complexity in energy systems optimization. *Adv. Appl. Energy* **2021**, *4*, 100063. [CrossRef]
18. Ervural, B.C.; Evren, R.; Delen, D. A multi-objective decision-making approach for sustainable energy investment planning. *Renew. Energy* **2018**, *126*, 387–402. [CrossRef]
19. Thompson, S. Strategic analysis of the renewable electricity transition: Power to the world without carbon emissions? *Energies* **2023**, *16*, 6183. [CrossRef]
20. Rozhkov, A. Applying graph theory to find key leverage points in the transition toward urban renewable energy systems. *Appl. Energy* **2024**, *361*, 122854. [CrossRef]
21. Adanma, U.M.; Ogunbiyi, E.O. Assessing the economic and environmental impacts of renewable energy adoption across different global regions. *Eng. Sci. Technol. J.* **2024**, *5*, 1767–1793. [CrossRef]
22. Moriarty, P.; Honnery, D. Feasibility of a 100% global renewable energy system. *Energies* **2020**, *13*, 5543. [CrossRef]
23. Akpan, J.; Olanrewaju, O. Towards a common methodology and modelling tool for 100% renewable energy analysis: A review. *Energies* **2023**, *16*, 6598. [CrossRef]
24. Elazab, R.; Dahab, A.A.; Adma, M.A.; Hassan, H.A. Reviewing the frontier: Modeling and energy management strategies for sustainable 100% renewable microgrids. *Discov. Appl. Sci.* **2024**, *6*, 168. [CrossRef]
25. Meschede, H.; Bertheau, P.; Khalili, S.; Breyer, C. A review of 100% renewable energy scenarios on islands. *Wiley Interdiscip. Rev. Energy Environ.* **2022**, *11*, e450. [CrossRef]
26. Marocco, P.; Novo, R.; Lanzini, A.; Mattiazzo, G.; Santarelli, M. Towards 100% renewable energy systems: The role of hydrogen and batteries. *J. Energy Storage* **2023**, *57*, 106306. [CrossRef]
27. O'malley, M. Towards 100% renewable energy system. *IEEE Trans. Power Syst.* **2022**, *37*, 3187–3189. [CrossRef]
28. Dahab, A.A.; Elazab, R.; Adma, M.A.A.; Hassan, H.F. Global Challenges and Economic Feasibility in Achieving 100% Renewable Energy. *Adv. Electr. Electron. Eng.* 2024, *in press*. [CrossRef]

29. Sahoo, G.S.; Mathur, M.; Zaidi, T.; Sharma, R. Comprehensive Assessment of Land Requirements for 100% Solar Energy Transition in Smart Cities. In Proceedings of the E3S Web of Conferences, EDP Sciences, Singapore, 7–9 June 2024; Volume 540, p. 04006.
30. Ulpiani, G.; Vetters, N.; Shtjefni, D.; Kakoulaki, G.; Taylor, N. Let's hear it from the cities: On the role of renewable energy in reaching climate neutrality in urban Europe. *Renew. Sustain. Energy Rev.* **2023**, *183*, 113444. [CrossRef]
31. Li, L.; Wang, J.; Zhong, X.; Lin, J.; Wu, N.; Zhang, Z.; Meng, C.; Wang, X.; Shah, N.; Brandon, N.; et al. Combined multi-objective optimization and agent-based modeling for a 100% renewable island energy system considering power-to-gas technology and extreme weather conditions. *Appl. Energy* **2022**, *308*, 118376. [CrossRef]
32. Tian, X.; Zhou, Y.; Morris, B.; You, F. Sustainable design of Cornell University campus energy systems toward climate neutrality and 100% renewables. *Renew. Sustain. Energy Rev.* **2022**, *161*, 112383. [CrossRef]
33. Cong, R.; Fujiyama, A.; Matsumoto, T. An Optimal Scheme Assists the Municipalities in Fukuoka, Japan in Achieving their Goal of 100% Renewable Energy Supply and Future Decarbonization. *Energy Nexus* **2024**, *13*, 100277. [CrossRef]
34. Liu, S.; Duffy, A.H.; Whitfield, R.I.; Boyle, I.M. Integration of decision support systems to improve decision support performance. *Knowl. Inf. Syst.* **2010**, *22*, 261–286. [CrossRef]
35. Prina, M.G.; Johannsen, R.M.; Sparber, W.; Østergaard, P.A. Evaluating near-optimal scenarios with EnergyPLAN to support policy makers. *Smart Energy* **2023**, *10*, 100100. [CrossRef]
36. Prina, M.G.; Lionetti, M.; Manzolini, G.; Sparber, W.; Moser, D. Transition pathways optimization methodology through EnergyPLAN software for long-term energy planning. *Appl. Energy* **2019**, *235*, 356–368. [CrossRef]
37. Ren, J. *Waste-to-Energy: Multi-Criteria Decision Analysis for Sustainability Assessment and Ranking*; Academic Press: Cambridge, MA, USA, 2020.
38. Brodny, J.; Tutak, M. Assessing the energy security of European Union countries from two perspectives–A new integrated approach based on MCDM methods. *Appl. Energy* **2023**, *347*, 121443. [CrossRef]
39. Kumar, A.; Sah, B.; Singh, A.R.; Deng, Y.; He, X.; Kumar, P.; Bansal, R.C. A review of multi criteria decision making (MCDM) towards sustainable renewable energy development. *Renew. Sustain. Energy Rev.* **2017**, *69*, 596–609. [CrossRef]
40. Kandakoglu, M.; Walther, G.; Ben Amor, S. The use of multi-criteria decision-making methods in project portfolio selection: A literature review and future research directions. *Ann. Oper. Res.* **2024**, *332*, 807–830. [CrossRef]
41. Manoj, V.; Pilla, R.; Kumar, Y.N.; Sinha, C.; Prasad, S.V.; Chakravarthi, M.K.; Bhogi, K.K. Towards Efficient Energy Solutions: MCDA-Driven Selection of Hybrid Renewable Energy Systems. *Int. J. Electr. Electron. Eng. Telecommun* **2024**, *13*, 98–111. [CrossRef]
42. Islam, M.R.; Aziz, M.T.; Alauddin, M.; Kader, Z.; Islam, M.R. Site suitability assessment for solar power plants in Bangladesh: A GIS-based analytical hierarchy process (AHP) and multi-criteria decision analysis (MCDA) approach. *Renew. Energy* **2024**, *220*, 119595. [CrossRef]
43. Halder, B.; Bandyopadhyay, J.; Sandhyaki, S. Impact assessment of environmental disturbances triggering aquaculture land suitability mapping using AHP and MCDA techniques. *Aquac. Int.* **2024**, *32*, 2039–2075. [CrossRef]
44. Więckowski, J.; Kizielewicz, B.; Sałabun, W. A multi-dimensional sensitivity analysis approach for evaluating the robustness of renewable energy sources in European countries. *J. Clean. Prod.* **2024**, 143225. [CrossRef]
45. Benítez, J.; Carpitella, S.; Certa, A.; Izquierdo, J. Management of uncertain pairwise comparisons in AHP through probabilistic concepts. *Appl. Soft Comput.* **2019**, *78*, 274–285. [CrossRef]
46. Aljburi, M.T.; Albahri, A.; Albahri, O.; Alamoodi, A.; Mohammed, S.M.; Deveci, M.; Tomášková, H. Exploring decision-making techniques for evaluation and benchmarking of energy system integration frameworks for achieving a sustainable energy future. *Energy Strategy Rev.* **2024**, *51*, 101251. [CrossRef]
47. Alamoodi, A.H.; Garfan, S.; Al-Zuhairi, O.; Zaidan, B.; Zaidan, A.; Albahri, O.S.; Ahmaro, I.Y.; Albahri, A.S.; Yussof, S.; Magableh, A.A. Exploring the integration of multi criteria decision analysis in the clean energy biodiesels applications: A systematic review and gap analysis. *Eng. Appl. Artif. Intell.* **2024**, *133*, 108023. [CrossRef]
48. Anchieta, T.; Meirelles, G.; Carpitella, S.; Brentan, B.; Izquierdo, J. Water distribution network expansion: An evaluation from the perspective of complex networks and hydraulic criteria. *J. Hydroinform.* **2023**, *25*, 628–644. [CrossRef]
49. Brentan, B.; Carpitella, S.; Zanfei, A.; Gabriel Souza, R.; Menapace, A.; Meirelles, G.; Righetti, M.; Izquierdo, J. Multi-criteria analysis applied to ranking rehabilitation strategies of water distribution networks. *Math. Methods Appl. Sci.* **2023**. [CrossRef]
50. Chatterjee, S.; Chakraborty, S. A study on the effects of objective weighting methods on TOPSIS-based parametric optimization of non-traditional machining processes. *Decis. Anal. J.* **2024**, *11*, 100451. [CrossRef]
51. Yazo-Cabuya, E.J.; Ibeas, A.; Herrera-Cuartas, J.A. Integration of Sustainability in Risk Management and Operational Excellence through the VIKOR Method Considering Comparisons between Multi-Criteria Decision-Making Methods. *Sustainability* **2024**, *16*, 4585. [CrossRef]
52. Menapace, A.; Thellufsen, J.Z.; Pernigotto, G.; Roberti, F.; Gasparella, A.; Righetti, M.; Baratieri, M.; Lund, H. The design of 100% renewable smart urb an energy systems: the case of Bozen-Bolzano. *Energy* **2020**, *207*, 118198. [CrossRef]
53. Battini, F.; Menapace, A.; Stradiotti, G.; Zanfei, A.; Nicolosi, F.F.; Dalla Torre, D.; Renzi, M.; Pernigotto, G.; Ravazzolo, F.; Righetti, M.; et al. Technical, economic and environmental optimization of renewable urban energy systems in a climate change scenario. *Smart Energy* 2024, *in press*.
54. ISTAT. Popolazione Residente al 1 Gennaio: Provincia Autonoma Bolzano/Bozen. 2023. Available online: http://dati.istat.it (accessed on 26 August 2024).

55. Grazieschi, G.; Zubaryeva, A.; Sparber, W. Energy and greenhouse gases life cycle assessment of electric and hydrogen buses: A real-world case study in Bolzano Italy. *Energy Rep.* **2023**, *9*, 6295–6310. [CrossRef]
56. Pinamonti, M.; Prada, A.; Baggio, P. Rule-based control strategy to increase photovoltaic self-consumption of a modulating heat pump using water storages and building mass activation. *Energies* **2020**, *13*, 6282. [CrossRef]
57. Hunter, G.W.; Vettorato, D.; Sagoe, G. Creating smart energy cities for sustainability through project implementation: A case study of Bolzano, Italy. *Sustainability* **2018**, *10*, 2167. [CrossRef]
58. Prina, M.G.; Cozzini, M.; Garegnani, G.; Moser, D.; Oberegger, U.F.; Vaccaro, R.; Sparber, W. Smart energy systems applied at urban level: The case of the municipality of Bressanone-Brixen. *Int. J. Sustain. Energy Plan. Manag.* **2016**, *10*, 33–52.
59. Fedak, W.; Anweiler, S.; Ulbrich, R.; Jarosz, B. The concept of autonomous power supply system fed with renewable energy sources. *J. Sustain. Dev. Energy, Water Environ. Syst.* **2017**, *5*, 579–589. [CrossRef]
60. Østergaard, P.A. Reviewing EnergyPLAN simulations and performance indicator applications in EnergyPLAN simulations. *Appl. Energy* **2015**, *154*, 921–933. [CrossRef]
61. Østergaard, P.A.; Lund, H.; Thellufsen, J.Z.; Sorknæs, P.; Mathiesen, B.V. Review and validation of EnergyPLAN. *Renew. Sustain. Energy Rev.* **2022**, *168*, 112724. [CrossRef]
62. Akpahou, R.; Mensah, L.D.; Quansah, D.A.; Kemausuor, F. Energy planning and modeling tools for sustainable development: A systematic literature review. *Energy Rep.* **2024**, *11*, 830–845. [CrossRef]
63. Denholm, P.; Margolis, R.M. Evaluating the limits of solar photovoltaics (PV) in traditional electric power systems. *Energy Policy* **2007**, *35*, 2852–2861. [CrossRef]
64. Kwon, P.S.; Østergaard, P.A. Priority order in using biomass resources–Energy systems analyses of future scenarios for Denmark. *Energy* **2013**, *63*, 86–94. [CrossRef]
65. Lund, H.; Marszal, A.; Heiselberg, P. Zero energy buildings and mismatch compensation factors. *Energy Build.* **2011**, *43*, 1646–1654. [CrossRef]

energies

MDPI

Article

Resilient Operation Strategies for Integrated Power-Gas Systems

Behdad Faridpak [1] and Petr Musilek [1,2,*]

[1] Department of Electrical and Computer Engineering, University of Alberta, Edmonton, AB T6G 1H9, Canada; faridpak@ualberta.ca
[2] Department of Applied Cybernetics, University of Hradec Králové, 500 03 Hradec Králové, Czech Republic
* Correspondence: pmusilek@ualberta.ca

Abstract: This article presents a novel methodology for analyzing the resilience of an active distribution system (ADS) integrated with an urban gas network (UGN). It demonstrates that the secure adoption of gas turbines with optimal capacity and allocation can enhance the resilience of the ADS during high-impact, low-probability (HILP) events. A two-level tri-layer resilience problem is formulated to minimize load shedding as the resilience index during post-event outages. The challenge of unpredictability is addressed by an adaptive distributionally robust optimization strategy based on multi-cut Benders decomposition. The uncertainties of HILP events are modeled by different moment-based probability distributions. In this regard, considering the nature of each uncertain variable, a different probabilistic method is utilized. For instance, to account for the influence of power generated from renewable energy sources on the decision-making process, a diurnal version of the long-term short-term memory network is developed to forecast day-ahead weather. In comparison with standard LSTM, the proposed approach reduces the mean absolute error and root mean squared error by approximately 47% and 71% for wind speed, as well as 76% and 77% for solar irradiance network. Finally, the optimal operating framework for improving power grid resilience is validated using the IEEE 33-bus ADS and 7-node UGN.

Keywords: multi-energy system; optimal operation; resilient power system; adaptive distributionally robust optimization

1. Introduction

Extreme weather conditions and man-made attacks have contributed to catastrophic power grid failures, leading to a growing global need for improvements in power system resilience. In this context, resilience refers to the ability of the power grid to meet an acceptable level of electricity demand during HILP events, where most power outages are limited to a few hours [1]. However, even short-time power outages can have significant consequences, especially when CLs are involved. Therefore, it is important to improve the resilience of ADS with economically efficient solutions for post-event load restoration.

A possible solution to improve system resilience is to connect a UGN and other local sources, such as RES, and coordinate their operation with the ADS. Compared to installing new components, using existing durable energy infrastructure is an effective approach, especially under common geographical and techno-economical limitations. The underground infrastructure is also better protected from HILP events [2]. However, the coordination of multiple energy systems for resilience enhancement is rarely considered. In addition, to the best of our knowledge, existing strategies only couple ADS and UGN on a predefined common bus. In the industrial practice of North American and European electric power utilities, ADS and UGN are usually coordinated by optimizing the location and capacity of GTs supplied with interruptible gas services [3–5]. Energy transfer is achieved through day-ahead gas-to-power contracts, which are more economical and practical than real-time contracts [6].

There are two main challenges in the area of resilience-oriented integration of multi-energy systems: (1) Modeling the intensity of the HILP events and their influence on ADS;

and (2) Optimizing the operation of ADS with weather-related power generated from RES, to determine appropriate locations and capacities of the coupling points. To address these challenges, there is a need for innovative solutions to improve the resilience of multi-energy systems and enhance the hardiness and stability of the power grid.

2. Literature Review

Optimization techniques for improving the resilience of the power grid have become a new field of research and development, with numerous studies focusing on short- and long-term improvements. Kwasinski [7] addressed the availability of power supply using a risk assessment method to improve resilience to hurricanes. Ahmed et al. [8] proposed to enhance resilience against false data injection attacks by equal power sharing in multi-area power systems. Xu et al. [9] used a stochastic integer program to minimize the average time that customers are without power. Similarly, Trakas and Hatziargyriou [10] described a stochastic programming solution to increase the resilience of the distribution system and minimize l^{sh} during a wildfire. The authors considered uncertainties associated with s, v, and the direction of the wind. According to Sharma et al. [11], distributed generators improved ADS resilience during islanding by mitigating the risks associated with uncertainties in load demand and RES generation. Wang et al. [12] studied the restoration of CLs in the presence of distributed generators and ESS using microgrid formation methods. Xu et al. [13] considered a priority for supplying CLs in demand response to enhance the resilience of distribution networks under extreme scenarios. Wang et al. [14] used an iterative algorithm considering an unbalanced three-phase power flow to develop an optimal decision-making method for serving CLs after blackouts. Robust tri-level planning for distributed energy resources was proposed by Samani and Aminifar [15] to solve the problem of enhancing resilience using column and constraint generation with BD. The use of electric vehicles as a backup power resource and their positive impact on resilience were discussed by Hussain and Musilek [16].

In addition to approaches based on locally distributed RES, the resilience of ADS can also be improved through the integration of other energy systems, decreasing the probability of load shedding. The technical and economic benefits of integrated energy systems were previously discussed by the authors [17–19]. According to [17], UGN is one of the most appropriate and readily available energy systems to integrate with ADS. There have been several studies conducted to optimize IDGS operation, including different strategies and coordination levels for coupling ADS and UGN. In general, IDGS optimization approaches can be categorized as sequential or simultaneous. In sequential optimization, the energy cost of ADS and UGN is minimized by defining two distinct objective functions. In contrast, simultaneous optimization strategies use a fully coupled IDGS operated by a single entity [20].

Due to the interdependence of multiple energy systems, a disruption in any of them can compromise the energy supply of the other ones. Ravadanegh et al. [2] modeled the effects of weather-related HILPs with various levels using multiphase performance response curves. In this way, they determined the time-dependent performance levels of the IDGS. Using the onsite supply strategy of the energy storage systems and demand response, Darvish et al. [3] enhanced the resilience of IDGS considering reserve scheduling and pre-event responses. Saravi et al. [21] introduced a resilience-oriented aggregator–agent splitting framework for IDGS. Cong et al. [22] presented a robust three-stage optimization model for the resilient operation of IDGS to minimize load curtailments imposed by attacks. Manshadi and Khodayar [23] described a bi-level optimization problem to address the optimal operation of multi-energy microgrids while considering the security and resilience of the system. Correa-Posada and Śanchez-Martın [24] presented a unique MILP approach for the power and gas flow considering security constraints for both normal and contingency scenarios. Wang et al. [25] increased IDGS resilience by protecting its most vulnerable components using tri-level MILP optimization. Sawas et al. [26] proposed a cyberattack-resistant scheduling model using supervised and unsupervised false data detection techniques.

Due to the high penetration rate of RES in ADS and their impact on resilience, accurate weather forecasting is crucial to formulate operating strategies. This can be accomplished using data-driven algorithms that evaluate time series of meteorological data and predict the power produced by RESs. To cope with the complex behavior of the weather and related uncertainties, machine learning algorithms (such as support vector machines, linear regression, or tree-based models) have often been used for this task [27]. However, these algorithms are not easily scaled to large data sets. Shoaei et al. [28] conducted a comprehensive study on the applications of artificial intelligence in renewable energy systems. A possible solution is to implement a robust technique with multivariate mapping capabilities. To take into account the time-series nature of weather data, recurrent neural networks allow feedback connections between their hidden layers. However, their applicability is limited due to the vanishing gradient problem. This and other issues have been alleviated by LSTM neural network models which have been successfully used for weather-dependent time series [29]. For example, Abdel-Nasser et al. [30] used LSTM to predict solar irradiation from input data aggregated by combining the Choquet integral using a fuzzy measure. Wang et al. [31] implemented an LSTM with a novel least absolute shrinkage and selection operator to increase the accuracy of short-term predictions. Zhou et al. [32] proposed a technique to improve the accuracy of LSTM by extracting significant features from the input. Two-dimensional convolutional neural networks and bidirectional LSTM units were used to predict wind power by Dolatabadi et al. [33]. In addition, Li et al. [34] used mathematical morphology to improve the accuracy of LSTM for wind speed forecasts.

This article deals with the optimal operation of ADS. The main objective is to improve the resilience of the electricity grid by coupling it with the UGN through GTs. To determine the capacities and locations of the GTs, we formulate a two-level tri-layer resilience-oriented optimal operation problem. Due to the unknown post-event status of the system, the problem is solved by a set of distribution functions. As a result, the uncertain consequences of HILP events are modeled by a probabilistic framework solved by an ADRO approach.

The required capacity of GTs depends on load specifications and generation from local RESs, influenced by extreme weather conditions. Therefore, to model these uncertainties, we construct moment-based probability distributions. Furthermore, by considering historical weather data, we propose an enhanced forecasting method based on a modified LSTM algorithm to predict expected weather-dependent RES generation.

To provide proper context, existing studies are compared in Table 1 along with the approach described in this article. Previous studies have addressed resilience in power systems using various optimization methods such as MINLP [1] and MILP [2]. Nevertheless, these methods cannot effectively handle the complex and uncertain nature of HILP events when RESs are involved. Although probabilistic approaches such as ARO [35] and RO [36] have been employed for uncertainty modeling, they lack the ability to address ambiguity sets created by more advanced uncertainty handling methods. In contrast, the proposed approach utilizes ADRO in combination with ambiguity sets, offering a more robust framework to handle both proactive and restoration processes. In addition, the solution methodology requires an efficient approach to capture these complexities. For instance, approaches like stochastic [10] and relaxation [12] methods are not efficient due to their high computational cost and inability to properly incorporate the impact of uncertainties. The proposed MCBD method provides a cost-effective and scalable solution to overcome IDGS resilience improvement challenges.

The major contributions are as follows.

1. A new approach to enhancing the resilience of ADS with the following properties:
    (a) A novel formulation of the resilience enhancement problem: Distribution functions of random post-event consequences are used in the calculations to address the probabilistic status of ADS. The performance of the proposed

model is evaluated by extensive comparative simulations considering different levels of robustness and resilience indices.

(b) A modified multi-cut decomposition method: To analyze the various effects of each probabilistic consequence and to address the complexity of the problem, a modified version of the decomposition approach is implemented. The multi-cut decomposition method captures more detailed information about the system, which is critical for managing the random post-event status of HILP events.

(c) Optimal use of the potential of existing natural gas infrastructure: The proposed approach determines the most resilient, secure and cost-efficient coordination points between ADS and UGN.

2. A new momentum-based approach with the following properties:

   (a) A learning-based approach to calculate the statistics of momentum: Training does not require complex filtering or pre-processing methods. The data presentation is modified to ensure smooth domain variation, transforming the prediction method from an hourly domain into a diurnal domain, and executing hourly predictive subtasks in parallel to construct probability distributions.

   (b) Improved accuracy in comparison with conventional LSTM network: The proposed diurnal learning approach is validated using multiple accuracy indicators and compared with the conventional approach.

Table 1. Comparative analysis of resilience-oriented methods.

Ref.	Optimization Method	Solution Method	Enhancement Method	Process Time		Uncertainty Handling	RSC	IDGS
				Proactive	Restoration			
[1]	MINLP	-	RES	✓	-	-	-	-
[2]	MILP	-	EH	-	✓	-	-	✓
[36]	MILP	RO	GT	-	✓	US	-	✓
[6]	MILP	NC&CG	GT	✓	✓	RBN		✓
[10]	MINLP	Stochastic	Hardening	-	✓	MC	-	-
[12]	MISDP	Relaxation	ES and DG	-	✓	-	-	-
[26]	-	Classification	BI	-	✓	NN	-	✓
[35]	ARO	C&CG	EH	✓	-	TI		-
[37]	RO	BD and C&CG	EH	-	✓	US		-
[38]	MINLP	MCB	-	-	✓	-	-	-
Proposed	ADRO	MCBD	GT	✓	✓	AS	✓	✓

MISDP: Mixed-integer semi-definite programming, NC&CG: nested column and constraint generation, MCB: modified combinatorial Benders, EH: emergency handling, DG: distributed generation, RBN: randomized binary numbers, BI: bidirectional interconnection, TI: three intervals, US: uncertainty set, NN: neural network, AS: ambiguity set, RSC: resilient and secure coordination.

3. Problem Formulation

To analyze the resilience enhancement of the ADS, two phases of system operation are defined: the pre-event and post-event, denoted by superscripts of 0 and 1, respectively. The main assumptions of the proposed approach are:

- The load is considered event-independent. In other words, the load is the same pre-event and post-event. However, since the specifications of an extreme event are unpredictable, the necessity of load supply (i.e., criticality of the load, CL/NCL) is randomly changed in the post-event phase.

- Towers and lines of the analyzed system are assumed to have similar fragility functions. Their restoration time is greater than the total optimization time interval T.
- Optimal AC power flow is used as a dispatch tool, assuming a reactive power balance among buses.
- After a HILP event, damage on the supply side and isolation from different zones and/or upstream power grid force reliance on local sources, such as RESs and ESSs.
- Interruption of any power source causes load shedding; the coordination with the UGN through GTs provides improved resilience against load shedding in the ADS.

3.1. Pre-Event

Under normal operation, without any faults in the system, the primary objective is to supply the load with the minimum power consumption.

$$\underset{P^{\mathrm{P}}}{\mathrm{Min}} \sum_{t=1}^{T} C_t^{\mathrm{P}} P_t^{\mathrm{P},0}, \tag{1}$$

where C^{P} is the power price, P^{P} and Q^{P} are the active and reactive power received from the upstream grid, respectively, T is the total operation time, and D is the total number of days in the data set. Symbols K, N, C, and Z express, respectively, the total number of buses, nodes, coordination points, and zones. Superscript $(.)^{0/1}$ is used to indicate pre-/post-event notation, and subscript t is the time index.

The objective function (1) is subject to the following constraints:

$$P_t^{\mathrm{P},0} + P_{k,t}^{\mathrm{WT}} + P_{k,t}^{\mathrm{PV}} + P_{k,t}^{\mathrm{dch},0} = \sum_{k,j}^{B} P_{kj,t}^{0} + r_{kj} I_{kj,t}^{0} + P_{k,t}^{\mathrm{L}} + P_{k,t}^{\mathrm{ch},0} \qquad \forall t, \forall k \tag{2}$$

$$Q_t^{\mathrm{P},0} + Q_{k,t}^{\mathrm{WT}} = \sum_{k,j}^{B} Q_{kj,t}^{0} + x_{kj} I_{kj,t}^{0} + Q_{k,t}^{\mathrm{L}} \qquad \forall t, \forall k \tag{3}$$

$$(P_{kj,t}^{0})^2 + (Q_{kj,t}^{0})^2 \le I_{kj,t}^{0} U_{k,t}^{0} \qquad \forall t, \forall k \tag{4}$$

$$U_{k,t}^{0} - U_{j,t}^{0} = 2P_{kj,t}^{0} r_{ij} + 2Q_{kj,t}^{0} x_{ij} - I_{kj,t}^{0}\left(r_{kj}^{2} + x_{kj}^{2}\right) \qquad \forall t, \forall k \tag{5}$$

$$P^{\mathrm{P,min}} \le P_t^{\mathrm{P},0} \le P^{\mathrm{P,max}} \qquad \forall t \tag{6}$$

$$Q^{\mathrm{P,min}} \le Q_t^{\mathrm{P},0} \le Q^{\mathrm{P,max}} \qquad \forall t \tag{7}$$

$$U_k^{\mathrm{min}} \le U_{k,t}^{0} \le U_k^{\mathrm{max}} \qquad \forall t, \forall k \tag{8}$$

$$0 \le I_{kj,t}^{0} \le I_{kj}^{\mathrm{max}} \qquad \forall t, \forall k \tag{9}$$

where P^{WT}, Q^{WT}, and $/P^{PV}$ are the amounts of power generated from RESs, P^{L} and Q^{L} are active and reactive load demands, and P^{ch} and P^{dch} are the charging and discharging powers of ESS, respectively. Symbols r and x mark resistance and inductance, while I and U represent the squared magnitudes of current and voltage, respectively. The superscripts $(.)^{\mathrm{max}}$ and $(.)^{\mathrm{min}}$ are the maximum and minimum limitations.

Equations (2) and (3) describe the constraints on the active and reactive power, respectively. They state that the amounts of power generated by different suppliers must be balanced with the amounts consumed. Branch flow is calculated using (4) and (5). Finally, the constraints on the active and reactive power, voltage and current received from the upstream grid are specified in (6)–(9).

The power generated from the RESs is formulated as follows [19].

$$P_{t,k}^{\mathrm{WT}} = \begin{cases} 0, & 0 \le v_t \le v_{\mathrm{ci}} \text{ or } v_t \ge v_{\mathrm{co}} \\ \frac{P_k^{\mathrm{WT,max}}(v_t - v_{\mathrm{ci}})}{(v_{\mathrm{r}} - v_{\mathrm{ci}})}, & v_{\mathrm{ci}} \le v_t \le v_{\mathrm{r}} \\ P_k^{\mathrm{WT,max}}, & v_r \le v_t \le v_{co} \end{cases} \qquad \forall t, \forall k \tag{10}$$

$$0 \le P_{t,k}^{\mathrm{WT}} \le P_k^{\mathrm{WT,max}} \qquad \forall t, \forall k \tag{11}$$

$$P_{t,k}^{\mathrm{PV}} = \begin{cases} \frac{P_k^{\mathrm{PV,max}} \times s_t}{s_{\mathrm{r}}}, & 0 \le s_t \le s_{\mathrm{r}} \\ P_k^{\mathrm{PV,max}}, & s_{\mathrm{r}} \le s_t \end{cases} \qquad \forall t, \forall k \tag{12}$$

$$0 \le P_{t,k}^{\mathrm{PV}} \le P_k^{\mathrm{PV,max}} \qquad \forall t, \forall k, \tag{13}$$

where parameters v_{ci}, v_{co}, and v_r are cut-in, cut-out, and rated wind speed. Additionally, the rated solar radiation, solar radiation, and wind speed are shown by s_r, s, and v, respectively.

ESSs are commonly used to mitigate the unpredictable nature of nondispatchable, intermittent RESs. The technical operating constraints of ESSs can be modeled as follows.

$$E_{k,t+1}^{0} = E_{k,t}^{0} + [P_{k,t}^{\mathrm{ch},0}.\eta - P_{k,t}^{\mathrm{dch},0}/\eta].\Delta t \qquad \forall t, \forall k \tag{14}$$

$$0 \le P_{k,t}^{\mathrm{ch},0} \le I_{k,t}^{\mathrm{ch},0}.P_k^{\mathrm{ch,max}} \qquad \forall t, \forall k, I_{k,t}^{\mathrm{ch},0} \in \{0,1\} \tag{15}$$

$$0 \le P_{k,t}^{\mathrm{dch},0} \le (1 - I_{k,t}^{\mathrm{ch},0}).P_k^{\mathrm{dch,max}}.\beta \qquad \forall t, \forall k, I_{k,t}^{\mathrm{ch},0} \in \{0,1\} \tag{16}$$

$$E^{\min} \le E_{k,t}^{0} \le E^{\max} \qquad \forall t, \forall k \tag{17}$$

$$E_{t=0}^{0} = E_{t=T}^{0} \qquad \forall t, \forall k \tag{18}$$

E and η are state of energy and efficiency. Also, I^{ch} and I^{dch} are the charging and discharging status of ESSs, respectively.

The state of energy at time $t+1$ depends on the state of energy and the charging/discharging rate of the ESS at the previous time slot (14). The constraints of ESS include charging (15) and discharging (16) rates, state of energy limitation (17), and the requirement of equality between the initial and final state of energy (18).

3.2. Post-Event

An extreme event, such as a weather-related disaster, can cause interruptions on the supply side. As a result, the electric load cannot be served efficiently. In most cases, locally distributed generators (such as PV, WT, and ESS) cannot entirely meet the load demands. The use of other available energy infrastructure, such as the UGN, is a possible approach to solve the problem of inadequate power supply. For example, power sources, including the upstream power grid or RESs, may be interrupted after a HILP event. In such cases, to avoid load shedding and substitute power shortage, GTs should be allocated to appropriate buses. Coordination between the ADS and the UGN can provide an effective approach to improve resilience.

Assuming C secure coordination points, the following objective function can be defined using natural gas price C^{G}, natural gas consumed by GTs G^{GT}, and penalty of load shedding ρ. The goal is to minimize the operational costs of the IDGS

$$\underset{P^{\mathrm{P}}, G^{\mathrm{GT}}, l^{\mathrm{sh}}}{\mathrm{Min}} \sum_{t=1}^{T} \left\{ C_t^{\mathrm{P}} P_t^{\mathrm{P},1} + \sum_{c=1}^{C} C_t^{\mathrm{G}} G_{t,c}^{\mathrm{GT},1} + \sum_{k=1}^{K} \rho l_{t,k}^{\mathrm{sh},1} \right\}, \tag{19}$$

subject to the following constraints

$$P_t^{P,1} + P_{k,t}^{WT} + P_{k,t}^{PV} + P_{k,t}^{dch,1} + l_{k,t}^{sh,1} = \sum_{k,j}^{B} P_{kj,t}^1 + r_{kj} I_{kj,t}^1 + P_{k,t}^L + P_{k,t}^{ch,1} \quad \forall t, \forall k \tag{20}$$

$$Q_t^{P,1} + Q_{k,t}^{WT} = \sum_{k,j}^{B} Q_{kj,t}^1 + x_{kj} I_{kj,t}^1 + Q_{k,t}^L \quad \forall t, \forall k \tag{21}$$

$$(P_{kj,t}^1)^2 + (Q_{kj,t}^1)^2 \leq I_{kj,t}^1 U_{k,t}^1 \quad \forall t, \forall k \tag{22}$$

$$U_{k,t}^1 - U_{j,t}^1 = \quad 2P_{kj,t}^1 r_{ij} + 2Q_{kj,t}^1 x_{ij} - I_{kj,t}^1 \left(r_{kj}^2 + x_{kj}^2 \right) \quad \forall t, \forall k \tag{23}$$

$$P^{P,\min} \leq P_t^{P,1} \leq P^{P,\max} \quad \forall t \tag{24}$$

$$Q^{P,\min} \leq Q_t^{P,1} \leq Q^{P,\max} \quad \forall t \tag{25}$$

$$U_k^{\min} \leq U_{k,t}^1 \leq U_k^{\max} \quad \forall t, \forall k \tag{26}$$

$$0 \leq I_{kj,t}^1 \leq I_{kj}^{\max} \quad \forall t, \forall k \tag{27}$$

$$0 \leq l_{k,t}^{sh,1} \leq P_{k,t}^L \quad \forall t, \forall k \tag{28}$$

$$E_{k,t+1}^1 = E_{k,t}^1 + [P_{k,t}^{ch,1}.\eta - P_{k,t}^{dch,1}/\eta] \quad \forall t, \forall k \tag{29}$$

$$0 \leq P_{k,t}^{ch,1} \leq I_{k,t}^{ch,1}.P_k^{ch,\max} \quad \forall t, \forall k, I_{k,t}^{ch,1} \in \{0,1\} \tag{30}$$

$$0 \leq P_{k,t}^{dch,1} \leq (1 - I_{k,t}^{ch,1}).P_k^{dch,\max}.\beta \quad \forall t, \forall k, I_{k,t}^{ch,1} \in \{0,1\} \tag{31}$$

$$E^{\min} \leq E_{k,t}^1 \leq E^{\max} \quad \forall t, \forall k \tag{32}$$

$$E_{t=0}^1 = E_{t=T}^1 \quad \forall t, \forall k \tag{33}$$

where β is depth of discharge

Note that constraints (20)–(33) are similar to the formulations introduced in the previous subsection; however, load shedding is incorporated in the post-event model (20) as a resilience index.

4. Resilience-Oriented Model

HILP incidents can result in supply-side disruptions that cause load loss l_t^{sh}. Coordination among multiple sources of energy can be used to enhance the resilience of the system [14]. The proposed model addresses the optimal operational-planning strategy that coordinates multiple energy sources, including the ADS and UGN, to minimize l_t^{sh}. First, the model incorporates the uncertainty of HILP events to capture potential damage. Then, it optimizes the operation of the ADS. Finally, it plans the optimal capacity and location for coordination. In addition to enhancing resilience during HILP events, this coordination process allocates limited GT resources to supply critical loads for longer periods and improves the handling of RES uncertainties.

For the system considered in this contribution, the resilience of the ADS can be enhanced by adding GTs to appropriate buses, as a backup. Since ADS has the priority to supply CLs, their configuration, location, and ratio in the post-event situation must be modeled. The optimal resilience-oriented operation problem can be mathematically formulated using a tri-layer two-level approach. Level 1 includes determining the maximum value of l_t^{sh} as the worst-case (Layer 1) and identifying the best coordination points (Layer 2). Layer 3 then checks the security validation of the IDGS plan in the Level 2 process. Details of the proposed model are described in the following subsections.

This multi-layer, multi-stage approach requires significant computational resources, particularly when dealing with large-scale systems or highly uncertain HILP events. Nonetheless, this complexity is unavoidable to capture the full extent of uncertainties and to enhance the resilience of the system in a realistic manner. While this may pose practical challenges, especially for systems with limited computational capabilities, modern computational tools coupled with parallel processing techniques can efficiently manage the computational load.

4.1. Level 1: Operation

The complex optimization problem at Level 1 of the proposed model is divided into two layers: Layer 1 and Layer 2. In Layer 1, various scenarios are defined. The combination of these scenarios generate multiple probability distributions. Then, the values of l_t^{sh} is calculated for the generated probability distributions. Finally, the maximum expected and feasible value of l_t^{sh} is found among all probability distributions and designated as the worst consequence for the resilience of the system.

To differentiate between CLs and NCLs, a penalty factor ρ is implemented with a different value for the two types of load such that $\rho^{\mathrm{cl}} \gg \rho^{\mathrm{ncl}}$. Hence, the last term of (19) can be reformulated by replacing the scenario-based penalty of load shedding

$$\underset{P^{\mathrm{P}},G^{\mathrm{GT}},l^{\mathrm{sh}}}{Min} \sum_{t=1}^{T} \left\{ C_t^{\mathrm{P}} P_t^{\mathrm{P},1} + \sum_{c=1}^{C} C_t^{\mathrm{G}} G_{t,c}^{\mathrm{GT},1} + \sum_{k=1}^{B} \sum_{\omega=1}^{W} \pi_\omega (\rho^{\mathrm{cl}} l_{t,k,\omega}^{\mathrm{sh|cl}} + \rho^{\mathrm{ncl}} l_{t,k,\omega}^{\mathrm{sh|ncl}}) \right\}. \tag{34}$$

where $l^{\mathrm{sh|cl}}$ and $l^{\mathrm{sh|ncl}}$ are load shedding of critical and non-critical loads.

4.2. Level 2: Secure Integration

The Level 2 equations correspond to the security check of the IDGS coordination points. To avoid the secure operational challenges of the UGN, associated with the interconnection of ADS and GTs, the following constraints of the gas system must be satisfied:

$$\tilde{q}_{mn,t}^2 = C_{mn}^2 \left(p_{m,t}^2 - p_{n,t}^2 \right) \quad \forall t, \forall m, \tag{35}$$

$$\tilde{q}_{mn,t} = (q_{mn,t} - q_{nm,t})/2 \quad \forall t, \forall m, \tag{36}$$

$$p_{mn}^{\min} \leq p_{mn,t} \leq p_{mn}^{\max} \quad \forall t, \forall m, \tag{37}$$

$$LP_{mn,t} = (q_{mn,t} + q_{nm,t})\Delta t + LP_{mn,t-1} \quad \forall t, \forall m, \tag{38}$$

$$\sum_{m,n}^{N} LP_{mn,t=0} = \sum_{m,n}^{N} LP_{mn,t=T} \geq LP^{\min} \quad \forall t, \forall m, \tag{39}$$

$$P_{k,t}^{\mathrm{GT}} = \alpha G_{m,t}^{\mathrm{GT}} \quad \forall t, \forall m, \tag{40}$$

$$P_k^{GT,\min} \leq P_{k,t}^{\mathrm{GT}} \leq P_k^{GT,\max} \quad \forall t, \forall m, \tag{41}$$

$$P_{m,t}^{\mathrm{GT}} - RD_m^{\mathrm{GT}} \leq P_{m,t+\Delta t}^{\mathrm{GT}} \leq P_{m,t}^{\mathrm{GT}} + RU_m^{\mathrm{GT}} \quad \forall t, \forall m, \tag{42}$$

where C_{mn}, p_{mn}, and LP are the Waymouth constant, node pressure of the natural gas network, and line-pack between nodes, respectively. In addition, α, $\tilde{q}(t)$, and $q_{mn,t}$ are gas-to-power conversion factor, average gas flow, and transmitted gas flow in a pipeline, respectively. Furthermore, RU^{GT} and RD^{GT} are the ramp-up and ramp-down rates of GT, respectively.

Formula (35) is the Weymouth equation of gas flow, expression (36) calculates the average value of gas transmitted through pipelines, and the nodal pressure and line-pack limits are described by (37)–(39). The GTs must be allocated at the coordination points considering the constraints (40)–(42).

For the proposed IDGS, the ADS Level 1 power balance (20) is modified as follows.

$$P_t^{\mathrm{P},1} + P_{k,t}^{\mathrm{GT}} + P_{k,t}^{\mathrm{WT}} + P_{k,t}^{\mathrm{PV}} + P_{k,t}^{\mathrm{ch},1} + l_{k,t}^{\mathrm{sh},1} = \sum_{k,j}^{B} P_{kj,t}^{1} + r_{kj} I_{kj,t}^{1} + P_{k,t}^{\mathrm{L}} + P_{k,t}^{\mathrm{dch},1} \quad \forall t, \forall k. \tag{43}$$

5. Handling of Uncertainties

The uncertainty challenges of the proposed scheduling problem are related to factors that have a substantial impact on scheduling decisions, but cannot be controlled by the system operator. There are two major sources of uncertainty considered: the power generated by the RESs and the load specifications.

5.1. Load Specifications and Extreme Events

Although historical data on load allocation in real-world scenarios exist, the accuracy of this data may be compromised by various factors, including missing or unavailable data, high costs associated with data acquisition, and other sources of uncertainty. To address these issues, we utilize a moment-based ambiguity set for **EC** .

First, to generate multiple realization scenarios of load types at each bus, the randomly variable elements of the **EC** matrix in (44) are created using a Monte-Carlo approach [10].

$$\mathbf{EC} = \overbrace{\begin{pmatrix} \mathrm{EC}_{11} & \cdots & \mathrm{EC}_{1W_1} \\ \vdots & \ddots & \vdots \\ \mathrm{EC}_{K1} & \cdots & \mathrm{EC}_{KW_1} \end{pmatrix}}^{W_1} \Bigg\} K \quad \forall \omega_1, \forall k, \mathrm{EC}_{k,\omega_1} \in \mathcal{S}_1 \tag{44}$$

where W_1 and ω_1 are the total scenario number and sample scenarios of EC.

The elements of **EC** are factors corresponding to the CL rate that have supplying priority at bus k for scenario ω_1, and $\mathcal{S}_1 \subseteq [0,1]^{\omega_1 \times k}$ is a convex support set. For instance, if IC_{k,ω_1} is x, then x% of P_k^L is CL, while the remaining $(1-x)$% is NCL.

Second, the moment information of the uncertain variable **EC** can be defined based on the mean and covariance as follows

$$\mathbf{EC}_{\mathcal{M}}\left(\mathcal{S}_1, \mu_1, \mathbf{\Sigma}_1, \varsigma_1^1, \varsigma_1^2\right) = \left\{ \begin{array}{c|c} \sum\limits_{\omega_{p1}=1}^{W} \pi_{\omega_{p1}} = 1 & \\ \omega_{p1} & \begin{array}{l} \left(\mathbb{E}_{\omega_{p1}}[\mathrm{EC}_{k,\omega_1}] - \mu_1\right)^T \mathbf{\Sigma}_1^{-1} \left(\mathbb{E}_{\omega_{p1}}[\mathrm{EC}_{k,\omega_1}] - \mu_1\right) < \varsigma_1^1 \\ \mathbb{E}_{\omega_{p1}}\left[(\mathrm{EC}_{k,\omega_1} - \mu_1)(\mathrm{EC}_{k,\omega_1} - \mu_1)^T \leq \varsigma_1^2 \mathbf{\Sigma}_1\right] \end{array} \end{array} \right\}$$
$$\forall \mathrm{EC}_{k,\omega_1} \in \mathcal{S}_1, \forall k \tag{45}$$

where the probability factor and momentum-based ambiguity set are represented by π and $\mathcal{M}$, respectively. The ambiguity set $\mathbf{EC}_{\mathcal{M}}$ represents a range of possible distributions, and the optimization problem is formulated to be robust against the worst-case distribution within this set. For each bus, the total probability of distribution functions is one (45). Moreover, the mean of any distribution should be within an ellipsoidal distance limited by ς_1, and the covariance matrix should be contained within a positive semi-definite cone restricted by $\varsigma_2 \mathbf{\Sigma}$. The calculations of parameters μ, $\mathbf{\Sigma}$, ς_1, and ς_2 are expressed in (46)–(51) using the results of [39,40].

$$\mu_1 = \frac{1}{\omega_1} \sum_{\omega_1=1}^{W_1} \mathrm{EC}_{k,\omega_1} \quad \forall \mathrm{EC}_{k,\omega_1} \in \mathcal{S}_1, \forall k \tag{46}$$

$$\mathbf{\Sigma}_1 = \frac{1}{\omega_1} \sum_{\omega_1=1}^{W_1} \left(\mathrm{EC}_{k,\omega_1} - \mu_1\right)\left(\mathrm{EC}_{k,\omega_1} - \mu_1\right)^T \quad \forall \mathrm{EC}_{k,\omega_1} \in \mathcal{S}_1, \forall k \tag{47}$$

$$\varsigma_1^1 = A_1 \tag{48}$$

$$\varsigma_1^2 = 1 + A_1 \tag{49}$$

$$A_1 = \frac{B_1^2}{\omega_1} \frac{\left(2+\sqrt{2ln\left(\frac{4}{1-\sqrt{1-\kappa}}\right)}\right)^2}{1-\frac{B_1^2}{\sqrt{\omega_1}}\left(\sqrt{1-\frac{\omega_1}{B_1^4}}+\sqrt{ln\left(\frac{4}{1-\sqrt{1-\kappa}}\right)}\right)-\left(\frac{B_1^2}{\omega_1}\right)\left(2+\sqrt{2ln\left(\frac{4}{1-\sqrt{1-\kappa}}\right)}\right)^2} \tag{50}$$

$$B_1 = \underset{\mathrm{EC}_{k,\omega_1}}{\mathrm{Max}} \left\| \frac{1}{\left(\mathrm{EC}_{k,\omega_1}-\mu_1\right)} \right\|_2 \tag{51}$$

The mean and covariance matrix of randomly generated arrays of **EC** are calculated in Equations (46) and (47), respectively. Controller parameters, $\varsigma_1 \geq 0$ in (48) and $\varsigma_2 \geq 1$ in (49) depend on the generated data to specify the size of the ambiguity set and the conservatism of optimal solutions. The auxiliary parameter A, introduced to simplify the equation, is presented in Equation (50), where κ represents the confidence level of uncertainty. Furthermore, the auxiliary variable B (51) denotes a radius region of $\mathcal{S}\mathbf{1}$ that includes EC_{k,ω_1}; here, $\|.\|_2$ indicates the second norm.

5.2. Renewable Energies and Extreme Events

To model renewable energies, the matrix **ER** is defined in (52). Based on historical data, the values in the arrays ER_{k,ω_2} can vary within an interval of 0 to the maximum power generated from RESs for each bus. Therefore, the convex support set of **ER** can be defined as $\mathcal{S}_\mathbf{2} \subseteq [0, (P_k^{\mathrm{WT,max}} + P_k^{\mathrm{PV,max}})]^{\omega_2 \times k}$.

$$\mathbf{ER} = \left.\overbrace{\begin{pmatrix} ER_{11} & \dots & ER_{1W_2} \\ \vdots & \ddots & \vdots \\ ER_{K1} & \cdots & ER_{KW_2} \end{pmatrix}}^{W_2}\right\}K \quad \forall\omega_2, \forall k, \mathrm{ER}_{k,\omega_2} \in \mathcal{S}_\mathbf{2} \tag{52}$$

where W_2 and ω_2 are the total scenario number and sample scenarios of ER.

Similar to the **EC**, the ambiguity set $\mathbf{ER}_\mathcal{M}$ can be modeled by (53).

$$\mathbf{ER}_\mathcal{M}\left(\mathcal{S}_\mathbf{2}, \mu_2, \mathbf{\Sigma_2}, \varsigma_2^1, \varsigma_2^2\right) = \left\{ \omega_{p2} \left| \begin{array}{l} \sum\limits_{\omega_{p2}=1}^{W} \pi_{\omega_{p2}} = 1 \\ \left(\mathbb{E}_{\omega_{p2}}[\mathrm{ER}_{k,\omega_2}] - \mu_2\right)^T \mathbf{\Sigma_2}^{-1}\left(\mathbb{E}_{\omega_{p2}}[\mathrm{ER}_{k,\omega_2}] - \mu_2\right) \leq \varsigma_2^2 \\ \mathbb{E}_{\omega_{p2}}\left[(\mathrm{ER}_{k,\omega_2} - \mu_2)(\mathrm{ER}_{k,\omega_2} - \mu_2)^T \leq \varsigma_2^2\mathbf{\Sigma_2}\right] \end{array} \right. \right\}$$
$$\forall \mathrm{ER}_{k,\omega_2} \in \mathcal{S}_\mathbf{2}, \forall k \tag{53}$$

To compute the mean μ_2, we use historical data to account for the uncertainty in the RES output power. The power generation from WTs and PVs is influenced by the weather conditions, which are represented by the time series of v and s, respectively. We use the compact notation $\mathbf{u}$ to denote the real datasets of v and s. Using these real-world datasets, we aim to accurately forecast the value of μ_2.

Several studies have used deep learning methods for time-series forecasting. LSTM, in particular, has shown great promise due to its advanced units and network topology [41]. LSTM's memory units enable it to capture time-series correlations and grasp the long-term behavior of underlying systems. In this study, we use an LSTM-based forecasting approach that leverages diurnal model to calculate the values of μ_2. This approach is described in more detail below.

The conventional use of LSTM from Figure 1a is modified as shown in Figure 1b. This allows us to capture rare events and increase the overall accuracy of the model. Since pattern recognition at each time interval t is assessed on a daily basis, this modified version of the network is qualified as diurnal and named DLSTM. Considering uncertain variable $u_{t,d}$, a daily prediction approach is used instead of solving the hourly time series prediction problem. Thus, the vertical analyses of time-series data can be transformed into a matrix-based prediction with T rows and D columns corresponding to time intervals t and days d, respectively. As shown in Figure 1b, the estimated data for each row with the same time slot t are collected as the DA forecast data $\bar{u}_{t,D+1}$.

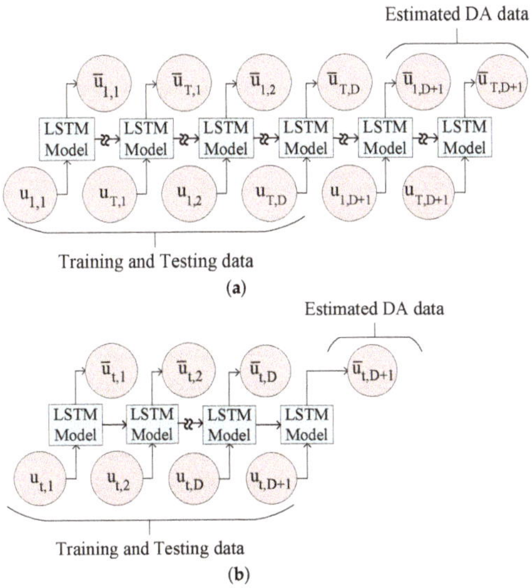

Figure 1. Implementation of the LSTM. (**a**) Conventional approach. (**b**) Proposed DLSTM approach.

The operation of the DLSTM model can be illustrated as follows: the variable v is random and non-stationary; the high randomness and fast variation of wind speed make it difficult to estimate wind speed data directly using the conventional method shown in Figure 1a. As shown in Figure 1b, a pre-processing model based on the DLSTM splits the entire data set into several time intervals for each day. This allows more effective forecasting and mitigation of the adverse effects of the stochasticity of v and s.

As shown in Figure 2, the cell sends and receives information at random intervals and the gates f_1, f_2, f_3 and f_4 follow the data flow from the cell's input to output. Furthermore, a DLSTM network's nodal formulations considering the sigmoid (σ) and the hyperbolic tangent (tanh) as activation functions are shown in [42].

$$\mathrm{f}_1 = \sigma(\mathrm{w}_1[h_{t,d-1}, u_t] + \mathrm{b}_1) \quad (54)$$

$$\mathrm{f}_2 = \sigma(\mathrm{w}_2[h_{t,d-1}, u_t] + \mathrm{b}_2) \quad (55)$$

$$\mathrm{f}_3 = \tanh(\mathrm{w}_3[h_{t,d-1}, u_t] + \mathrm{b}_3) \quad (56)$$

$$\mathrm{f}_4 = \sigma(\mathrm{w}_4[h_{t,d-1}, u_t] + \mathrm{b}_4) \quad (57)$$

$$\mathrm{f}_{t,d} = \mathrm{f}_1 \times \mathrm{f}_{t,d-1} + \mathrm{f}_2 \times \mathrm{f}_3 \quad (58)$$

$$h_{t,d} = \mathrm{f}_4 \tanh(\mathrm{f}_{t,d}) \quad (59)$$

Training the historical data using the DLSTM structure, as illustrated in Figure 1 and based on the cells depicted in Figure 2, yields deterministic DA forecasts for each time slot. However, in the context of extreme weather conditions during a HILP event, these forecasted values are designated as μ_2.

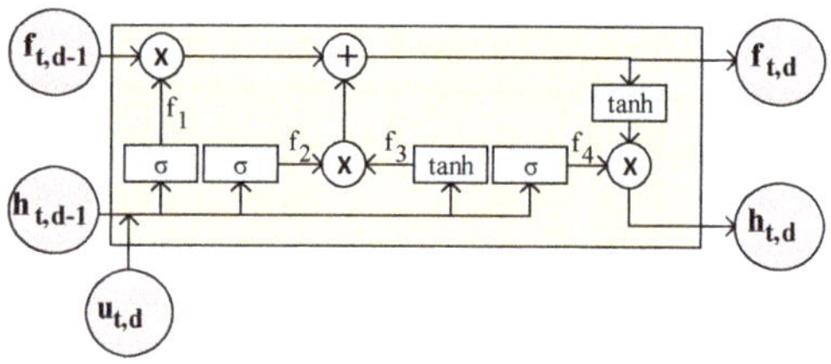

Figure 2. The structure of a DLSTM model.

The calculations of parameters $\mathbf{EC}_{\mathcal{M}}$ in (47)–(51) are rewritten for $\mathbf{ER}_{\mathcal{M}}$ in (60)–(64).

$$\boldsymbol{\Sigma}_2 = \frac{1}{\omega_2} \sum_{\omega_1=1}^{W_1} \left(\mathrm{ER}_{k,\omega_2} - \mu_2\right)\left(\mathrm{ER}_{k,\omega_2} - \mu_2\right)^T \quad \forall \mathrm{ER}_{k,\omega_2} \in \mathcal{S}_2, \forall k \tag{60}$$

$$\varsigma_2^1 = A_2 \tag{61}$$

$$\varsigma_2^2 = 1 + A_2 \tag{62}$$

$$A_2 = \frac{B_2^2}{\omega_2} \frac{\left(2 + \sqrt{2ln\left(\frac{4}{1-\sqrt{1-\kappa}}\right)}\right)^2}{1 - \frac{B_2^2}{\sqrt{\omega_2}}\left(\sqrt{1 - \frac{\omega_2}{B_2^4}} + \sqrt{ln\left(\frac{4}{1-\sqrt{1-\kappa}}\right)}\right) - \left(\frac{B_2^2}{\omega_2}\right)\left(2 + \sqrt{2ln\left(\frac{4}{1-\sqrt{1-\kappa}}\right)}\right)^2} \tag{63}$$

$$B_2 = \underset{\mathrm{ER}_{k,\omega_2}}{\mathrm{Max}} \left\| \frac{1}{\left(\mathrm{ER}_{k,\omega_2} - \mu_2\right)} \right\|_2 \tag{64}$$

6. Solution Strategy

After modeling the uncertainties, an MCBD is adapted with ADRO to solve the proposed problem. The flowchart in Figure 3 summarizes the proposed solution. The detailed methodology is explained in the following subsections.

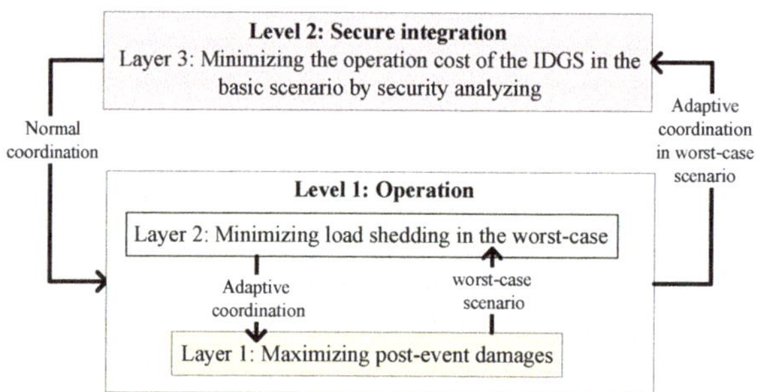

Figure 3. Resilience enhancement of the ADS by making coordination with UGN through the proposed two-level tri-layer ADRO model.

6.1. Modeling Equivalence

To facilitate analysis, the proposed problem can be expressed in a compact form. The following formulation is a simplified version of the tri-layer two-level model for each time interval.

The deterministic mixed-integer nonlinear problem (19) can be formulated as follows

$$C(\mathbf{x},\mathbf{z}) = \underset{\mathbf{x},\mathbf{z}}{\text{Min}}\ \mathbf{c_1^T x} + \mathbf{c_2^T z} \tag{65}$$

$$\text{s.t. } \mathbf{Ax} \leq \mathbf{a} \tag{66}$$

$$\mathbf{Bz} \leq \mathbf{b} \tag{67}$$

$$\mathbf{D_1 x} + \mathbf{D_2 z} \leq \mathbf{d}, \tag{68}$$

where **A**, **B**, **D** are matrices of auxiliary coefficients and **a**, **b**, **d** are vectors of auxiliary parameters of the compact model. The objective function (65) is a compact matrix representation of minimizing the cost function C in the post-event situation. For notational brevity, **x** and **z** are, respectively, vectors of decision variables to model the energy consumed by ADS (i.e., power and gas) and load shedding. These decision variables are minimized, considering constraints (66)–(68). Constraints (66) and (67) are related to the capacity limitations (10)–(13), (21)–(33), and (35)–(42). The last inequality (68) models constraints (20) and (43).

6.2. Final Model Formulation

The solution method must be adopted to post-event operational requirements for various uncertainties and be able to deal with practical scheduling plans for HILP events. The proposed tri-layer two-level problem can be solved using an adaptive mixed-integer distributionally robust scheduling model considering the uncertainties raised from EC, EZ, and ER. Correspondingly, to model the solution process the following adaptive robust optimization model can be obtained from (65).

$$C(\mathbf{x},\mathbf{z}) = \underset{\mathbf{x},\mathbf{z}}{\text{Min}} \left[\mathbf{c_1^T x} + \underset{\mathbf{z}|\mathbf{u}}{\text{Max}}\ \underset{\mathbf{z}}{\text{Min}}\ \mathbf{c_2^T z}\right]. \tag{69}$$

The resilience enhancement factor **z** is closely related to the values of **u**. Hence, the Level 1 solver finds the worst-case operation point of ADS to specify the most robust interconnection with UGN considering various realization distributions. This is achieved by a resilience-oriented max − min function. Then, a robust scheduling of supplied power is achieved by incorporating the energy costs of IDGS at Level 2. The general form of the tri-layer two-level ADRO model is presented as follows for each uncertain probability distribution ω_p constructed from sampled data of ω.

$$C^{(\omega_p)}(\mathbf{x},\mathbf{z}^{(\omega_p)}) = \underset{\mathbf{x},\mathbf{z}}{\text{Min}}[\mathbf{a_1^T x} + \underset{\pi_1,\ldots,\pi_W \geq 0}{\text{Max}}\ \underset{\mathbf{z}_{\omega_p} \in Z^{(\omega_p)}(\mathbf{x},\mathbf{u})}{\text{Min}} \sum_{\omega_p=1}^{W} \pi_{\omega_p} \mathbf{a_2^T z}_{\omega_p}] \quad \forall \omega_p,$$

subject to

$$Z^{(\omega_p)}(\mathbf{x},\mathbf{u}) = \left\{ z_{\omega_p} \left| \begin{array}{c} (66)\text{–}(68) \\ \sum_{\omega_p=1}^{W} \pi_{\omega_p} = 1 \\ \sum_{\omega_p=1}^{W} \left|\pi_{\omega_p} - \hat{\pi}_{\omega_p}\right| \leq (1-\Gamma) \end{array} \right. \right\} \quad \forall \omega_p, \tag{70}$$

As shown in (70), the resilience factor **z** is not known exactly in advance. Thus, a set of random distributions with different probabilities π is castrated, considering the vector $Z^{(\omega_p)}$. The value of **z** in each distribution depends on the values of **x** and **u**. By identifying the worst-case through the max − min function, the distributionally robust energy scheduling of the proposed IDGS can be achieved. The set of constraints (70) states that the limits of the decision variables must be considered for each distribution of the ADRO problem and that the sum of all probability factors must be equal to 1. Since the proposed stochastic framework has a large sample space, the variation in the probability π_ω of realized distributions with respect to the probability $\hat{\pi}_\omega$ derived from the data is limited to Γ.

The proposed resilience-oriented ADRO problem is a complex two-level tri-layer optimization program shown in Figure 3. The objective function in (70) is different from the compact representation (65) because the ambiguity set of uncertainties are modeled by distributions in the full model. It also differs from (69), because the full model specifies a set of various probabilities for each random distributions. Moreover, while conventional adaptive robust optimization is associated with a single worst-case scenario, ADRO is driven by the statistical characterization of distributions developed at Level 1 of the model. In this way, ADRO (70) can take into account ambiguous variations from the expected values and simultaneously cover both the results obtained by solving (65) and (69).

6.3. The Proposed Methodology

To allow for the solution of the proposed problem using commercial software packages and to mitigate computational issues, we develop a MCBD algorithm considering post-event ω_p. The MCBD approach incorporates multiple cuts into the master problem. A similar method for a stochastic (scenario-based) approach was proposed in [43]. In contrast to [43], this paper generates multiple cuts for each distribution rather than individually for each scenario. Consequently, the master problem acquires more realistic and detailed information, which definitely contributes to a more accurate outcome.

(1) Initialization: The BD algorithm requires a linear sub-problem. The power flow constraint (22) and the Weymouth gas flow (35) cause the problem to become nonlinear. In this paper, both constraints are linearized using an approach similar to [44]. First, the starting points of variables $\bar{\mathbf{x}}$ are calculated using the first-order Taylor expansion. Then, an approximate canonical form is obtained by defining coefficients $\mathbf{H}, \mathbf{h}$ and $\mathbf{M}, \mathbf{m}$. Finally, based on the second-order cone approach, a convex form of the nonlinear constraints is constructed, as shown below.

$$||\mathbf{Hx}||_2 \leq \mathbf{hx}, \tag{71}$$

$$||\mathbf{M_1}(\bar{\mathbf{x}})\mathbf{x} + \mathbf{M_2}(\bar{\mathbf{x}})||_2 \leq \mathbf{m_1}(\bar{\mathbf{x}})\mathbf{x} + \mathbf{m_2}(\bar{\mathbf{x}}). \tag{72}$$

In addition, at the first iteration, the initial decision variables of Level 1 are set as feasible solutions. These initial values are obtained by solving (65)–(68). The lower and upper bounds are considered as $\pm\infty$. Furthermore, the convergence tolerance for Level 2 is specified by $\epsilon > 0$.

(2) Master Problem: To find the optimal Level 2 decision for the worst-case expected cost, master problem (73) is minimized, under constraints (74)–(77).

$$\underset{\mathbf{x},\mathbf{\Phi}^{(i)}}{\text{Min}}\ \mathbf{c_1}^T\mathbf{x}^i + \sum_{\omega_p=1}^{W} \pi_{\omega_p}\mathbf{\Phi}^{(i)} \qquad \forall\omega_p, \forall i, \tag{73}$$

s.t.

$$A\mathbf{x}^i_{\omega_p} \leq a \qquad \forall\omega_p, \forall i, \tag{74}$$

$$(||\mathbf{Hx}||_2)_{\omega_p} \leq \boldsymbol{h}\boldsymbol{x}_{\omega_p} \qquad \forall\omega_p, \forall i, \tag{75}$$

$$\left(\left\|\boldsymbol{M_1}(\bar{\boldsymbol{x}})\boldsymbol{x}_{\omega_p} + \mathbf{M_2}(\bar{\mathbf{x}})\right\|_2\right)_{\omega_p} \leq \boldsymbol{m_1}(\bar{\boldsymbol{x}})\boldsymbol{x}_{\omega_p} + \mathbf{m_2}(\bar{\mathbf{x}}) \qquad \forall\omega_p, \forall i, \tag{76}$$

$$\Phi^{(i)}_{\omega_p} \geq \gamma^{(i-1)}[e - E\mathbf{x}^i_{\omega_p}] \qquad \forall\omega_p, \forall i, \tag{77}$$

where $\Phi^{(0)}$ is the set of initialized decision variables and $\Phi^{(i)}$ is an approximation for the cost of Level 1 in the ith iteration. In (73), the results of the master problem determine the lower bound of the proposed ADRO problem for each scenario. Inequalities (74)–(76) indicate the convex formulation (71) and (72) should be applied for each ω_p. For constraint (68), including complicating variable $\mathbf{z}$, dual variable γ is defined in (77) to generate Benders' cuts for each iteration.

(3) Sub-problem: In the proposed MCBD compared to the conventional BD, the sub-problem generates multi cuts to analyze the ADRO problem in more detail. The following model defines the sub problem $SP^{(\omega_p)}$.

$$\underset{\pi_1,\ldots,\pi_W \geq 0}{\text{Max}} \; \underset{\mathbf{z}^i_{\omega_p}}{\text{Min}} \sum_{\omega_p=1}^{W} \pi_{\omega_p} \mathbf{a}_2^T \mathbf{z}^i_{\omega_p} \qquad \forall \omega_p, \forall i, \tag{78}$$

$$\mathbf{z}^i_{\omega_p} \leq e - E\mathbf{x}^{*,i}_{\omega_\mathbf{p}} \qquad \forall \omega_p, \forall i, \qquad : \gamma^{(i)} \tag{79}$$

$$D\mathbf{z}^i_{\omega_p} \leq d \qquad \forall \omega_p, \forall i, \tag{80}$$

To calculate $SP^{(\omega_p)}$, W scenarios of each probability distributions ω_{p1} and ω_{p2} are sampled. Since the nature of ω_{p1} and ω_{p2} are different, to achieve an efficient sampling method, two different approaches that utilized random sampling and cluster sampling were used.Since ω_{p1} is constructed based on the synthetic data, equal chance of being selected is considered for every individual array using simple random sampling. On the other hand, ω_{p2} is based on real historical data depending on extreme weather. Therefore, a cluster sampling method is employed. Firstly, the data sets of s and v are divided into the various clusters (c) modeling the intensity of situations. Then, samples are chosen systematically to ensure representation from each cluster using LHS. The total number of selected sample is W. Finally, the dual sub-problem is degenerated feasible W cuts for the same MP.

The sub-problems are solved in parallel and each sub-problem generates a cut added to the master problem. Indeed, the multiple-cut sub-problem version ($SP^{(\omega_p)}$) of BD is generated from all sampled scenarios, which is different from the single-cut one (SP). By implementing MCBD to solve the ARDO problem, the optimized value (maximum) of (78) among multi-cuts returned to MP in each iteration.

7. Numerical Results

7.1. Case Study and Input Data

The efficiency of the proposed resilience enhancement strategy is demonstrated using the test systems depicted in Figure 4. The network parameters are adopted from [45]. The configuration of wind and solar, and the energy storage capacity have been selected such that the IDGS operates under normal (pre-event) conditions, i.e., with no load shedding. The configuration of wind and solar, and the energy storage capacity have been selected such that the IDGS operates under a pre-event (normal) conditions, i.e., with no load shedding. To properly implement the proposed optimization approach, the test ADS is sectionalized into four distinct zones. These zones include RESs and ESs of various sizes.

Real-world data are used to train the model and calculate the power generated by RESs. The two data sets include 4 years (D = 1462) of hourly values v (at 10 m) and s (at 2 m) collected on the South Campus of the University of Alberta, Edmonton, Canada, from 2018 to 2022. This data is recorded by weather stations operating in the province of Alberta and provided by the Alberta Climate Information Service [46]. These time series are processed using the proposed DLSTM method. Considering the hourly-based DA optimization (T = 24), the raw time series data are distributed among 24 subtasks that operate in parallel. Therefore, the proposed matrix-based prediction method has dimensions of 24 rows and 1462 columns. To achieve more practical results, avoid the need for filtering, and capture rare events, each distributed data set is classified into three clusters using the k-means algorithm. Thus, for each data set, the model is run 72 times to predict three variables in 24 time slots. Finally, as a case study, the predicted v and s data for 15 February 2022 (chosen arbitrarily) is used as the input to calculate the power generated by RESs. Theses predicted values are μ_2 used to construct matrix ER.

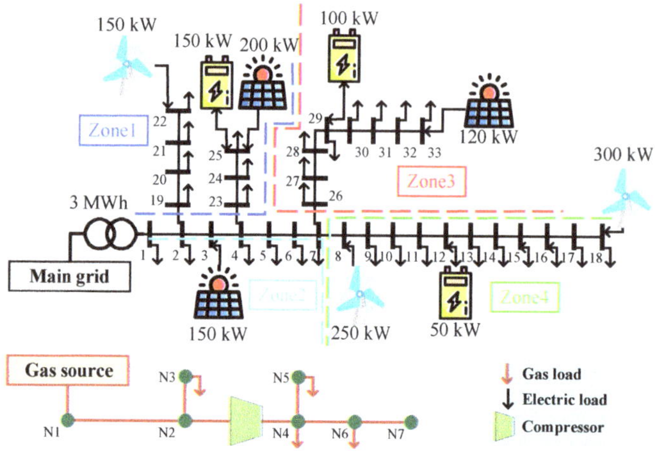

Figure 4. Topology of the modified IEEE 33-bus ADS and seven-node UGN.

The proposed diurnal learning approach is implemented in Python 3.8.15 using package Keras 2.7.0 with TensorFlow 2.7.1 backend. For each hour (subtask), the data is divided into training (75%), validation (15%), and testing (10%) subsets [33]. The DLSTM network layout is identical for all hours using (54)–(59). It is a sequential model with 64 hidden layers including 32 LSTM blocks/neurons with *ReLU* activation function, and a dense output layer. MSE is used as a loss function to optimize the learning rate, epoch size, and time step by Adam optimizer for each hour. The developed software is run on a virtual server with NVIDIA GPU and 32 GB of RAM [47]. Figure 5 shows the generated input data for handling of uncertainties in the post-event situation. Figure 5a depicts DA values of v and s obtained from all subtasks at each hour. This figure includes results of k-means clustering for W = 3 (cf. the three semi-transparent areas of different colors) and DLSTM results (solid lines) for each cluster as μ_2. The weight of each data point is assigned by evaluating the standard deviation and cluster density [48]. Figure 5b presents realization scenarios of CLs' factor at each bus generated by the MC method. Finally, using (44)–(53) and (60)–(64), the results illustrated in Figure 5a,b are utilized to construct ω_{p1} and ω_{p2}, respectively.

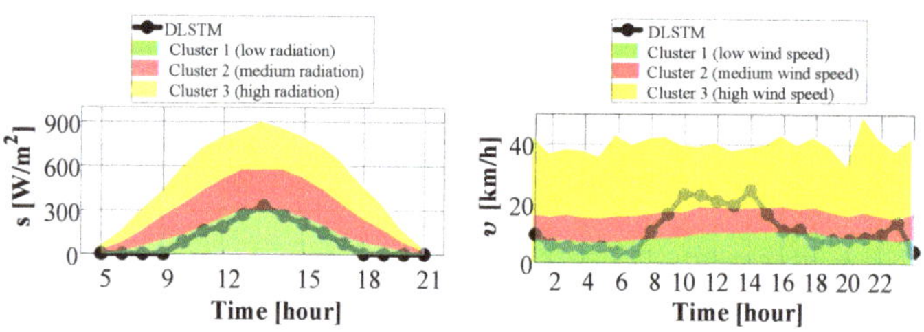

(a)

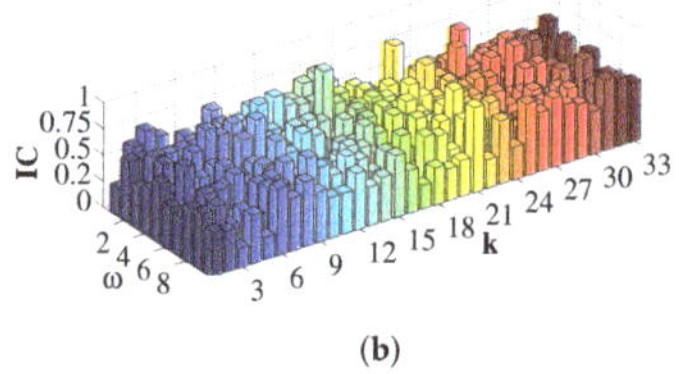

(b)

Figure 5. Data preparation (**a**) predicted wind speed and solar radiation by the proposed DLSTM method, (**b**) critical load factors in each bus for 10 scenarios.

7.2. Results and Discussion

After modeling the uncertainties the proposed solution methodology is applied using MCBD. The proposed methodology was simulated using CPLEX solver in GAMS 24.1.2 [49] and executed on a PC with an Intel Core i7, 1.8 GHz CPU, and 8 GB of memory.

(1) Resilience enhancement:

The expected optimal problem-solving outcomes are illustrated for three distinct phases to allow for a detailed comparison. In addition to the pre-event and post-event phases, the base IDGS case (when both the integrated ADS and the UGN are operating normally) is also considered.

To keep the presentations simple, the operation of the ESs is only depicted for the pre-event phase, as shown in Figure 6. To prevent load shedding, the system has been modified so that ADS exchanges power with ESs using their maximum capacities most of the time. Consequently, this system is more vulnerable and should be made resilient against potential outages in any zone.

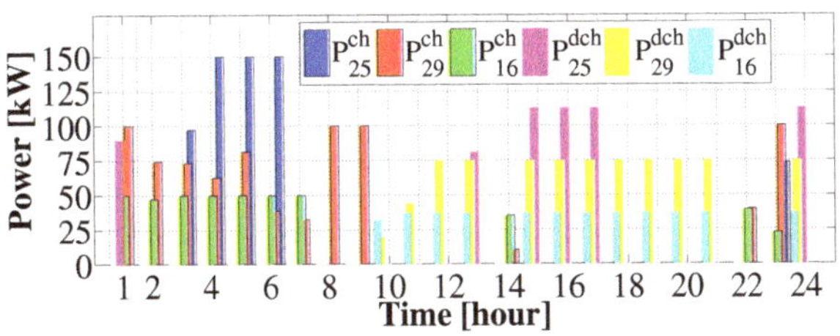

Figure 6. Optimal status of ESs.

By applying the solution in Layer 1, the worst-case scenario includes the outage of lines that disconnect Zone 1 from the system (lines between buses 2–19 and 4–23). Taking this outage into account, Layer 2 recognizes the appropriate buses for the installation of GTs considering the l^{sh} of each bus. This way, buses 19 and 24 are identified as the most vulnerable. Finally, the results of Layer 3 determine the appropriate capacities and gas nodes to coordinate between ADS and UGN so that the resilience of the system is enhanced. This results in nodes N2 and N3 supplying the GTs in buses 19 and 24, with capacities of approximately 750 kW and 800 kW, respectively.

Figures 7 and 8 show the results of the proposed two-level approach for P^{P} and P^{GT}, respectively. As can be seen in Figure 7, the pattern of power received from the upstream grid during normal pre-event operation of ADS follows the price of power, RESs generation, and load profile.

Due to outages, the loads in Zone 1 are supplied by local suppliers. At the same time, the secure power flow constraints of the ADS limit the power received from the upstream feeder. Consequently, during the post-event phase, the DA values of P^{P} are relatively low compared to the pre-event phase.

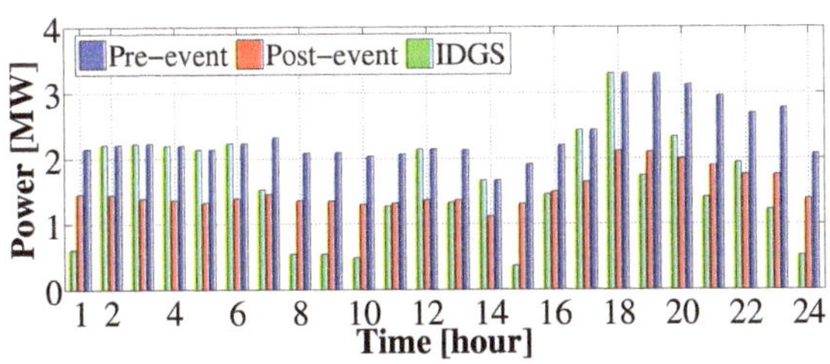

Figure 7. Optimal power received from upstream grid.

To eliminate l^{sh} and enhance the resilience of the ADS, GTs use the pattern of P^{P} which is almost the same as post-event in Figure 8. As shown in the same figure, the installed GTs can be alternatively used as backup suppliers during the normal operation of the IDGS. As a result, for peak load and high power price hours, the operation of GTs is cost-efficient as they consume relatively inexpensive gas from UGN (G^{GT}) to generate power for ADS (P^{GT}).

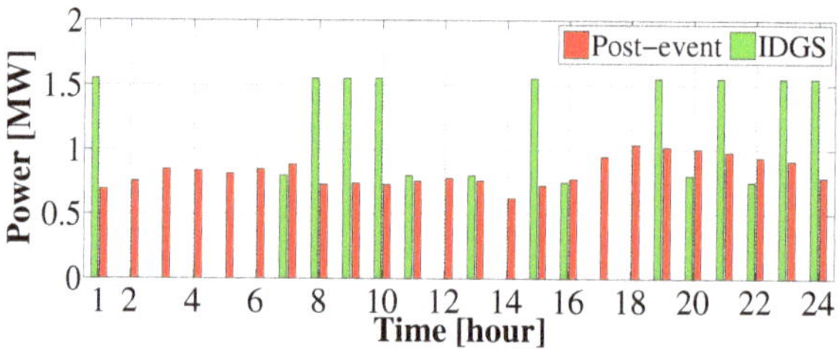

Figure 8. Optimal power generated by GTs.

Although the proposed solution for integrating ADS and UGN is the optimal method for enhancing resilience, it is unlikely that the GT capacity determined this way would be considered in a practical coordination scheme. In practice, additional operational variables, such as the vulnerability of UGN and variation in gas load, should be included in the security analysis. When the GTs are installed, the voltage stress of the ADS buses increases significantly. In addition, the node pressure of the connected GTs drops to the minimum level. The effect of different GT penetration rates on l_{sh}, for the worst-case scenario, is shown in Figure 9. According to the figure, the level of commitment between ADS and UGN operators has a direct impact on the total DA load shedding and resilience of the ADS. However, due to the operational limitations for higher penetration rate of GTs, the amount of critical load shedding is higher than the non-critical load shedding in the proposed decision-making approach. For the IDGS, the best commitment of GT capacity for coordination can be selected using the results shown in Figure 9.

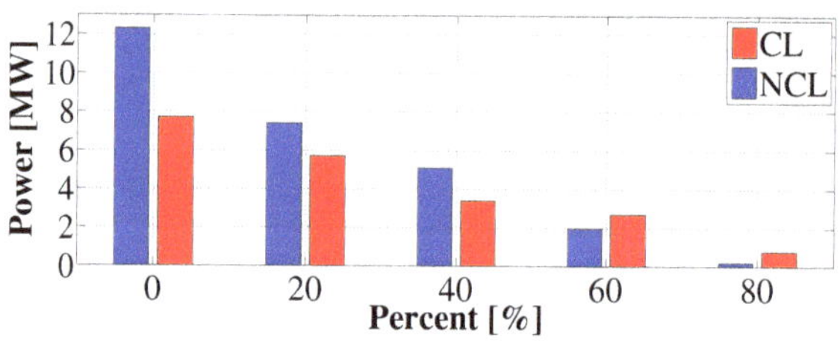

Figure 9. Total day ahead l_{sh} vs. GTs penetration rate considering two types of loads for the worst-case scenario.

The optimization results of the proposed system depend on the degree of uncertainty at Level 1. Table 2 compares the results obtained with two deterministic methods (MILP and MINLP) using a sample scenario, and two robust models (ARO [35] and ADRO) using a set of realization scenarios. It can be seen that linearization (MILP) results in lower l_{sh} compared to the basic nonlinear formulation (MINLP). The remaining two robust scenario-based models demonstrate the effect of various uncertainty parameters on resilience metric calculations. When uncertainties are incorporated into both optimization methods (ARO and ADRO), the value of l_{sh} increases. However, the results of the ADRO approach are more realistic and practical than those of ARO for different levels of uncertainties. In other words, in ARO, for different uncertainty budget percentages (Γ^*), the operation costs of Level 1 are higher than for the corresponding percentages Γ in ARDO. Indeed, the ARO method constructs a deterministic calculation based on the most likely realization scenario as the worst-case. In contrast, the proposed ADRO method considers the occurrence probabilities, or the symmetry in the uncertainty distributions. As a consequence, considering the uncertainty distribution results in a smaller gap between the real-time and DA analyses.

Table 2. Resilience comparison of different optimization methods during post-event for Level 1.

MINLP					MILP				
l^{sh}(MW)	18.21				l^{sh}(MW)	17.51			
ARO [35]					ADRO				
Γ^*(%)	95	90	85	80	Γ(%)	95	90	85	80
l^{sh}(MW)	22.19	22.85	23.91	24.89	l^{sh}(MW)	19.80	20.01	20.78	21.23

(2) Learning method:

Figure 10a and Figure 10b depict, respectively, the results of the DA forecasts for the wind speed and solar irradiance test sets. For a fair comparison, the parameters for LSTM and DLSTM neural networks are identical. The results shown in the figures confirm that the new forecasting model can predict both variables effectively and with high accuracy. This is due to the fact that the DLSTM network can decrease the sample variation by extracting the hourly features leading to bias reduction.

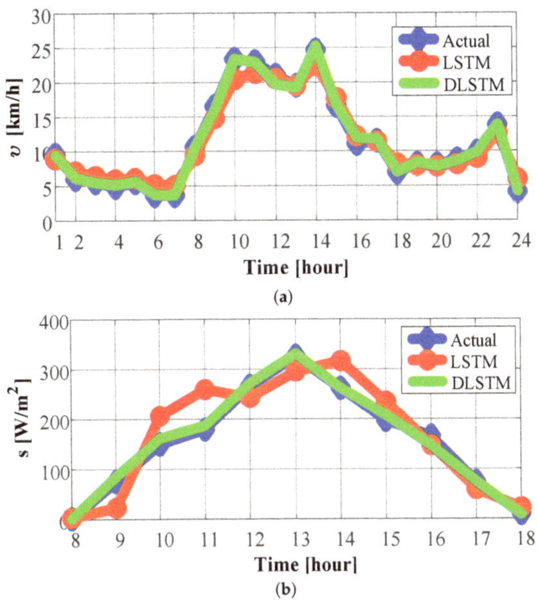

Figure 10. Comparison results of the proposed DLSTM for 15 February 2022. (**a**) wind speed, (**b**) solar radiation.

Prediction accuracy is also evaluated using two commonly used error metrics: MAE and MAPE. Forecast errors are listed in Table 3. Using DLSTM, the MAE and RMSE for v are reduced by about 47% and 71%, respectively, compared to the standard LSTM. For s, the reductions are 76% and 77% for MAE and RMSE, respectively. This confirms that the proposed DLSTM neural network can capture deep generalizations of various time-series data by distributing them across several forecasting tasks. Thus, neither normalization nor filtering is required because the data for each subtask are within its own logical range. This allows DLSTM to generate accurate forecasts for complex time series.

Table 3. Prediction error of conventional LSTM and proposed DLSTM.

Forecasting Method	Wind Speed		Solar Radiation	
	RMSE (km/h)	MAE (km/h)	RMSE (W/m^2)	MAE (W/m^2)
LSTM	3.22	2.34	45.87	30.14
DLSTM	1.71	0.68	11.13	6.92

The illustrations in Figure 11, which demonstrate the error frequency of 24 subtasks, confirm the results presented in Table 3. According to the entire data set of v, the RMSE and MAE values for the subtasks are improved between the ranges of $[34\%, 52\%]$ and $[67\%, 77\%]$. Similarly, each subtask of the s data set achieves RMSE and MAE improvement ranges of $[34\%, 93\%]$ and $[57\%, 95\%]$, respectively. As indicated previously, to ensure a fair comparison, the fitting model for both the LSTM and DLSTM approaches is the same. However, the values in Figure 11 can be further improved by optimizing the hyperparameters of the neural network in each subtask.

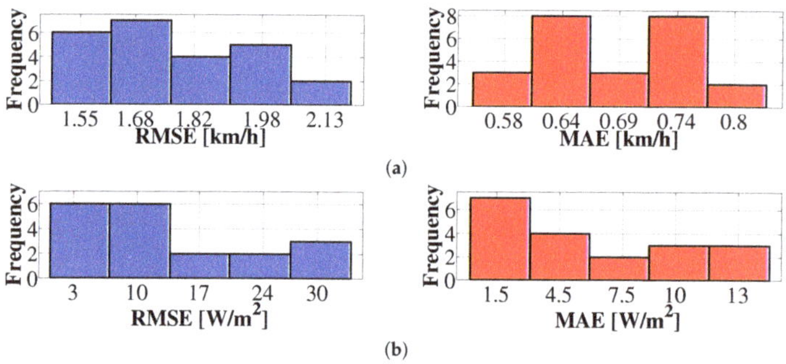

Figure 11. Error distributions of subtasks: (**a**) Wind speed, (**b**) solar radiation.

8. Conclusions

This article proposes a two-level tri-layer problem to improve the resilience of ADS by coupling it with UGN. The complex resilience-oriented problem is solved using an ADRO approach based on the MCBD method. Compared to ARO, the proposed approach results in approximately 11% and 15% reduction in load shedding for higher and lower uncertainty budget percentages, respectively. To handle the uncertainty of the post-event structural consequences, realization scenarios are modeled using the MC method. The results confirm that the use of GTs with optimal capacity and location enhances the resilience of the system against HILP events. Accordingly, optimizing the location and capacity for coordination between ADS and UGN can reduce load shedding by approximately 94% compared to uncoordinated systems. In summary, the most effective solution for the proposed resilience enhancement approach is a learning-based optimization method for IDGS. An additional contribution is the newly proposed diurnal learning method to address the uncertainties of weather-related power generated from RESs. This method, DLSTM, achieves more accurate and practical results than conventional LSTM. In future work, we plan to extend the proposed solution to cover additional resilience enhancement mechanisms such as demand response and electric vehicles.

The proposed approach has several limitations which offer opportunities for its further development. First, it may be possible to develop a more streamlined version of the model that balances the complexity of the resilience-oriented model with computational efficiency. The model also relies on accurate data to effectively capture uncertainties. In scenarios where adequate data are not available, the performance of the model may be compromised, leading to suboptimal resilience outcomes. In addition, the presented study focuses mainly on the technical aspects of enhancing power system resilience, not considering social and economic aspects such as costs, the regulatory environment, and public acceptance. These considerations are critical for the real-world adoption of the proposed resilience enhancement solutions and can be addressed in future work.

Supplementary Materials: The following supporting information can be downloaded at: https://www.mdpi.com/article/10.3390/en17246270/s1.

Author Contributions: Conceptualization, B.F. and P.M.; methodology, B.F. and P.M.; investigation, B.F.; resources, P.M.; writing—original draft preparation, B.F.; writing—review and editing, P.M.; supervision, P.M.; project administration, P.M.; funding acquisition, P.M. All authors have read and agreed to the published version of the manuscript.

Funding: This research has been supported by the Natural Sciences and Engineering Research Council (NSERC) of Canada grant (award number 2024-04565), and by the U.S.-Canada Center on Climate-Resilient Western Interconnected Grid (NSF WIRED Global Center) funded jointly by the National Science Foundation (award number 2330582) and NSERC (award number 2023-585094).

Data Availability Statement: Data are available on request due to restrictions (large data size). Minimal data representing the original were provided at submission and are available in supplementary file.

Conflicts of Interest: The authors declare no conflict of interest.

Nomenclature

HILP	High-impact low-probability
CL	Critical load
NCL	Non-critical load
UGN	Urban gas network
RES	Renewable energy source
ADS	Active distribution system
GT	Gas turbine
ESS	Energy storage system
IDGS	Integrated distribution and gas system
MILP	Mixed integer linear programming
DLSTM	Diurnal long short-term memory
ADRO	Adaptive distributionally robust optimization
WT	Wind turbines
PV	Photovoltaic panels
DA	Day ahead
MAE	Mean absolute error
MAPE	Mean absolute percentage error
MC	Monte-Carlo
C_{mn}	Waymouth constant
p_{mn}	Node pressure of natural gas network [bar]
$\bar{q}(t)/q_{mn,t}$	Average/Transmitted Gas flow in pipeline [Mm^3]
LP	Line-pack between nodes [kcf]
α	Gas to power conversion factor [kW/kcf]
RU^{GT}/RD^{GT}	Ramp-up/Ramp-down rate of GT [kW]
ω	ambiguity set
ω_1/ω_2	Sample scenarios of EC/ ER
$\mathbf{EC}$	Encoding matrix of CLs
$\mathcal{M}$	Momentum-based ambiguity set
$\mathbf{ER}$	Encoding matrix of generated power from RESs
π	Probability factor
$\mathcal{S}$	Support
μ	Mean
$\mathbf{\Sigma}$	Covariance

ς_1/ς_2	Controller of ambiguity set
f	Forgetting gate output
h	Hidden layer output
w	Connection weight
ω_p	Probability distributions
ϵ	Convergence tolerance of MCBD
W	Total probability distributions
d	Daily time intervals
i	Iteration of MCBD
C^{P}	Power price [USD/kw]
$(.)^{0/1}$	Pre-/Post-event notation
$P^{\mathrm{P}}/Q^{\mathrm{P}}$	Active/reactive power received from upstream grid [kW]
t	Time [h]
T	Total operation time [24 h]
D	Total day interval of data set
$K/N/C/Z$	Total bus/node/coordination point/zone
$p^{ch/dch}$	ESS's charging/discharging power [kW]
I_{kj}	Squared current magnitude from k to j [A]
U_k/U_j	Squared voltage magnitude of bus k/j [V]
r/x	resistance/inductance [Ω]
$v_{ci}/\ v_{co}/\ v_r$	Cut-in/ cut-out/ rated wind speed [m/s]
s_r	Rated solar radiation [W/m^2]
s	Solar radiation [W/m^2]
v	Wind speed [m/s]
$(.)^{\max/\min}$	Maximum/minimum limitation
E	State of energy [kWh]
$I^{ch/dch}$	ESS's charging/discharging status [binary]
η	Efficiency of ESS [%]
C^{G}	Natural gas price [USD/kcf]
G^{GT}	Consumed natural gas by GTs [kcf]
ρ	Penalty of load shedding [USD/kW]
W_1/W_2	Total scenario number of EC and ER
β	Depth of discharge [%]
$l^{\mathrm{sh}\vert\mathrm{cl}}/l^{\mathrm{sh}\vert\mathrm{ncl}}$	Load shedding of CL/NCL [kW]

References

1. Abdin, I.; Fang, Y.P.; Zio, E. A modeling and optimization framework for power systems design with operational flexibility and resilience against extreme heat waves and drought events. *Renew. Sustain. Energy Rev.* **2019**, *112*, 706–719. [CrossRef]
2. Ravadanegh, S.N.; Jamali, S.; Vaniar, A.M. Multi-infrastructure energy systems resiliency assessment in the presence of multi-hazards disasters. *Sustain. Cities Soc.* **2022**, *79*, 103687. [CrossRef]
3. Darvish, S.M.; Derakhshan, G.; Hakimi, S.M. Enhancing resilience of the smart urban electrical and gas grids considering reserve scheduling and pre-event responses via the onsite supply strategy of the energy storage systems and demand response. *J. Energy Storage* **2024**, *83*, 110633. [CrossRef]
4. Jia, K.; Liu, C.; Li, S.; Jiang, D. Modeling and optimization of a hybrid renewable energy system integrated with gas turbine and energy storage. *Energy Convers. Manag.* **2023**, *279*, 116763. [CrossRef]
5. Manesh, M.K.; Firouzi, P.; Kabiri, S.; Blanco-Marigorta, A.M. Evaluation of power and freshwater production based on integrated gas turbine, S-CO2, and ORC cycles with RO desalination unit. *Energy Convers. Manag.* **2021**, *228*, 113607. [CrossRef]
6. Sayed, A.R.; Wang, C.; Bi, T. Resilient operational strategies for power systems considering the interactions with natural gas systems. *Appl. Energy* **2019**, *241*, 548–566. [CrossRef]
7. Kwasinski, A. Technology planning for electric power supply in critical events considering a bulk grid, backup power plants, and micro-grids. *IEEE Syst. J.* **2010**, *4*, 167–178. [CrossRef]
8. Ahmed, Z.; Nasir, M.; Alsekait, D.M.; Shah, M.Z.H.; AbdElminaam, D.S.; Ahmad, F. Control Conditions for Equal Power Sharing in Multi-Area Power Systems for Resilience Against False Data Injection Attacks. *Energies* **2024**, *17*, 5757. [CrossRef]
9. Xu, N.; Guikema, S.D.; Davidson, R.A.; Nozick, L.K.; Çağnan, Z.; Vaziri, K. Optimizing scheduling of post-earthquake electric power restoration tasks. *Earthq. Eng. Struct. Dyn.* **2007**, *36*, 265–284. [CrossRef]
10. Trakas, D.N.; Hatziargyriou, N.D. Optimal distribution system operation for enhancing resilience against wildfires. *IEEE Trans. Power Syst.* **2017**, *33*, 2260–2271. [CrossRef]

11. Sharma, A.; Srinivasan, D.; Trivedi, A. A decentralized multi-agent approach for service restoration in uncertain environment. *IEEE Trans. Smart Grid* **2016**, *9*, 3394–3405. [CrossRef]
12. Wang, Z.; Wang, J.; Chen, C. A three-phase microgrid restoration model considering unbalanced operation of distributed generation. *IEEE Trans. Smart Grid* **2016**, *9*, 3594–3604. [CrossRef]
13. Xu, G.; Guo, Z. Resilience enhancement of distribution networks based on demand response under extreme scenarios. *IET Renew. Power Gener.* **2024**, *18*, 48–59. [CrossRef]
14. Wang, Y.; Xu, Y.; He, J.; Liu, C.C.; Schneider, K.P.; Hong, M.; Ton, D.T. Coordinating multiple sources for service restoration to enhance resilience of distribution systems. *IEEE Trans. Smart Grid* **2019**, *10*, 5781–5793. [CrossRef]
15. Samani, E.; Aminifar, F. Tri-level robust investment planning of DERs in distribution networks with AC constraints. *IEEE Trans. Power Syst.* **2019**, *34*, 3749–3757. [CrossRef]
16. Hussain, A.; Musilek, P. Resilience Enhancement Strategies For and Through Electric Vehicles. *Sustain. Cities Soc.* **2022**, *80*, 103788. [CrossRef]
17. Faridpak, B.; Farrokhifar, M.; Murzakhanov, I.; Safari, A. A series multi-step approach for operation Co-optimization of integrated power and natural gas systems. *Energy* **2020**, *204*, 117897. [CrossRef]
18. Faridpak, B.; Farrokhifar, M.; Alahyari, A.; Marzband, M. A mixed epistemic-aleatory stochastic framework for the optimal operation of hybrid fuel stations. *IEEE Trans. Veh. Technol.* **2021**, *70*, 9764–9774. [CrossRef]
19. Faridpak, B.; Alahyari, A.; Farrokhifar, M.; Momeni, H. Toward Small Scale Renewable Energy Hub-Based Hybrid Fuel Stations: Appraising Structure and Scheduling. *IEEE Trans. Transp. Electrif.* **2020**, *6*, 267–277. [CrossRef]
20. Raheli, E.; Wu, Q.; Zhang, M.; Wen, C. Optimal coordinated operation of integrated natural gas and electric power systems: A review of modeling and solution methods. *Renew. Sustain. Energy Rev.* **2021**, *145*, 111134. [CrossRef]
21. Saravi, V.S.; Kalantar, M.; Anvari-Moghaddam, A. A cooperative resilience-oriented planning framework for integrated distribution energy systems and multi-carrier energy microgrids considering energy trading. *Sustain. Cities Soc.* **2024**, *100*, 105039. [CrossRef]
22. Cong, H.; He, Y.; Wang, X.; Jiang, C. Robust optimization for improving resilience of integrated energy systems with electricity and natural gas infrastructures. *J. Mod. Power Syst. Clean Energy* **2018**, *6*, 1066–1078. [CrossRef]
23. Manshadi, S.D.; Khodayar, M.E. Resilient operation of multiple energy carrier microgrids. *IEEE Trans. Smart Grid* **2015**, *6*, 2283–2292. [CrossRef]
24. Correa-Posada, C.M.; Sánchez-Martın, P. Security-constrained optimal power and natural-gas flow. *IEEE Trans. Power Syst.* **2014**, *29*, 1780–1787. [CrossRef]
25. Wang, C.; Wei, W.; Wang, J.; Liu, F.; Qiu, F.; Correa-Posada, C.M.; Mei, S. Robust defense strategy for gas–electric systems against malicious attacks. *IEEE Trans. Power Syst.* **2016**, *32*, 2953–2965. [CrossRef]
26. Sawas, A.M.; Khani, H.; Farag, H.E. On the resiliency of power and gas integration resources against cyber attacks. *IEEE Trans. Ind. Inform.* **2020**, *17*, 3099–3110. [CrossRef]
27. Kor, Y.; Reformat, M.Z.; Musilek, P. Predicting weather-related power outages in distribution grid. In Proceedings of the 2020 IEEE Power & Energy Society General Meeting (PESGM), IEEE, Montreal, QC, Canada, 2–6 August 2020; pp. 1–5.
28. Shoaei, M.; Noorollahi, Y.; Hajinezhad, A.; Moosavian, S.F. A review of the applications of artificial intelligence in renewable energy systems: An approach-based study. *Energy Convers. Manag.* **2024**, *306*, 118207. [CrossRef]
29. Li, Q.; Zhao, Y.; Yu, F. A novel multichannel long short-term memory method with time series for soil temperature modeling. *IEEE Access* **2020**, *8*, 182026–182043. [CrossRef]
30. Abdel-Nasser, M.; Mahmoud, K.; Lehtonen, M. Reliable solar irradiance forecasting approach based on choquet integral and deep LSTMs. *IEEE Trans. Ind. Informatics* **2020**, *17*, 1873–1881. [CrossRef]
31. Wang, Y.; Shen, Y.; Mao, S.; Chen, X.; Zou, H. LASSO and LSTM integrated temporal model for short-term solar intensity forecasting. *IEEE Internet Things J.* **2018**, *6*, 2933–2944. [CrossRef]
32. Zhou, H.; Zhang, Y.; Yang, L.; Liu, Q.; Yan, K.; Du, Y. Short-term photovoltaic power forecasting based on long short term memory neural network and attention mechanism. *IEEE Access* **2019**, *7*, 78063–78074. [CrossRef]
33. Dolatabadi, A.; Abdeltawab, H.; Mohamed, Y.A.R.I. Deep Spatial-Temporal 2-D CNN-BLSTM Model for Ultrashort-Term LiDAR-Assisted Wind Turbine's Power and Fatigue Load Forecasting. *IEEE Trans. Ind. Informatics* **2021**, *18*, 2342–2353. [CrossRef]
34. Li, M.; Zhang, Z.; Ji, T.; Wu, Q. Ultra-short term wind speed prediction using mathematical morphology decomposition and long short-term memory. *CSEE J. Power Energy Syst.* **2020**, *6*, 890–900.
35. Gholami, A.; Shekari, T.; Grijalva, S. Proactive management of microgrids for resiliency enhancement: An adaptive robust approach. *IEEE Trans. Sustain. Energy* **2017**, *10*, 470–480. [CrossRef]
36. Shao, C.; Shahidehpour, M.; Wang, X.; Wang, X.; Wang, B. Integrated planning of electricity and natural gas transportation systems for enhancing the power grid resilience. *IEEE Trans. Power Syst.* **2017**, *32*, 4418–4429. [CrossRef]
37. Yan, M.; He, Y.; Shahidehpour, M.; Ai, X.; Li, Z.; Wen, J. Coordinated regional-district operation of integrated energy systems for resilience enhancement in natural disasters. *IEEE Trans. Smart Grid* **2018**, *10*, 4881–4892. [CrossRef]
38. Sekhavatmanesh, H.; Cherkaoui, R. A novel decomposition solution approach for the restoration problem in distribution networks. *IEEE Trans. Power Syst.* **2020**, *35*, 3810–3824. [CrossRef]
39. Jiang, J.; Peng, S. Mathematical programs with distributionally robust chance constraints: Statistical robustness, discretization and reformulation. *Eur. J. Oper. Res.* **2024**, *313*, 616–627. [CrossRef]

40. Luo, Z.; Yin, Y.; Wang, D.; Cheng, T.; Wu, C.C. Wasserstein distributionally robust chance-constrained program with moment information. *Comput. Oper. Res.* **2023**, *152*, 106150. [CrossRef]
41. Korstanje, J. *Advanced Forecasting with Pyton*; Apress: Berkeley, CA, USA, 2021.
42. Ren, L.; Dong, J.; Wang, X.; Meng, Z.; Zhao, L.; Deen, M.J. A data-driven auto-CNN-LSTM prediction model for lithium-ion battery remaining useful life. *IEEE Trans. Ind. Informatics* **2020**, *17*, 3478–3487. [CrossRef]
43. Ge, X.; Jin, Y.; Fu, Y.; Ma, Y.; Xia, S. Multiple-cut benders decomposition for Wind-Hydro-Thermal optimal scheduling with quantifying various types of reserves. *IEEE Trans. Sustain. Energy* **2019**, *11*, 1358–1369. [CrossRef]
44. Sayed, A.R.; Wang, C.; Zhao, J.; Bi, T. Distribution-level robust energy management of power systems considering bidirectional interactions with gas systems. *IEEE Trans. Smart Grid* **2019**, *11*, 2092–2105. [CrossRef]
45. Khani, H.; Farag, H.E.Z. Optimal day-ahead scheduling of power-to-gas energy storage and gas load management in wholesale electricity and gas markets. *IEEE Trans. Sustain. Energy* **2017**, *9*, 940–951. [CrossRef]
46. Alberta Climate Information Service (ACIS), Current and Historical Alberta Weather Station Data Viewer, Government of Alberta. 2022. Available online: https://acis.alberta.ca/ (accessed on 13 May 2022).
47. ISAIC: Industry Sandbox for AI Computing, 2022. Available online: https://www.isaic.ca/ (accessed on 11 June 2022).
48. Sagi, O.; Rokach, L. Ensemble learning: A survey. *Wiley Interdiscip. Rev. Data Min. Knowl. Discov.* **2018**, *8*, e1249. [CrossRef]
49. Bussieck, M.R.; Meeraus, A. General Algebraic Modeling System (GAMS). In *Modeling Languages in Mathematical Optimization*; Kallrath, J., Ed.; Springer: Boston, MA, USA, 2004; pp. 137–157. [CrossRef]

energies

MDPI

Article

Economic–Energy–Environmental Optimization of a Multi-Energy System in a University District

Luca Bacci, Enrico Dal Cin, Gianluca Carraro, Sergio Rech * and Andrea Lazzaretto

Department of Industrial Engineering, University of Padova, Via Venezia 1, 35131 Padova, Italy; luca.bacci.1@studenti.unipd.it (L.B.); enrico.dalcin@phd.unipd.it (E.D.C.); gianluca.carraro@unipd.it (G.C.); andrea.lazzaretto@unipd.it (A.L.)
* Correspondence: sergio.rech@unipd.it

Abstract: The integration of energy generation and consumption is one of the most effective ways to reduce energy-system-related waste, costs, and emissions in cities. This paper considers a university district consisting of 32 buildings where electrical demand is currently met by the national grid, and 31% of thermal demand is supplied by a centralized heating station through a district heating network; the remainder is covered by small, dedicated boilers. Starting from the present system, the goal is to identify "retrofit" design solutions to reduce cost, environmental impact, and the primary energy consumption of the district. To this end, three new configurations of the multi-energy system (MES) of the district are proposed considering (i) the installation of new energy conversion and storage units, (ii) the enlargement of the existing district heating network, and (iii) the inclusion of new branches of the electrical and heating network. The configurations differ in increasing levels of integration through the energy networks. The results show that the installation of cogeneration engines leads to significant benefits in both economic (up to −12.3% of total annual costs) and energy (up to −10.2% of the primary energy consumption) terms; these benefits increase as the level of integration increases. On the other hand, the limited availability of space for photovoltaics results in increased CO_2 emissions when only total cost minimization is considered. However, by accepting a cost increase of 8.4% over the least expensive solution, a significant reduction in CO_2 (−23.9%) can be achieved while still keeping total costs lower than the existing MES.

Keywords: multi-energy system (MES); university district; MILP; economic–energy–environmental optimization; decarbonization; primary energy saving

1. Introduction

The urgent need to reduce carbon emissions is driving the energy sector toward harnessing distributed resources, integrating generation and consumption at the local level, and exploiting synergies between end-use sectors [1]. In fact, more interconnected energy infrastructures allow increasing efficiency, reducing primary energy consumption and increasing the penetration of renewable sources [2]. Accordingly, multi-energy systems are emerging as promising configurations due to their holistic approach in decarbonizing power, thermal, and other sectors while minimizing economic impacts [3]. These benefits can be promoted by the increased flexibility deriving from the active role of end users in energy management, including active participation in demand-side management programs [4].

In general terms, a multi-energy system (MES) is an energy system of any geographical extension, from a single building to a national system, that combines multiple energy vectors (as, for instance, electricity, heat, and fuels) and fulfils the end users' demand for different forms of energy [5]. The design and operation of an MES is a challenging task that requires proper optimization methods, due to the necessity of contextually considering energy conversion, storage, delivery, and utilization [6]. The goal is to properly evaluate the number, type, size, and management over time of the energy conversion and storage units meeting the demand of the end users, as well as the type, layout, and capacity of the energy networks needed for interconnection. This can be formulated, in general, as a mixed-integer non-linear programming (MINLP) problem due to the coexistence of both binary and continuous decision variables (as the existence of a component and its size, respectively) and the presence of non-linearities (e.g., off-design maps) [7]. The large number of decision variables involved can easily make the problem unsolvable with currently available computational technologies [8]. However, linearization approaches that transform the problem into a mixed-integer linear programming (MILP) one allow simplifying the optimization process, thereby drastically reducing computational requirements [9]. Rech [10] proposed a rigorous methodology for modeling the components of an energy system within an MILP problem. This methodology is used in this work as well.

The aforementioned design and operation optimization of an MES is an inherently integrated problem that requires being solved in "one shot". However, the common approach in the literature is to consider the optimization of energy conversion and storage systems separately from the optimization of energy networks. This may lead to solutions that are not optimal for the system as a whole. For instance, in a cost minimization problem, the optimal solution for energy conversion may not be the global optimum because of the failure to consider the interconnection costs with end users.

Few examples are provided of works dealing with the optimization of energy conversion and storage units while neglecting the design of the networks. Dal Cin et al. [11] solved the multi-objective design and operation optimization of an MES serving the electricity and heating demand of a renewable energy community, with the goal of minimizing life-cycle costs and CO_2 emissions. They considered photovoltaic (PV) panels, cogeneration units, and storage systems (both electrical and thermal) but did not model the energy networks. Rech et al. [12] minimized the total cost of an MES providing electricity and heat to a small mountain town by means of renewable plants. Heat is distributed to the users through a district heating network (DHN) that is assumed ideal and modeled as a "black box" connecting generators, storage systems, and consumers. Wirtz [13] proposed a web tool for the optimal conceptual design of the generation mix in a district MES, to define the optimal size of the energy conversion and storage plants available in an energy hub that fulfils electricity, heating, and cooling demands. Heating and cooling networks can also be considered and optimized in terms of operation but not in terms of design (i.e., the capacity of a line is assumed to be sufficient to transport all the energy required). Mashayekh et al. [14] developed an optimization method for the design of the energy conversion and storage plants supplying a multi-energy microgrid that involves both electricity and heating networks. The optimization addresses the generation mix selection and sizing, the resource siting and allocation, and the operation scheduling. However, the design of the networks is provided as input and cannot be improved.

On the other hand, works that optimize the topology and size (i.e., rated power that can flow in each branch) of energy networks usually consider energy conversion systems as input data, thereby neglecting their design. For instance, Pizzolato et al. [15] optimized the topology and capacity of a DHN in Turin, Italy, and analyzed the inclusion of network loops to increase the reliability of the system. They considered an existing heat generation facility to provide the required thermal energy. Röder et al. [16] optimized the design of a DHN serving a mixed residential–commercial district in Germany by searching for the cost-optimal solution with the inclusion of distributed thermal storage fixing in advance of the position, type, and size of the energy conversion plants and storage systems.

Works in the literature that considered the design and operation optimization of both the energy conversion and storage systems and energy networks of an MES as a single problem applied many simplifications to ensure the achievement of a solution in a reasonable time. For instance, Keirstead et al. [17] optimized the design and operation of the electricity, heating, and gas networks, as well as conversion systems (mainly cogeneration units), of an urban MES. The one-day operation of the MES is aggregated into two representative time intervals that allow solving the problem in an acceptable time (the yearly operation, composed of few representative days, is still described by a small number of time intervals). The drawback is that the coarse representation of the simulated time frame is not sufficient to properly model the short-term variability of time-dependent quantities (e.g., availability of renewable sources, changes in energy demand). Sidnell et al. [18] optimized the design and operation of a neighborhood MES integrating electricity and heating networks with renewable plants (PV) and cogeneration units. They considered a simplified network representation in which network lines are modeled as linear segments that connect adjacent buildings, regardless of the real geographical layout of the MES. Lerbinger et al. [19] provided a more detailed model of the DHN in an MES by constraining heating network lines to follow only predetermined paths (such as the streets of a neighborhood), thus avoiding unfeasible connections between nodes. Morvaj et al. [20] highlighted that a "street-following approach" for the optimal design of energy networks guarantees more realistic solutions than a "green-field approach". Dal Cin et al. [21] adopted the street-following approach to optimize the design and operation of a district MES including both electricity and heating networks, as well as PV plants, cogeneration engines, heat pumps, and both thermal and electrical storage systems. Moreover, they made use of typical days obtained by K-medoids clustering to reduce computational complexity while ensuring a sufficient accuracy of the simulated time frame. It should be emphasized that in the mentioned paper, the authors also introduced the concept of "retrofit design", which aims at adding new components or additional installed capacities to an existing system in order to improve its initial layout. Finally, the same authors presented in [22] a general method for the integrated optimization of an MES considered as a whole, i.e., by considering energy conversion units, storage units, and the energy network in the same synthesis, design, and operation (SDO) optimization problem. However, the latter works [21,22] focus on the optimization methodology, the validity of which is demonstrated through hypothetical case studies. They lack, instead, implementations in real-case studies that further strengthen the potential of the proposed method in real applications.

Among the several studies on MESs available in the literature, this paper focuses on those related to university districts, where proposed MES systems have proven to be by far superior solutions in terms of reducing cost and environmental impact compared to the conventional systems based on boiler and electricity withdrawal from the grid. Moreover, the analysis of this type of MES is interesting because the associated studies are often based on measured demand data rather than simulated or standard consumption profiles, thus providing realistic solutions to practical problems. Gabrielli et al. [23] optimized the MES operation of a university facility in Zurich, Switzerland, by focusing on geothermal storage. It turned out that CO_2 emissions can be reduced by 87% compared to a conventional system based on centralized heating and cooling. Martelli et al. [24] studied the subsidies and taxes required to achieve specified decarbonization targets in MESs. Considering a university campus in Parma, Italy, and optimizing both the design and operation of the associated MES, they found that a carbon tax of about 160 EUR/ton favors the installation of PV systems, thereby reducing CO_2 emissions by 25%. Venturini et al. [25] also included a life-cycle assessment in the design and operation optimization problem. The analysis of a university facility in Parma, Italy, showed that the optimal sizing and operation of cogeneration systems results in primary energy savings of about 15% and total cost reductions of about 12% over the life cycle of the MES. Testi et al. [26] developed a multi-objective stochastic optimization methodology to evaluate the integrated optimal sizing and operation of an MES under uncertainties in climate, space occupancy, energy loads, and fuel costs. They considered a university campus in Trieste, Italy, as a test case by focusing on the integration of heat pumps, which can increase the share of renewable energy, reduce the operating cost of the system, and moderate the investment risk. Dos Santos et al. [27] focused on the optimal design of the electrical microgrid serving a university campus in São Paulo, Brazil. The authors of the study optimized the size and location of distributed energy conversion and storage units (mainly photovoltaic plants and battery storage systems) and also the capacity of the cables constituting the microgrid. Finally, Comodi et al. [28] modeled the MES of a university facility in Singapore, with the goal of defining the optimal mix of energy conversion and storage systems and energy network infrastructure needed to meet electricity and cooling demands. They found that a district cooling network can reduce the total installed capacity of electric chillers, which results in capital cost savings. It is worth noting that the aforementioned design optimization studies modeled the MES of the university districts "from scratch" regardless of the pre-existing system configurations, thus achieving optimal layouts that are substantially alternative to the available ones. How to apply "retrofit design" optimization to this type of MES in order to improve existing system configurations (i.e., by starting with the components already available and replacing or resizing them) requires further research.

This paper considers a university district in Padova, northern Italy, and explores new and "smart" solutions to reduce the cost, primary energy consumption, and environmental impact of meeting its electricity and heating demand. This study takes the existing system configuration as a starting point and applies the retrofit design approach proposed in [21] for the optimization of the associated MES. In the existing system layout, the heating and summer-cooling demands of buildings are mostly covered by autonomous small boilers fired by natural gas and compression refrigeration systems, respectively. The buildings belonging to a part of the district are already interconnected through a district heating network (DHN) and a local electrical distribution network (EDN). A centralized thermal power generation system covers via the DHN the heat demand of this part, corresponding to 31% of the total heating demand of the district. The goal of this study is to identify the capacity and operation of the new plants to be installed, including photovoltaic (PV) panels, air–water heat pumps (HPs), and gas-fired cogeneration internal combustion engines (CHP ICEs), as well as the additional capacity and possible expansion of the available DHN and EDN. More specifically, three new energy system layouts are proposed and compared to determine the best solution for the district in terms of cost, primary energy consumption, and environmental impact. The first layout maintains the same structure and capacity of the existing DHN and EDN. The second and third layouts consider instead the expansion of the DHN and EDN. The optimization approach proposed in [21] is then applied to the three layouts to find the most cost-effective design and operation of the district MES (Cost Minimization scenario, CM). It is worth emphasizing that the optimization problem is based on electricity and natural gas consumption data measured over multiple years, from which annual electricity and heating demands are derived with an hourly resolution. This allows the optimization method proposed in [21] to be tested on a real-case study (conversely, the original study made use of standardized demand curves). Two additional scenarios are finally identified by imposing target reduction values on carbon emissions (Low Carbon Emissions scenario, LCE) and primary energy consumption (Primary Energy

Saving scenario, PES) as secondary objectives ("epsilon-constraints") in the optimization procedure. The goal is to identify the suitable trade-off between cost, energy efficiency, and environmental impact.

What is new compared to the literature on MESs in university districts is the implementation of the retrofit design problem to improve the economic and environmental performance of the existing system and its components. This makes it possible to take into account the energy infrastructure already available and to decide on new interventions in the current system. This approach contrasts with that commonly used in the literature, which is based on the design "from scratch" of new system configurations regardless of the components already available, thus providing MES configurations that are completely alternative to the existing ones. In addition, through the consideration of a real-case study based on measured data, this paper provides a practical validation to the optimization methodology proposed in [22], which has been used for "retrofit design" in other contexts but has been applied so far only to hypothetical test cases.

The rest of this paper is organized as follows. Section 2 describes in detail the considered university district with the existing energy infrastructure and outlines the three new MES configurations proposed to improve the existing configuration. Section 3 concerns the methodology and focuses on the input data to be provided to the optimization problem, as well as on the mathematical formulation of the optimization problem. Section 4 shows the results in terms of installed capacities, investment and operational costs, primary energy consumption, and CO_2 emissions for the three new MES configurations in the considered scenarios (CM, LCE, and PES). Section 5 discusses the results in order to determine the best solution for the university district among those proposed. Finally, Section 6 draws conclusions.

2. System Description

This section describes the university district and the proposed configurations considered in the optimization problem in order to explore new solutions to reduce the costs, primary energy consumption, and environmental impact of the district.

2.1. Reference Case

The multi-energy system (MES) considered as a case study is a university district located in the northeast of Padova, Italy. The size of the district area is 422,000 m^2, consisting of 32 university buildings placed in a residential neighborhood. The entire district, shown in Figure 1a, is crossed by a river that separates the north and south areas.

Currently, the yearly electrical demand of 22.5 GWh is covered by withdrawing electricity from 14 PODs connected to the national grid, all referring to the same primary substation. The yearly heat demand of 12.6 GWh is covered only by natural-gas-fired boilers. The north area accounts for 50% of the electrical demand and 40% of the heating demand, fulfilled mostly by a single POD and a centralized boiler, exploiting the existing local EDN and DHN (see blue/red line connecting node 6 to 9/8 in Figure 1b). The energy system of the south area presents a more fragmented situation; the electrical energy is withdrawn by 11 PODs, and each building is equipped with small boilers.

Hourly data of the electrical consumption are available for each POD, starting from 2019 to 2022. For the heat demand, the only data available are the cumulative monthly gas consumptions measured at each of the 21 gas delivery points (GDPs) during 2019 and 2020.

The MES outlined in Figure 1b is modeled as a multi-nodal system composed of N = 20 nodes subdivided into two types, i.e., "aggregation" and "connection" nodes. The former represent energy withdrawal points associated with aggregated buildings, whereas the latter shape both existing and new energy networks by identifying key points where network branches divide or join. Table 1 provides a comprehensive description of each node in the MES, including the existing installed capacity of gas boilers (GBs). The current total capacity of GBs, most of which are condensing boilers, is 20.5 MW (including backup units). "Aggregation" nodes are nodes 0 to 5, node 7, and nodes 9 to 12. The criteria adopted for the aggregation of buildings are as follows:

- Sharing: buildings connected to the same POD, conversion unit, or pre-existing distribution network;
- Proximity: nearby buildings, not separated by public streets;
- Space availability: availability of existing indoor rooms or usable outdoor spaces to be used for technical rooms.

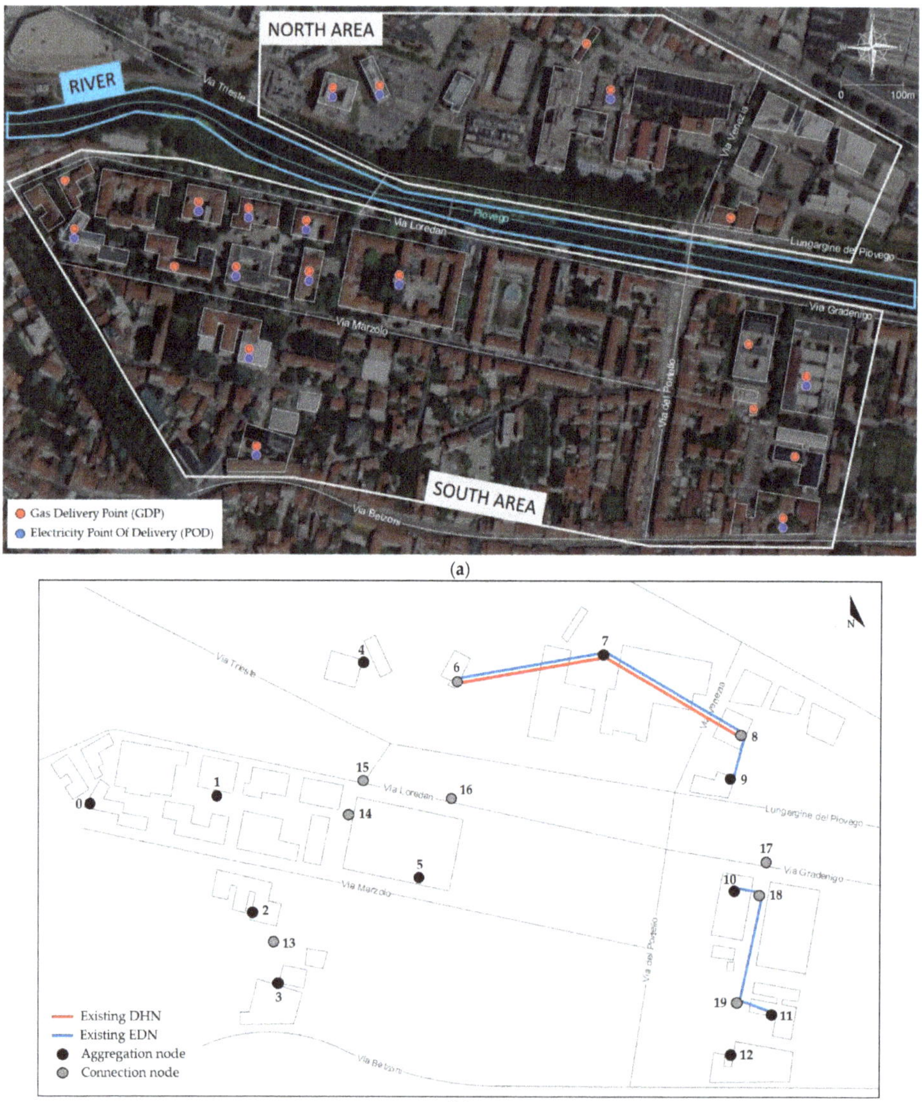

Figure 1. University district: (**a**) satellite image identifying university buildings and the locations of PODs and GDPs; (**b**) multi-nodal representation of the MES (nodes are identified by black or gray dots and numbered).

Table 1. Characterization of the nodes of the starting MES layout.

Node	Type	POD	Installed GB Capacity [kWth]
0	Aggregation node	✓	2427
1	Aggregation node	✓	2274
2	Aggregation node	✓	540
3	Aggregation node	✓	750
4	Aggregation node	✓	1291
5	Aggregation node	✓	1368
6	Connection node	-	-
7	Aggregation node	✓	9000
8	Connection node	-	-
9	Aggregation node	-	220
10	Aggregation node	✓	1980
11	Aggregation node	-	380
12	Aggregation node	✓	313
13	Connection node	-	-
14	Connection node	-	-
15	Connection node	-	-
16	Connection node	-	-
17	Connection node	-	-
18	Connection node	-	-
19	Connection node	-	-

"Connection" nodes are all the remaining nodes (6, 8, and 13 to 20). Their location is chosen to follow the paths of the existing network branches (nodes 6 and 8 identify the existing EDN and DHN together with aggregation nodes 7 and 9, see Figure 1b) and to draw possible new network branches, considering all constraints associated with their installation in accordance with a street-following approach.

Despite its central location (1 km from the city center), the district presents a high spatial density of energy consumption compared with residential neighborhoods; thus, it represents a good test case for the study of energy intervention strategies, combining the advantage of working on a relatively small area with the opportunity to affect a substantial amount of consumptions and related emissions. However, the historical-interest restriction on many of the buildings, the limited roof availability for PV systems, and the absence of other renewable sources strongly limit the renewable share and possible range of interventions.

2.2. Proposed Configurations

Three different MES configurations are considered in the optimization problem. Given the limitations of the case study in terms of renewable energy share, the proposed configurations consider the extension of energy networks that may favor centralized systems over autonomous plants and lead to reduced costs while achieving a higher average conversion efficiency. An increasing number of potential new network branches characterizes the three MES configurations. This may lead to an increasing integration of the MES, intended as the degree of interconnection between nodes through the energy networks and the consequent flexibility in energy distribution:

A. Existing networks: only the pre-existing DHN and EDN are considered (Figure 1b);
B. Medium integration: the north, south-east, and south-west areas constitute three independent DHNs and EDNs, each with a sufficient availability of technical space to accommodate all types of energy conversion and storage units (Figure 2a);
C. High integration: a single DHN and EDN extend throughout the district (Figure 2b).

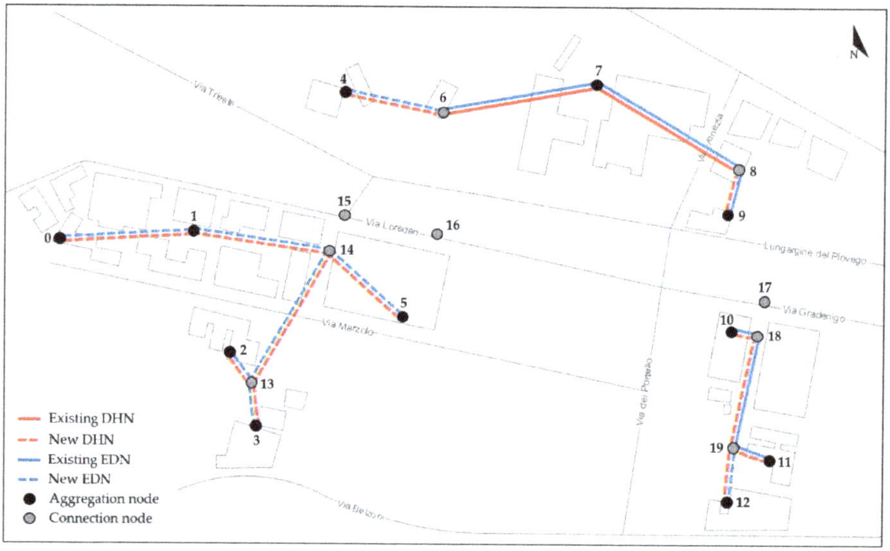

(a)

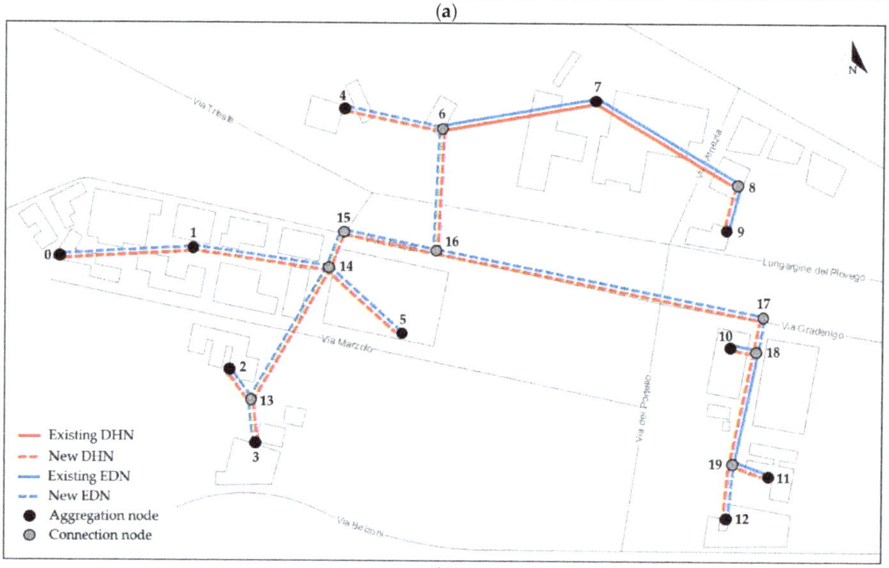

(b)

Figure 2. Proposed new configurations: (**a**) medium integration; (**b**) high integration.

The design of each configuration starts with the current MES assets (i.e., existing boilers, PODs, DHN, and EDN) and evaluates the installation of new energy conversion and storage units and additional network infrastructure. In particular, the proposed solutions combine the installation of PV systems, where possible, and HPs with an effective use of the available fossil resources (natural gas). To this end, integration is promoted with the installation of centralized combined heat and power internal combustion engines (CHP ICEs). The possible installation of electric and thermal energy storage (EES and TES) units is also considered.

The configurations are all characterized by the same dataset, reported in Table 2, which defines the type and maximum size of the energy conversion and storage units that can be installed at the aggregation nodes. The possible locations and maximum sizes of new installations have been identified by considering the available roof area and technical rooms for PV systems (based on a feasibility study initiated by the University of Padova) and CHP ICEs, respectively; the size of existing GBs for HPs (which can be installed at any node associated with a heating demand); the installation of PV systems for small EESs; and the installation of ICEs for large EESs and TESs. In this regard, the maximum sizes of the EESs and TESs at nodes 1, 7, and 10 are intended as sufficiently large thresholds for exploring the optimization results.

Table 2. Candidate energy conversion and storage units to be included in the optimal MES configuration and their maximum capacities.

Node	POD	Max GB Capacity [kW_{th}]	Max PV Capacity [kW_{el}]	Max CHP ICE Capacity [kW_{el}]	Max HP Capacity [kW_{th}]	Max EES Capacity [kWh]	Max TES Capacity [kWh]
0	✓	3000	39	-	2500	250	-
1	✓	3000	-	4000	2500	10,000	20,000
2	✓	1000	-	-	600	-	-
3	✓	1000	58	-	800	250	-
4	✓	2000	-	-	1500	-	-
5	✓	2000	-	-	1500	-	-
7	✓	9000	272	4000	4000	10,000	20,000
9	-	500	-	-	300	-	-
10	✓	2500	-	4000	2000	10,000	20,000
11	-	500	116	-	600	600	-
12	✓	500	-	-	600	-	-

Finally, it is worth noting that the proposed optimization of the design considers not only the installation of new units and network branches but also the retrofit of the existing ones, i.e., the addition of new capacity to units and networks initially available in the reference case.

3. Methods

Figure 3 provides an overview of the procedure used to optimize the design and operation of the MES configurations in Section 2.2.

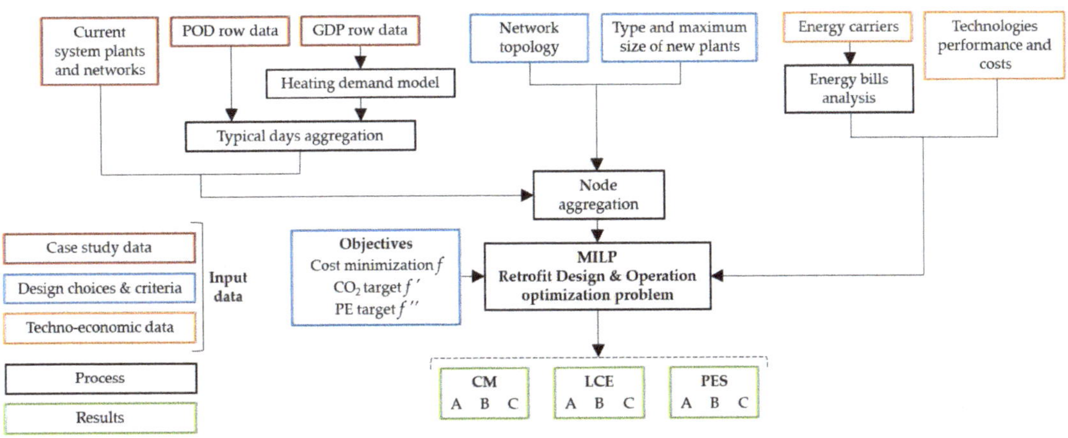

Figure 3. Flow chart of the procedure used to optimize the design and operation of the MES configurations. Different colored boxes are used to identify different types of steps in the procedure.

The "design choices" are already discussed for each configuration in Section 2.2; the "case study" and "techno-economic data" steps are described in Section 3.1. Finally,

Section 3.2 describes the mathematical formulation of the "MILP design and operation optimization problem" and the "criteria" adopted for each scenario.

3.1. Input Data

The optimization problem requires as input data the following time series with an hourly resolution:

- electricity demand of each node associated with an electrical load;
- heating demand of each node associated with a thermal load;
- global solar irradiance, required to calculate the PV generation (source PVGIS [29]);
- ambient temperature, required to assess the coefficient of performance (COP) of HPs (source PVGIS [29]).

The yearly time series of electricity demand has been obtained by averaging the hourly data available for the years 2019 and 2022, neglecting 2020 and 2021 due to the bias introduced by the COVID-19 pandemic. For the heating demand, only the cumulated monthly gas consumptions of each GDP were available, requiring some assumptions in order to build up the hourly time series needed to solve the optimization problem. A simple approach has been adopted, assuming a flat demand curve during the scheduled operating hours of the gas boilers on weekdays, Saturdays, and holidays, for each month of the year. The resulting curves are then normalized with respect to the measured monthly consumption, ensuring that the integral of the monthly demand equals the measured data. Finally, data from PODs and GDPs are stacked into node demands, according to the criteria adopted for the aggregation presented in Section 2 (energy demands of the aggregated buildings are summed hour by hour).

To reduce the computational effort required to solve the optimization problem, the yearly demand time series data are aggregated into K = 36 typical days, considering three days for each month of the year, representative of the average weekday, Saturday, and Sunday/holiday, respectively. Each typical day is defined as H = 24 hourly time steps and is associated with a weight that is equal to the number of represented days in the year (the sum of all weights is 365). Figure 4 shows the energy demand curves of two typical days at one of the aggregation nodes. The shape of the demand profile is homogeneous throughout the district and characterized by the simultaneity of the electrical and thermal demands during the heating period. The choice of 36 typical days allows accounting for both weekly and annual energy demand seasonality. The weekly seasonality is related to the different occupancy of buildings during weekdays, Saturdays, and Sunday/holidays, while seasonal seasonality is mainly related to different environmental conditions (e.g., ambient temperature and daylight hours). For instance, during the heating period (October 15–April 15), the existing heating system (gas boilers) operates from 7 to 19 on weekdays and from 7 to 17 on Saturdays from November to March; from 7 to 17 on weekdays and from 8 to 14 on Saturdays in October and April; and is always off during holidays. Seasonality can be observed in the electric and thermal energy balances (Figure 5), for each day of the year, in the reference case. The heat balance shows clearly the periods when the heating system is in operation (winter, weekdays, and Saturday, daytime); the electric balance shows that the electrical demand during summer is the highest due to the use of compression refrigeration systems. It is worth recalling that in the reference case, the entire heating demand is covered by gas boilers, while the entire electricity demand is met by the national electric grid.

In addition, the data related to the system topology must be provided. These data include the Cartesian position of each node with respect to two reference axes, the position and capacity of the available network branches (identified by the extreme nodes), and the position and capacity of each available conversion unit along with those new units that can be installed. To this end, the considered technologies are as follows:

- Photovoltaic (PV);
- Combined heat and power gas-fired internal combustion engine (CHP ICE);
- Gas-fired boiler (GB);
- Air–water heat pump (HP);
- Thermal energy storage (TES) based on hot water tanks;
- Electrical energy storage (EES) based on lithium batteries.

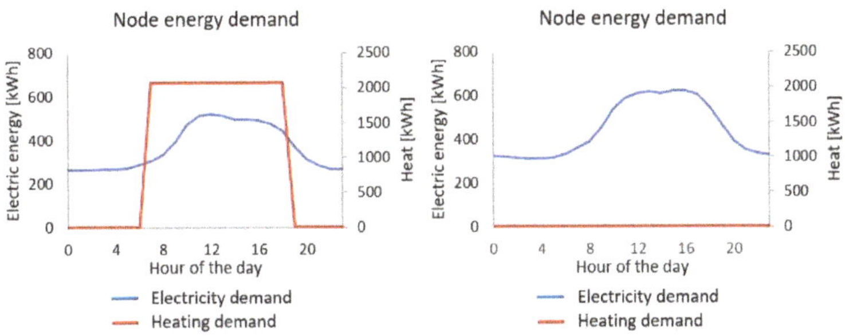

(a) (b)

Figure 4. Daily energy demand curves of node 0, corresponding to the following typical days: (**a**) 0 (weekday of January); (**b**) 18 (weekday of July).

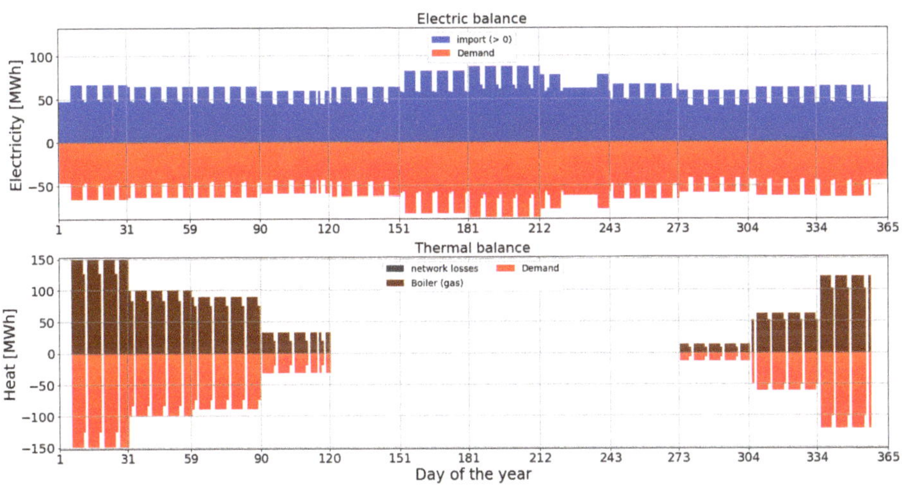

Figure 5. Electric and thermal energy balances in the reference case (each day of the year is replaced by the typical day representing it).

The other information required to solve the model is the techno-economic data of the energy conversion and storage technologies (e.g., lifetime of the units, investment costs), which are summarized in Table 3.

Finally, Table 4 summarizes the specific costs of each energy carrier (i.e., electricity and natural gas), which are evaluated by averaging the university bills collected from October 2022 to September 2023, and their CO_2 emission factors are also provided, which consider the Italian average generation mix of the electricity withdrawn from the national grid.

Table 3. Linearized investment cost and lifetime of the energy conversion and storage units and distribution networks.

Technology	Investment Cost (c_{inv})		Lifetime [Year]
	$c_{inv,var}$	$c_{inv,fix}$	
PV	1250 EUR/kW	-	20
CHP ICE	1398 EUR/kW_{el}	25.8 kEUR	20
GB	65 EUR/kW	1.6 kEUR	20
HP	343 EUR/kW_{th}	6.3 kEUR	20
EES	880 EUR/kWh	3.5 kEUR	20
TES	244 EUR/kWh	1.0 kEUR	20
DHN	0.2 EUR/kW_{th}/m	103 EUR/m	40
EDN	0.006 EUR/kW_{el}/m	34 EUR/m	40

Table 4. Specific cost and emissions of the considered energy carriers.

Carrier	Cost [EUR/MWh]	Emission Factor [kgCO_2/MWh]
Electricity from the grid	259	271
Natural gas	101	197

3.2. Optimization Problem

The design and operation optimization of the MES configurations in Section 2.2 is formulated as a mixed-integer linear programming (MILP) problem:

$$\begin{aligned} &\text{Find } \mathbf{x}_D^* \text{ and } \mathbf{x}_O^*(t) \in \mathbb{R}^n \text{ or } \mathbb{I}^m \text{ that maximize } f(\mathbf{x}_D, \mathbf{x}_O(t))_Y = \boldsymbol{a}^T\mathbf{x}_D + \boldsymbol{b}^T\mathbf{x}_O(t) \\ &\text{subject to } \boldsymbol{g}(\mathbf{x}_D, \mathbf{x}_O(t)) = \boldsymbol{C}\mathbf{x}_D + \boldsymbol{D}\mathbf{x}_O(t) \geq \boldsymbol{e} \end{aligned} \quad (1)$$

where f is the linear objective function; $\mathbf{x}_D$ and $\mathbf{x}_O(t)$ are the decision variables associated with the design (constant in the whole period Y) and operation (time-varying) of the MES, respectively; and g represents the linear equality/inequality constraints that make up the model of the MES. The variables and equations involved in the problem in Equation (1) are discussed in detail in the following.

The relationships g include (i) the electric and thermal balances of the MES, (ii) the characteristic equations describing the performance of the energy conversion and storage units, and (iii) the constraints related to the capacity of the energy networks lines.

The electric and thermal balances (relationships (i)) are defined for each node of the MES (see Figure 1a) and written in accordance with [22]. Each balance, as shown in Equation (2), imposes that the sum of all electric/thermal power flows that enter, are produced, or discharged in the node must be equal to the sum of the electric/thermal power flows that exit, are consumed, or charged in the node.

$$\begin{aligned} &\left(F_{imp}(t)\right) + \sum_i P_i(t) + \sum_j P_j(t) + \sum_z T_z(t)\cdot(1 - \lambda_z(t)\cdot l_z) \\ &\qquad = \left(P_{exp}(t)\right) + F_{dem}(t) + \sum_{i'} F_{i'}(t) + \sum_{j'} F_{j'}(t) + \sum_{z'} T_{z'}(t) \end{aligned} \quad (2)$$

where $F_{imp}(t)$ and $P_{exp}(t)$ are included in the electric balances only and represent the electric power flows that are imported and exported through the node POD, respectively; $P_i(t)$ is the power output of each energy conversion unit i in the node; $P_j(t)$ is the power flow discharged from each energy storage unit j in the node; $T_z(t)$ and $T_{z'}(t)$ are the power flows entering and exiting the node through the local EDN/DHN, respectively; $F_{dem}(t)$ is the power flow associated with the electrical/thermal demand; $F_{i'}(t)$ is the power input of each energy conversion unit i' in the node; and $F_{j'}(t)$ is the power flow charged in each energy storage unit j' in the node. In Equation (2), the subscripts i and j refer exclusively to units that produce or store the considered energy carrier (electricity or heat) in the node, while the subscripts i' and j' refer to units that consume or store the considered energy carrier in the node. Note that the power flow entering a node through the local EDN/DHN equals the power flow leaving the node at the other end of the branch, minus the losses ($\lambda_z(t)$ is the network-specific loss per unit of network length l_z).

Relationships (ii) are written in accordance with [10,22], subdividing the energy conversion units into dispatchable and non-dispatchable ones.

The input power (fuel of the unit, $F_i(t)$) of a dispatchable conversion unit i (i.e., CHP ICE, GB, HP) varies linearly with the power output (product of the unit, $P_i(t)$), as given in Equation (3), where $k_i(t)$ is a correction factor depending on the input data (e.g., ambient temperature in the case of HP), C_i and D_i are constant coefficients linearizing the off-design performance map of the unit, and $\delta_i(t)$ is a binary variable describing the on/off status of the unit.

$$F_i(t) = k_i(t) \cdot (C_i \cdot P_i(t) + D_i \cdot \delta_i(t)) \tag{3}$$

The power output is upper and lower bounded according to the unit rated capacity P_i^{max}, as shown in Equation (4), which includes the auxiliary variable $M_i(t)$ to avoid non-linearities [30].

$$\begin{gathered} P_i(t) \leq P_i^{max} \cdot \beta_i \\ M_i(t) \leq P_i^{max} \cdot \delta_i(t) \\ P_i(t) - M_i(t) \leq (1 - \delta_i(t)) \cdot P_i^{max} \\ m_i \cdot P_i^{max} \leq P_i(t) \leq P_i^{max} \end{gathered} \tag{4}$$

In Equation (4), β_i is a binary variable (constant in the total period) describing the existence of the unit i, and m_i is the minimum load of the unit i referred to its capacity P_i^{max}. Note that if $\beta_i = 0$, the power output $P_i(t)$ is zero for all the time steps in the total period (i.e., the unit i is excluded from the MES), whereas when $\beta_i = 1$, the power output $P_i(t)$ can vary between the minimum ($k_i \cdot P_i^{max}$) and maximum (P_i^{max}) load when $\delta_i(t) = 1$ (the unit i is on) or is equal to zero when $\delta_i(t) = 0$ (the unit i is off).

For CHP ICEs, only the additional linear relationship in Equation (5) is added to calculate the available thermal power that can be recovered (second product of the unit, Q_i) as a function of the unit load (P_i)

$$Q_i(t) \leq C_{2,i} \cdot P_i(t) + D_{2,i} \cdot \delta_i(t) \tag{5}$$

The power output of PV is calculated using Equation (6), where the correction factor $k_{PV}(t)$ is a function of the solar irradiance in the time step t.

$$P_{PV}(t) = k_{PV}(t) \cdot P_{PV}^{max} \tag{6}$$

The state of charge of a storage unit j (TES or EES) is modeled as the sum of the contribution from an intra-day state of charge (periodic within each typical day) and an inter-day state of charge (periodic within the total period of one year) [31].

The intra-day state of charge ($E_j(t)$) accounts for the daily storage and is calculated using the dynamic energy balance in Equation (7), where $k_j(t)$ is a correction factor depending on the input data and the relative coefficient of self-discharge losses of the unit, and $\eta_{char,j}$ and $\eta_{disc,j}$ are the charging and discharging efficiencies, respectively.

$$E_j(t) = k_j(t) \cdot E_j(t-1) + \eta_{char,j} \cdot F_j(t) - \frac{1}{\eta_{disc,j}} \cdot P_j(t) \tag{7}$$

$E_j(t)$ is upper and lower bounded according to the maximum capacity of the storage unit E_j^{max}, as shown in Equation (8). Moreover, it is assumed that each storage unit is completely empty at the beginning of each typical day (second relationship in Equation (8)).

$$\begin{gathered} m_j \cdot E_j^{max} \leq E_j(t) \leq E_j^{max} \\ E_j(t=0) = 0 \end{gathered} \tag{8}$$

The inter-day state of charge accounts for seasonal storage and is calculated as proposed in [31].

The constraints at point (iii) above related to the capacity of the energy networks (EDN and DHN) are taken from [22]. The capacity T_z^{max} of each energy network branch z to carry electricity (heat) between two connected nodes (see Figures 1 and 2) is upper bounded by a "big-M constraint", where $T_z(t)$ is the power flow in the branch z, and M_z is a "large enough" value, as given in Equation (9).

$$T_z^{max} \leq M_z \tag{9}$$

Finally, Equation (10) constrains the power flow $T_z(t)$ in each network branch to be lower than the branch capacity.

$$T_z(t) \leq T_z^{max} \tag{10}$$

For each dispatchable conversion unit i, the design decision variables ($\mathbf{x}_D$) include the unit rated capacity P_i^{max} and the binary variable β_i in Equation (4), the latter describing the existence of the unit i. For PV and TES/EES, the only design decision variable is the maximum unit capacity (P_{PV}^{max} and E_j^{max}), as well as for the energy network branches (T_z^{max}). A value of P_{PV}^{max}, E_j^{max}, or T_z^{max} equal to zero corresponds to the exclusion of the specific unit/network branch from the optimal MES design.

For all dispatchable energy conversion units, storage units, and energy network branches, the operation decision variables ($\mathbf{x}_O(t)$) include a power flow associated with unit/branch load ($P_i(t)$, $F_j(t)$ and $P_j(t)$, $T_z(t)$) for each of the 864 hourly time steps (24 h × 36 typical days). In addition to these, all binary variables $\delta_i(t)$ describing the on/off status of the dispatchable energy conversion units (Equation (3)), and the thermal power output $Q_i(t)$ for CHP ICEs only (Equation (5)) are included among the operation decision variables.

The objective function to be minimized (f in Equation (1)) is the annual cost of the MES, i.e., the annual levelized cost of investments C_{inv} plus the annual operating costs C_{oper}:

$$f(\mathbf{x}_D, \mathbf{x}_O(t))_Y = C_{inv} + C_{oper} = \sum\nolimits_{i,j,z} c_{inv_{i,j,z}} + C_{oper} \tag{11}$$

C_{inv} in Equation (11) is obtained as the sum of the annual levelized cost of investments of each unit ($c_{inv_{i,j,z}}$), which are calculated according to the following linear equations:

$$\begin{aligned} c_{inv_i} &= \left(c_{inv,var_i} \cdot P_i^{max} + c_{inv,fix_i}\right) \cdot F_D && \text{for energy conversion units} \\ c_{inv_j} &= \left(c_{inv,var_j} \cdot E_j^{max} + c_{inv,fix_j}\right) \cdot F_D && \text{for storage units} \\ c_{inv_z} &= \left(c_{inv,var_z} \cdot T_z^{max} + c_{inv,fix_z}\right) \cdot F_D && \text{for energy network branches} \end{aligned} \tag{12}$$

where $c_{inv,var_{i,j,z}}$ and $c_{inv,fix_{i,j,z}}$ are the variable and fixed linearization coefficients, respectively; P_i^{max}, E_j^{max}, and T_z^{max} are the unit capacities; and F_D is the discount factor based on a 5% interest rate and the expected lifetime of the specific technology.

Finally, C_{oper} is calculated as the sum of expenditures for the imported energy (gas and electricity, subscript imp) and revenues for the exported electricity (subscript exp):

$$C_{oper} = \sum_{Y} \left(F_{imp}(t) \cdot c_{el,imp}(t) + G_{imp}(t) \cdot c_{gas}(t) - P_{exp}(t) \cdot c_{el,exp}(t) \right) \cdot h \quad (13)$$

where $F_{imp}(t)$, $G_{imp}(t)$, and $P_{exp}(t)$ are the power flows associated with the imported electricity, imported natural gas, and exported electricity in each hourly time step, respectively; $c_{el,imp}$, c_{gas}, and $c_{el,exp}$ are their specific costs/prices; and h is the considered time step size (1 h).

Optimization results also include the assessment of the annual carbon emissions ($f\prime(\mathbf{x}_D, \mathbf{x}_O(t))_Y$ in Equation (14)) and the annual primary energy consumption of the MES ($f''(\mathbf{x}_D, \mathbf{x}_O(t))_Y$ in Equation (15)) due to the system operation.

$$f\prime(\mathbf{x}_D, \mathbf{x}_O(t))_Y = \sum_{Y} \left(\left(F_{imp}(t) - P_{exp}(t) \right) \cdot e_{el} + G_{imp}(t) \cdot e_{gas} \right) \cdot h \quad (14)$$

where e_{el} and e_{gas} are the average emission factors related to the electricity produced by the Italian electric system and the combustion of the natural gas, respectively.

$$f''(\mathbf{x}_D, \mathbf{x}_O(t))_Y = \sum_{Y} \left(\frac{\left(F_{imp}(t) - P_{exp}(t) \right)}{\eta_{el}} + G_{imp}(t) + P_{PV}(t) \right) \cdot h \quad (15)$$

where η_{el} is the average primary energy-to-electricity conversion efficiency of the Italian electricity system, and $P_{PV}(t)$ is the electric power produced by the PV. Note that in Equation (15), the contribution of photovoltaics to primary energy consumption is equal to the electricity produced by this technology, as it is proposed in [32].

The solution of the optimization problem in Equation (1) with the only objective of minimizing the cost of the MES (Equation (11)) falls under what is called the Cost Minimization (CM) scenario.

The fixed objective values of the reduction in carbon emissions (Equation (14)) and primary energy consumption (Equation (15)) compared to the CM scenario are imposed in specific additional runs of the optimization problem (Low Carbon Emissions—LCE scenario and Primary Energy Saving—PES scenario, respectively) to identify the optimal trade-off between cost, energy efficiency, and environmental impact.

4. Results

The optimization problem of the design and operation of the MES has been solved for each proposed configuration in the CM scenario. The following sections present the results considering the reference case as a term of comparison. The results are then compared with the other scenarios (LCE and PES) in which additional constraints are included to reduce both the primary energy consumption and CO_2 emissions.

All optimization runs were performed using the Gurobi (version 11.0.2) Python API [33] optimization software library on a standard laptop equipped with a 9th Gen i7 processor and 16 GB RAM. Execution times varied widely, ranging from 30 min to 4 days depending on the complexity of the configuration (the optimization runs involving configuration C are the most time-consuming), with tolerances in the solution on the order of 1%.

4.1. Reference Case

The reference case represents the current operation of the district. It considers all the electricity demand met by withdrawing electricity from the national grid and all the heating demand covered by gas-fired boilers.

The investment costs of the existing units and networks are not included because the capital cost of all the existing facilities is assumed to be already depreciated (all considered existing facilities are dated), thus the yearly cost of 7029.8 kEUR is only composed of the operation contribution. The electrical energy withdrawn by the national grid is

22,478 MWh/year, while the annual gas consumption is 11,960.5 MWh, resulting in a total primary energy consumption of 59,183.2 MWh/year and 8447.8 t/year of CO_2 emissions.

In the following, the electricity and heating demands of the reference case are named "reference electricity demand" and "reference heating demand", respectively, to provide homogeneous terms of comparison for the proposed new configurations. In fact, in each of the proposed configurations, the total electricity demand also includes the electricity used by HPs and the EDN losses associated with the specific configuration, while the total heating demand also includes the DHN losses associated with the specific configuration.

4.2. Configuration A: Existing Networks

The first configuration considers the design and operation optimization of the university district MES without modifications to the topology of the pre-existing DHN and EDN.

Figure 6 shows the resulting position and size of the energy conversion and storage units. The results (Table 5, column A) suggest that even if gas-fired boilers can completely cover the heating demand, it is convenient to install CHP ICEs and HPs to minimize the costs. ICEs and HPs contribute to meet the heating demand with 3027.7 MWh/year (24% of the reference heating demand) and 4465.3 MWh/year (35.5% of the reference heating demand), respectively. The installed PV capacity of 485 kW_p fully exploits the limited availability of roof area and contributes to the overall electrical generation with 704 MWh/year (3.1% of the reference electricity demand). The electricity generation from CHP ICEs instead is 9835.3 MWh/year and covers 43.8% of the reference electricity demand. No EES units are included in the solution due to the very limited PV capacity. In addition, the shape of the heating demand is well met by the installed CHP ICE and HP units, with no need for TES, while its simultaneity with the electrical demand favors cogeneration. Finally, the design optimization shows that an increase in the capacity of the existing DHN and EDN is not convenient (i.e., the existing energy networks' capacity is sufficient to distribute the flows produced by existing and new units in the optimal operation).

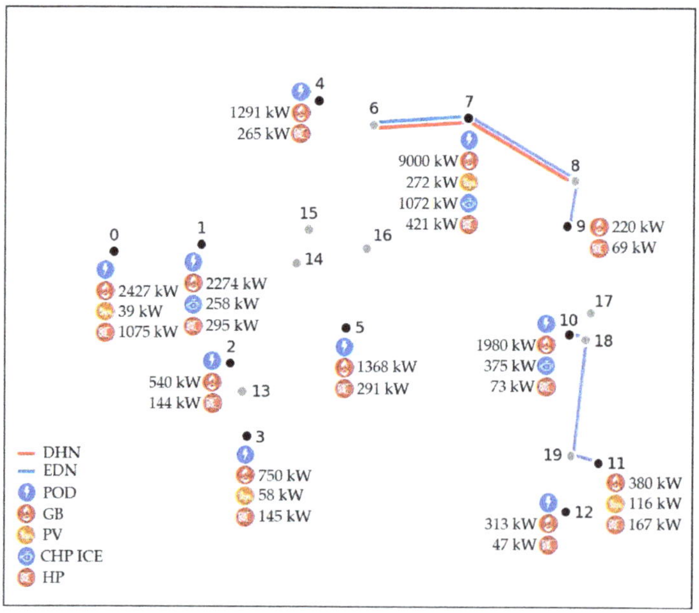

Figure 6. Configuration A, CM scenario: resulting network topology and size and position of the energy conversion units.

Table 5. Results of the design and operation optimization of the three configurations in the CM scenario.

	Configurations A	Configurations B	Configurations C	Reference
Total PV capacity [kW_p]	485	485	485	0
Total CHP ICE capacity [kW_{el}]	1705	2643	3723	0
Total GB capacity [kW_{th}]	20,543	20,543	20,543	20,543
Total HP capacity [kW_{th}]	2992	2448	3670	0
Overall yearly cost [kEUR/year]	**6609.2**	**6367.9**	**6163.3**	**7029.8**
Overall yearly cost [$\%_{ref}$]	**−6.0**	**−9.4**	**−12.3**	-
Operation costs [kEUR/year]	6213.3	5862.3	5467.8	7029.8
Investment costs [kEUR/year]	395.9	505.7	695.5	0.0
Imported electricity [MWh/year]	13,079.9	4371.6	1059.0	22,478.0
Exported electricity [MWh/year]	30.0	53.2	23.5	0.0
Ren [1] electricity generation [MWh/year]	704.0	704.0	704.0	0.0
NRen [2] electricity generation [MWh/year]	9835.5	18,337.6	21,936.7	0.0
HP heat generation [MWh/year]	4465.3	3487.4	4602.5	0.0
GB heat generation [MWh/year]	5125.3	4608.3	2557.3	12,594.5
CHP heat generation [MWh/year]	3027.7	4597.5	5582.2	0.0
CHP factor [%]	30.8	25.1	25.4	-
Usage of natural gas [MWh/year]	27,464.7	45,850.8	50,202.7	11,960.5
HP usage of electricity [MWh/year]	1117.0	879.2	1172.8	0.0
PE [3] consumption [MWh/year]	**55,647.4**	**55,738.8**	**53,131.4**	**59,183.2**
PE [3] consumption [$\%_{ref}$]	**−6.0**	**−5.8**	**−10.2**	-
CO_2 emissions [t/year]	**8955.2**	**10,217.3**	**10,176.9**	**8447.8**
CO_2 emissions [$\%_{ref}$]	**6.0**	**20.9**	**20.5**	-

[1] Ren: renewable; [2] NRen: non-renewable; [3] PE: primary energy.

The resulting overall cost, composed of the operation and investment costs, is 6609.2 kEUR/year, 6% less than the reference case also considering the amortization of investment costs for the new units, which is equal to 395.9 kEUR/year. This is because the operation costs in the optimized configuration A are 6213.3 kEUR/year, 11.6% less than the reference case. The primary energy consumption is 55,647.4 MWh/year, 6% less than the reference case, while the annual CO_2 emissions value of 9282.5 t is 6% higher than the reference case.

Higher emissions are due to the relatively low CHP factor, i.e., the ratio between the recovered waste heat and the electricity produced by the ICEs, which is equal to 30.8%. Table 5 summarizes the results.

4.3. *Configuration B: Medium Integration*

The second configuration allows a partial integration of the university district according to the possible network paths shown in Figure 2a. Figure 7 shows the resulting position and size of the energy conversion and storage units, and the topology of the DHNs and EDNs of the three distinct areas in which the system is divided.

The CHP ICEs' capacity of 2643 kW_{el} (Table 5, column B) contributes 18,337.6 MWh/year (81.6% of the reference electricity demand) and 4597.5 MWh/year (36.5% of the reference heating demand) to the electricity and heating production, respectively, with a CHP factor of 25.1%. The optimal HPs' capacity is concentrated into six nodes with an overall heat production of 3487.4 MWh/year (27.7% of the reference heating demand). The remaining 36.3% of the overall heat production, which includes also the DHN losses, is covered by the existing boilers, the capacity of which is not increased. The installed PV capacity is the maximum available. PV and CHP ICEs together cover 84.7% of the reference electricity demand. No energy storage capacities are included in the optimal solution.

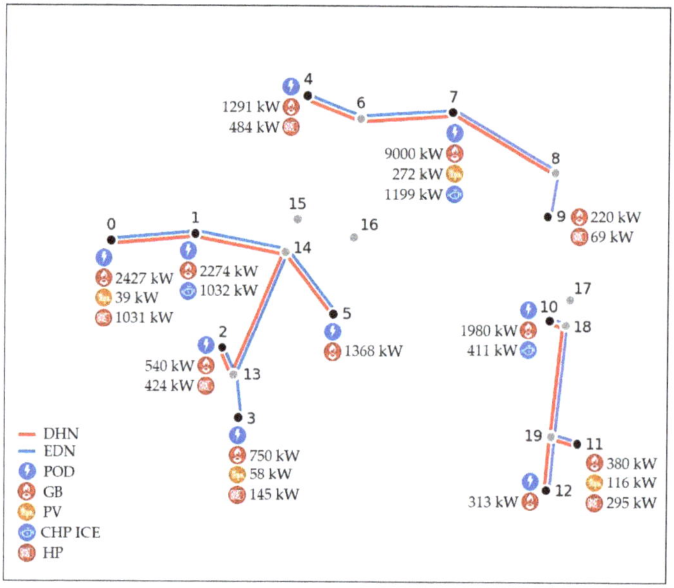

Figure 7. Configuration B, CM scenario: resulting network topology and size and position of the energy conversion units.

All nodes are connected to the EDN, while the results exclude nodes 3 and 9 from the DHN. As expected, the highest capacities of the DHN and EDN branches are concentrated in the nodes where the CHP ICEs are installed and in adjacent nodes.

Finally, the results of the optimization problem show an overall cost of 6367.9 kEUR/year (9.4% less than the reference case), with operation costs of 5862.3 kEUR/year (16.6% less than the reference case) and investment costs of 505.7 kEUR/year. The resulting primary energy consumption is 55,738.8 MWh/year, 5.8% lower than the reference case, and the annual CO_2 emissions are 10,217.3 t, 20.9% higher than the reference. Both results are mainly influenced by the large amount of electrical energy generated by the CHP ICEs without taking advantage of the total available waste heat.

4.4. *Configuration C: High Integration*

The third configuration represents the highest integrated solution, with the whole university district served by the DHN and EDN. Figure 8 shows the topology of the networks and the resulting position and size of the energy conversion and storage units.

CHP ICEs are installed at nodes 1 and 7, with an overall capacity of 3723 kW_{el} (Table 5, column C), while the total capacity of HPs is 3670 kW_{th}. Both capacities are the highest among the three considered configurations. CHP ICEs and HPs contribute 5582.2 MWh/year and 4602.5 MWh/year, respectively, to cover 80.9% of the reference heating demand. At 21,936.7 MWh/year, the electricity generation of ICEs covers 97.6% of the reference electricity demand; thus, in this configuration, the autonomous electricity generation of CHP ICEs and PV completely fulfills the reference electricity demand. Once again, the optimal solution excludes electrical and thermal energy storage units.

Figure 9 shows the progressive expansion of both the DHN and EDN across the three layouts resulting from the optimization in the CM scenario. In configuration C, the resulting capacities of branches 6–16, 15–16, and 16–17, laid out for both the DHN and EDN, allow the three main areas of the MES to be effectively interconnected. Only node 3 and branch 18–19 are excluded by the resulting DHN, while all the available links are installed in the EDN.

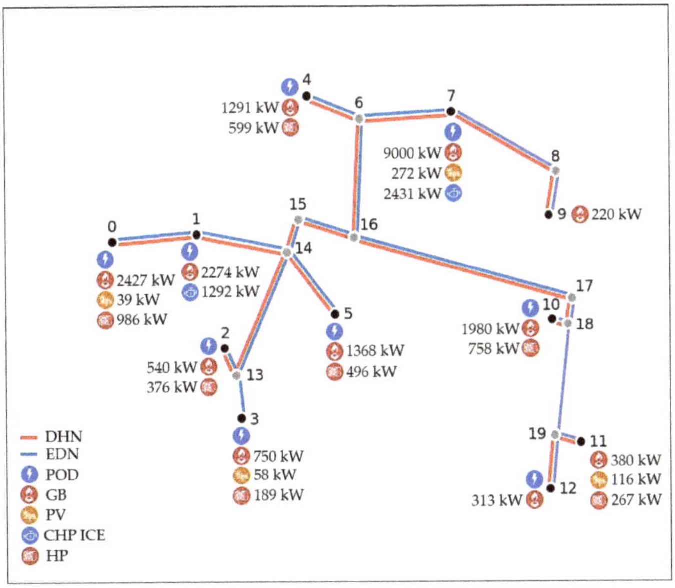

Figure 8. Configuration C, CM scenario: resulting network topology and size and position of the energy conversion units.

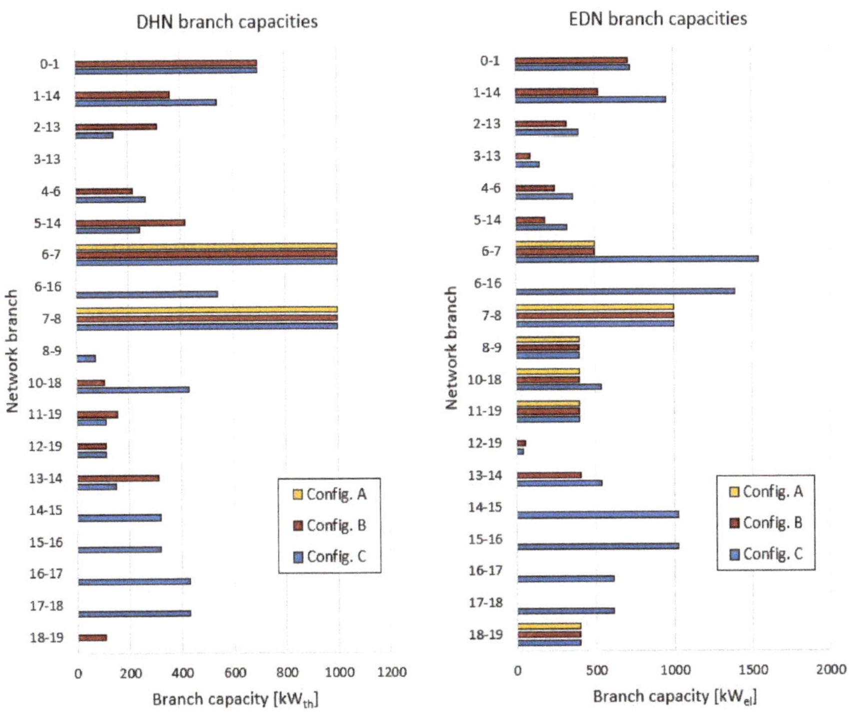

Figure 9. Resulting branch capacities in configurations A, B, and C of the CM scenario: (**a**) district heating network; (**b**) electrical distribution network. Network branches are identified by the pairs of connected nodes.

The resulting overall cost of 6163.3 kEUR/year is the lowest among the three configurations, 12.3% less than the reference case, with operation costs of 5467.8 kEUR/year, 22.2% less than the reference. In contrast, the investment costs of 695.5 kEUR/year are the highest among the proposed configurations, due to the higher capacity of CHP ICEs and HPs and the greater extent of the distribution networks. Finally, this configuration also achieves the best result in terms of primary energy savings, with a consumption of 53,131.4 MWh/year, 10.2% less than the reference case, while the annual CO_2 emissions are 10,176.9 t, about the same as configuration B.

4.5. Low Carbon Emissions and Primary Energy Saving Scenarios

In the CM scenario, an increased network integration (i.e., greater interconnection between nodes) promotes the energy generation from CHP ICEs, leading the proposed configurations to perform better than the reference case, both in terms of overall costs (up to −12.3%) and primary energy consumption (up to −10.2%). On the other hand, the installation of CHP ICEs, particularly with low CHP factors, increases the CO_2 emissions of the system, due to the lower share of the renewables of this "local" system compared to that of the national grid.

Starting from these considerations, the Cost Minimization (CM) scenario is compared with two different scenarios: Low Carbon Emissions (LCE) and Primary Energy Saving (PES). Cost minimization remains the objective in both the LCE and PES scenarios, but the former also considers a cap of −17% (referred to the reference case) on the maximum CO_2 emissions, while the latter imposes a cap of −11% (referred to the reference case) on the maximum primary energy consumption and constrains the minimum hourly utilization of CHP ICE waste heat to be equal to 10% to foster the CHP operation of ICEs. The caps are obtained by minimizing the CO_2 emissions and primary energy consumption of configuration A in two separate optimization problems. Table 6 shows the results of the three scenarios.

Table 6. Results of the optimization of the proposed configurations A, B, and C in the three scenarios: Cost Minimization (CM), Low Carbon Emissions (LCE), and Primary Energy Saving (PES).

	CM Scenario			LCE Scenario			PES Scenario			Reference
	A	B	C	A	B	C	A	B	C	
Total PV capacity [kW_p]	485	485	485	485	485	485	485	485	485	0
Total CHP ICE capacity [kW_{el}]	1705	2643	3723	387	0	0	2334	3318	3268	0
Total GB capacity [kW_{th}]	20,543	20,543	20,543	20,543	20,543	20,543	20,543	20,543	20,543	20,543
Total HP capacity [kW_{th}]	2992	2448	3670	9716	8136	7874	6682	3430	2864	0
Overall yearly cost [kEUR/year]	**6609.2**	**6367.9**	**6163.3**	**6952.4**	**6904.9**	**6908.2**	**6819.8**	**6733.2**	**6683.9**	**7029.8**
Overall yearly cost [$\%_{ref}$]	**−6.0**	**−9.4**	**−12.3**	**−1.1**	**−1.8**	**−1.7**	**−3.0**	**−4.2**	**−4.9**	-
Operation costs [kEUR/year]	6213.3	5862.3	5467.8	6478.0	6531.4	6533.7	6207.6	6099.3	6077.8	7029.8
Investment costs [kEUR/year]	395.9	505.7	695.5	474.4	373.6	374.5	612.2	633.9	606.1	0.0
Imported electricity [MWh/year]	13,079.9	4371.6	1059.0	23,993.0	24,436.2	24,443.9	20,070.2	17,447.8	16,938.9	22,478.0
Exported electricity [MWh/year]	30.0	53.2	23.5	0.0	0.0	0.0	14.6	26.4	1.7	0.0
Ren. electricity generation [MWh/year]	704.0	704.0	704.0	704.0	704.0	704.0	704.0	704.0	704.0	0.0
NRen. electricity generation [MWh/year]	9835.5	18,337.6	21,936.7	480.5	0.0	0.0	3672.6	5462.0	5805.4	0.0
HP heat generation [MWh/year]	4465.3	3487.4	4602.5	10,550.9	10,611.8	10,670.5	7685.0	4390.8	3826.3	0.0
GB heat generation [MWh/year]	5125.3	4608.3	2557.3	1480	2081.5	2070.4	1480.9	3287.7	3858.9	12,594.5
CHP heat generation [MWh/year]	3027.7	4597.5	5582.2	587.4	0.0	0.0	3452.4	5016.0	5056.2	0.0
CHP factor [%]	30.8	25.1	25.4	122.2	-	-	94.0	91.8	87.1	-
Usage of natural gas [MWh/year]	27,464.7	45,850.8	50202.7	2586.5	1976.7	1966.2	9804.7	15,313.9	16,383.2	11,960.5
HP usage of electricity [MWh/year]	1117.0	879.2	1172.8	2705.0	2667.8	2675.4	1959.8	1112.4	969.1	0.0
PE consumption [MWh/year]	**55,647.4**	**55,738.8**	**53,131.4**	**53,695.9**	**54,017.3**	**54,022.9**	**52,673.0**	**52,673.0**	**52,673.0**	**59,183.2**
PE consumption [$\%_{ref}$]	**−6.0**	**−5.8**	**−10.2**	**−9.3**	**−8.7**	**−8.7**	**−11.0 ***	**−11.0 ***	**−11.0 ***	-
CO_2 emissions [t/year]	**8955.2**	**10,217.3**	**10,176.9**	**7011.6**	**7011.6**	**7011.6**	**7370.6**	**7745.2**	**7817.9**	**8447.8**
CO_2 emissions [$\%_{ref}$]	**6.0**	**20.9**	**20.5**	**−17.0 ***	**−17.0 ***	**−17.0 ***	**−12.8**	**−8.3**	**−7.5**	-

* Fixed objective values in the optimization procedure.

In the LCE scenario, the installed capacity of CHP ICEs is reduced to zero (a marginal CHP ICE capacity is maintained only in configuration A) in favor of the HPs, whose capacity in configuration A is more than tripled compared to the CM scenario. Again, in this scenario, as well as in the PES scenario, PV capacity saturates roof availability (485 kW_p)

while the installation of EESs and TESs is not beneficial. It is worth noting that given the limitations of PV availability and the absence of other renewable sources, a large reduction in CO_2 emissions can only be achieved by using the renewable share of the national electric generation mix for heat generation as well, thus employing HPs. Installing HPs reduces investment costs but increases operation costs, so that the overall yearly costs are the highest among the three scenarios, although lower than the reference case (up to −1.8%). HPs cover 84% of the reference heating demand in all three configurations of the LCE scenario, which results in a lower primary energy consumption compared to the reference case (between −9.3% and −8.7%). Regarding CO_2 emissions, all the configurations reach the −17% cap imposed by the scenario.

The constraints imposed on the PES scenario result in high CHP factors: 94%, 91.8%, and 87.1% in configurations A, B, and C, respectively. The installed capacity of HPs is higher in configuration A (6682 kW_{th}), where a reduced network integration limits the amount of cogenerated heat. In fact, CHP ICE heat generation is higher in configurations B and C (up to 46.5% more in configuration C than in configuration A). CHP ICEs' operation results in lower overall yearly costs compared to the LCE scenario: 3%, 4.2%, and 4.9% less than the reference case in configurations A, B, and C, respectively. Primary energy consumption is 52,673 MWh/year in the three configurations, reaching the −11% cap imposed by the scenario. Finally, regarding CO_2 emissions, the best result within the PES scenario is obtained in configuration A (12.8% less than the reference case), due to the higher heat generation of HPs. In any case, compared with the CM scenario, the high CHP factors in the PES scenario entail a strong reduction in CO_2 emissions in all three configurations.

Finally, Figure 10 shows a comparison of the energy balances for configuration C of the district MES for the three considered scenarios. In the CM scenario (Figure 10a), the economic convenience of electricity generation from the CHP ICEs leads these conversion units in operation throughout the entire year to cover 97.6% of the electricity demand (green area in Figure 10a). In contrast, the operating conditions required to achieve the CO_2 emissions target of the LCE scenario (Figure 10b) result in a completely different solution, in which all the electricity is withdrawn from the national grid (except for the 3.1% produced by PV), and HPs cover 84.7% of the reference heating demand. The additional constraint on the minimum hourly utilization of CHP ICE waste heat (10%, PES scenario) leads to a CHP factor of 87.1%, and therefore, ICEs' operation is limited to the heating period.

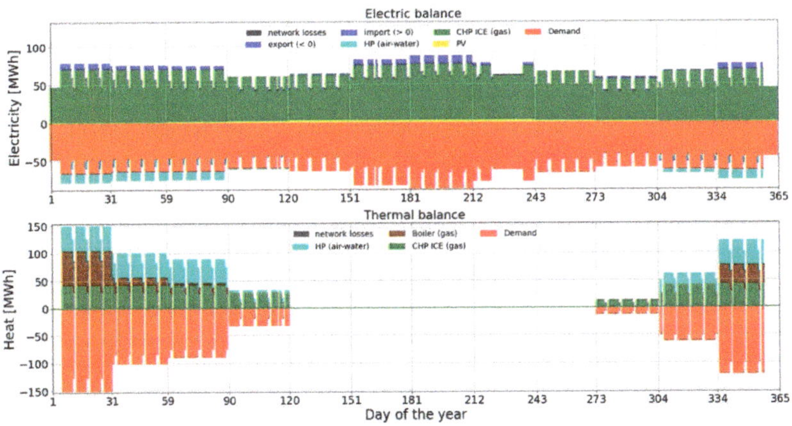

(a)

Figure 10. *Cont.*

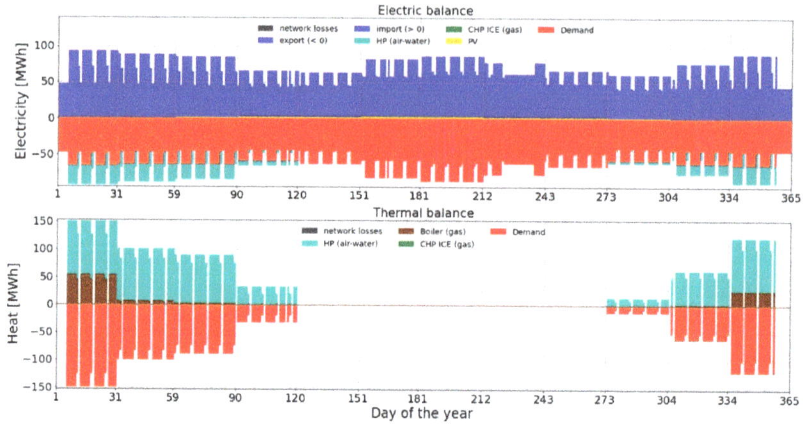

(**b**)

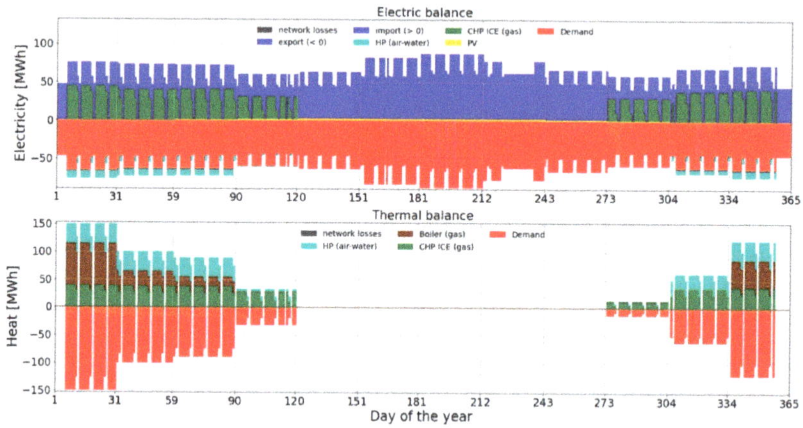

(**c**)

Figure 10. Electric and thermal energy balances of configuration C in the (**a**) Cost Minimization (CM) scenario; (**b**) Low Carbon Emissions (LCE) scenario; (**c**) Primary Energy Saving (PES) scenario.

5. Discussion

In terms of overall annual costs and primary energy consumption, all the proposed configurations in the three different scenarios perform better than the reference case. A higher integration (i.e., greater interconnection between nodes through energy networks) is beneficial from the economic point of view, with configuration C achieving the best result in both the PES scenario and, in particular, the CM scenario, with an overall yearly cost 12.3% lower than the reference case. In the CM scenario, network integration is also associated with lower primary energy consumptions (10.2% less than the reference case in configuration C). Both trends are related to the increased use of CHP ICEs promoted by a greater extension of the DHN and EDN (note that CHP ICEs can only be installed at nodes 1, 7, and 10), which allows for a better match with the district's energy demands. In particular, the reduction in primary energy consumption is associated with the increase in heat cogeneration: 84% more in configuration C than in configuration A (CM scenario), with only a 3% increase in HPs' heat generation.

Network integration and CHP are also associated with a cost-effective reduction in primary energy consumption. This is evident in the PES scenario, where for the same primary energy consumption, configuration C achieves an additional 1.9% reduction in overall yearly cost compared to configuration A, reducing both operation and investment costs.

On the other hand, when the goal is a strong reduction in CO_2 emissions, the results of the LCE scenario show that the installed capacity of CHP ICEs is reduced to zero in favor of HPs. In this scenario, DHN integration in configuration B allows the reduction of the total HPs' installed capacity (16.3% less than configuration A), resulting in an overall yearly cost reduction of 1.1% in configuration A and 1.8% in configuration B (referred to the reference case). Configuration C, however, does not benefit further from the extended network topology.

Nevertheless, a moderate reduction in CO_2 emissions is still feasible in a hybrid context of CHP ICEs and HPs, as shown by the three configurations in the PES scenario, which achieve 12.8%, 8.3%, and 7.5% reductions in CO_2 emissions compared with the reference case, respectively.

Summarizing, in the context of this case study and considering the limited roof availability for PV systems and the absence of other renewable sources, the extension of the DHN and EDN combined with the introduction of CHP ICE units is effective in terms of both overall yearly cost reduction and primary energy saving. The use of CHP ICEs, however, limits the reduction in CO_2 emissions otherwise achievable by installing only HP units. It is also true that the penalization of CHP in this sense is strongly related to the comparison with the electricity supply from the national grid coupled with heating electrification, which can rely on a much larger renewable share: 42% in 2020 (source IEA [34]) on a national basis versus 3.1% locally.

The use of gas boilers in the three different configurations deserves further consideration. Although design optimization never affects their size, in all the proposed solutions, a part of the heating demand is met by GBs (from 11.8% to 40.7% of the reference heating demand). The cost-effectiveness of heat generation from GBs is partly biased by the high installed capacity, which favors this solution over HPs; moreover, the absence of a minimum load threshold in the GB model (unlike CHP ICEs and HPs) makes the use of boilers necessary to fulfill the heating demand at very low load conditions. This is evidenced by the results of configuration A in the LCE scenario, where despite the constraint of minimizing CO_2 emissions, HPs' heat generation cannot completely replace GBs' heat generation (11.7% of total heat generation). In this sense, the optimization process would benefit from the implementation of a database of commercial energy conversion units, in order to consider the installation of units of discrete sizes (instead of considering size as a continuous variable), thus improving the prediction of the part-load performance of the MES.

It is worth noting that the obtained results can be easily extended also to other districts of different sizes and similar characteristics in terms of the limited availability of renewable sources (urban context) and use of the buildings (which results in an almost simultaneous demand for electricity and heating). In fact, in these cases, the inclusion of district CHP units and networks typically leads to reduced costs and primary energy consumption, as demonstrated in [25,28], and a trade-off between cost reduction and the massive installation of heat pumps must be found to limit emissions of CO_2.

Finally, cost assumptions on technologies and energy carriers play a crucial role in the optimization problem, driving the design choices and thus influencing the economic and environmental outcomes. In this regard, the data available from the bills used to calculate the average energy costs are affected by the increase in energy costs that occurred in Europe in winter 2022–2023 due to the Russo–Ukrainian war. To this end, the subsequent reduction in energy costs registered globally could lead to favorable economic conditions.

In conclusion, it is important to emphasize that the retrofit design and operation optimization approach applied in this paper is a completely general tool with broad applicability. While this study focuses on a specific case in terms of renewable energy availability, energy demands, and economic context, the approach can be applied to other districts or systems characterized by very different technical (e.g., high availability of different renewable sources) and economic constraints, yielding variable results tailored to each specific case.

6. Conclusions

Identifying effective retrofit solutions to improve existing energy systems is a key challenge to reduce energy-related costs and environmental impacts in cities. To this end, design and operation optimization tools that consider both energy conversion and storage units and energy distribution networks play a crucial role in choosing the best investment strategy.

This paper investigates the case study of a university district in Padova, proposing three new configurations of the multi-energy system (MES) with increasing possibilities for DHN and EDN interconnection and optimizes the design and operation of additional energy conversion units and energy distribution networks applying a new MILP optimization approach suggested by the same group of authors. Three retrofit design and operation optimizations are solved for each MES configuration, considering Cost Minimization (CM), Low Carbon Emissions (LCE), and Primary Energy Saving (PES) scenarios. Results are compared to search for a suitable trade-off between the economic, environmental, and energy efficiency objectives.

All the optimized configurations in the three different scenarios (CM, LCE, and PES) perform better than the reference case in terms of overall cost reduction and primary energy saving. To this end, the results show that DHN and EDN expansion combined with CHP installation are effective intervention strategies. In fact, the most integrated configuration (C), in both the CM and PES scenarios, has the highest CHP penetration (43.8% and 39.7% of cogenerated heat, respectively) and achieves the best economic results, with overall cost reductions of 12.3% and 4.9%, respectively, compared to the reference case. In addition, configuration C also achieves the best result in terms of primary energy consumption within the CM scenario, 10.2% less than the reference case. On the other hand, the operation of ICEs at low CHP factors results in high CO_2 emissions (up to 20.9% more than the reference case) in the CM scenario, and the LCE scenario shows that ICE installation is reduced to zero if a strong reduction in CO_2 emissions is required (17% less than the reference case), albeit with a negative impact on total costs that still remain lower than in the reference case (up to −1.8%). Nevertheless, a moderate reduction in CO_2 emissions is still feasible in hybrid solutions combining ICEs (operating at high CHP factors) and HPs, as evidenced by the 8.3% reduction in CO_2 emissions and the 39.5% of cogenerated heat achieved by configuration B in the PES scenario.

In conclusion, current needs for drastic emission reduction and consumption containment make the environmental objective of any intervention imperative. To this end, MES integration through expanded energy networks and cogenerating solutions that emerge from the PES scenario represent cost-effective intervention strategies capable of mitigating the environmental impact of the district while preserving the economic sustainability of the investment. Further reduction in CO_2 emissions is also possible at the expense of overall annual cost and primary energy consumption.

Future developments of this work will consider the addition of the "synthesis" problem to the current design and operation optimization problem, i.e., the addition of the variable associated with MES topology among the decision variables of the optimization problem. Moreover, the performance prediction of the energy conversion and storage units will be improved by implementing a database of commercial units of known sizes. Finally, a cost sensitivity analysis will be conducted to determine the influence of cost assumptions.

Author Contributions: Conceptualization, L.B., S.R. and A.L.; methodology, L.B., E.D.C. and S.R.; software, L.B., E.D.C. and G.C.; formal analysis, L.B. and S.R.; data curation, L.B.; writing—original draft preparation, L.B., E.D.C. and S.R.; writing—review and editing, G.C., S.R. and A.L.; visualization, L.B.; supervision, S.R. and A.L. All authors have read and agreed to the published version of the manuscript.

Funding: This research received no external funding.

Data Availability Statement: The original contributions presented in this study are included in the article. Further inquiries can be directed to the corresponding author.

Acknowledgments: The authors would like to thank Alessandro Mazzari, head of the energy sustainability sector office at the University of Padova, for providing access to the data related to the existing energy system and to the electricity and gas bills used in this research.

Conflicts of Interest: The authors declare no conflicts of interest.

References

1. Papadis, E.; Tsatsaronis, G. Challenges in the decarbonization of the energy sector. *Energy* **2020**, *205*, 118025. [CrossRef]
2. Guelpa, E.; Bischi, A.; Verda, V.; Chertkov, M.; Lund, H. Towards future infrastructures for sustainable multi-energy systems: A review. *Energy* **2019**, *184*, 2–21. [CrossRef]
3. Alabi, T.M.; Agbajor, F.D.; Yang, Z.; Lu, L.; Ogungbile, A.J. Strategic potential of multi-energy system towards carbon neutrality: A forward-looking overview. *Energy Built Environ.* **2023**, *4*, 689–708. [CrossRef]
4. Vahid-Ghavidel, M.; Javadi, M.S.; Gough, M.; Santos, S.F.; Shafie-khah, M.; Catalão, J.P.S. Demand Response Programs in Multi-Energy Systems: A Review. *Energies* **2020**, *13*, 4332. [CrossRef]
5. Mancarella, P. MES (multi-energy systems): An overview of concepts and evaluation models. *Energy* **2014**, *65*, 1–17. [CrossRef]
6. Xu, Y.; Yan, C.; Liu, H.; Wang, J.; Yang, Z.; Jiang, Y. Smart energy systems: A critical review on design and operation optimization. *Sustain. Cities Soc.* **2020**, *62*, 102369. [CrossRef]
7. Mancò, G.; Tesio, U.; Guelpa, E.; Verda, V. A review on multi energy systems modelling and optimization. *Appl. Therm. Eng.* **2024**, *236*, 121871. [CrossRef]
8. Frangopoulos, C.A. Recent developments and trends in optimization of energy systems. *Energy* **2018**, *164*, 1011–1020. [CrossRef]
9. Urbanucci, L. Limits and potentials of Mixed Integer Linear Programming methods for optimization of polygeneration energy systems. *Energy Procedia* **2018**, *148*, 1199–1205. [CrossRef]
10. Rech, S. Smart Energy Systems: Guidelines for Modelling and Optimizing a Fleet of Units of Different Configurations. *Energies* **2019**, *12*, 1320. [CrossRef]
11. Dal Cin, E.; Carraro, G.; Volpato, G.; Lazzaretto, A.; Danieli, P. A multi-criteria approach to optimize the design-operation of Energy Communities considering economic-environmental objectives and demand side management. *Energy Convers. Manag.* **2022**, *263*, 115677. [CrossRef]
12. Rech, S.; Lazzaretto, A. Smart rules and thermal, electric and hydro storages for the optimum operation of a renewable energy system. *Energy* **2018**, *147*, 742–756. [CrossRef]
13. Wirtz, M. nPro: A web-based planning tool for designing district energy systems and thermal networks. *Energy* **2023**, *268*, 126575. [CrossRef]
14. Mashayekh, S.; Stadler, M.; Cardoso, G.; Heleno, M. A mixed integer linear programming approach for optimal DER portfolio, sizing, and placement in multi-energy microgrids. *Appl. Energy* **2017**, *187*, 154–168. [CrossRef]
15. Pizzolato, A.; Sciacovelli, A.; Verda, V. Topology Optimization of Robust District Heating Networks. *J. Energy Resour. Technol.* **2017**, *140*, 020905. [CrossRef]
16. Röder, J.; Meyer, B.; Krien, U.; Zimmermann, J.; Stührmann, T.; Zondervan, E. Optimal Design of District Heating Networks with Distributed Thermal Energy Storages—Method and Case Study. *Int. J. Sustain. Energy Plan. Manag.* **2021**, *31*, 5–22. [CrossRef]
17. Keirstead, J.; Samsatli, N.; Shah, N.; Weber, C. The impact of CHP (combined heat and power) planning restrictions on the efficiency of urban energy systems. *Energy* **2012**, *41*, 93–103. [CrossRef]
18. Sidnell, T.; Clarke, F.; Dorneanu, B.; Mechleri, E.; Arellano-Garcia, H. Optimal design and operation of distributed energy resources systems for residential neighbourhoods. *Smart Energy* **2021**, *4*, 100049. [CrossRef]

19. Lerbinger, A.; Petkov, I.; Mavromatidis, G.; Knoeri, C. Optimal decarbonization strategies for existing districts considering energy systems and retrofits. *Appl. Energy* **2023**, *352*, 121863. [CrossRef]
20. Morvaj, B.; Evins, R.; Carmeliet, J. Optimising urban energy systems: Simultaneous system sizing, operation and district heating network layout. *Energy* **2016**, *116*, 619–636. [CrossRef]
21. Cin, E.D.; Carraro, G.; Lazzaretto, A.; Tsatsaronis, G. Optimizing the Retrofit Design and Operation of Multi-Energy Systems Integrated With Energy Networks. *J. Energy Resour. Technol.* **2024**, *146*, 042102. [CrossRef]
22. Dal Cin, E.; Carraro, G.; Volpato, G.; Lazzaretto, A.; Tsatsaronis, G. DOMES: A general optimization method for the integrated design of energy conversion, storage and networks in multi-energy systems. *Appl. Energy* **2025**, *377*, 124702. [CrossRef]
23. Gabrielli, P.; Acquilino, A.; Siri, S.; Bracco, S.; Sansavini, G.; Mazzotti, M. Optimization of low-carbon multi-energy systems with seasonal geothermal energy storage: The Anergy Grid of ETH Zurich. *Energy Convers. Manag. X* **2020**, *8*, 100052. [CrossRef]
24. Martelli, E.; Freschini, M.; Zatti, M. Optimization of renewable energy subsidy and carbon tax for multi energy systems using bilevel programming. *Appl. Energy* **2020**, *267*, 115089. [CrossRef]
25. Bahlawan, H.; Morini, M.; Spina, P.R.; Venturini, M. Inventory scaling, life cycle impact assessment and design optimization of distributed energy plants. *Appl. Energy* **2021**, *304*, 117701. [CrossRef]
26. Testi, D.; Urbanucci, L.; Giola, C.; Schito, E.; Conti, P. Stochastic optimal integration of decentralized heat pumps in a smart thermal and electric micro-grid. *Energy Convers. Manag.* **2020**, *210*, 112734. [CrossRef]
27. Dos Santos, K.V.; Higino Silva Santos, L.; Bañol Arias, N.; López, J.C.; Rider, M.J.; Da Silva, L.C.P. Optimal Sizing and Allocation of Distributed Energy Resources in Microgrids Considering Internal Network Reinforcements. *J. Control Autom. Electr. Syst.* **2023**, *34*, 106–119. [CrossRef]
28. Comodi, G.; Bartolini, A.; Carducci, F.; Nagaranjan, B.; Romagnoli, A. Achieving low carbon local energy communities in hot climates by exploiting networks synergies in multi energy systems. *Appl. Energy* **2019**, *256*, 113901. [CrossRef]
29. European Commission. Photovoltaic Geographical Information System (PVGIS). Available online: https://re.jrc.ec.europa.eu/pvg_tools/it/#MR (accessed on 1 June 2024).
30. Glover, F. Improved Linear Integer Programming Formulations of Nonlinear Integer Problems. *Manag. Sci.* **1975**, *22*, 455–460. [CrossRef]
31. Kotzur, L.; Markewitz, P.; Robinius, M.; Stolten, D. Time series aggregation for energy system design: Modeling seasonal storage. *Appl. Energy* **2018**, *213*, 123–135. [CrossRef]
32. International Energy Agency, World Energy Outlook. 2023. Available online: https://www.iea.org/reports/world-energy-outlook-2023 (accessed on 1 June 2024).
33. Gurobi Optimizer. Available online: https://www.gurobi.com (accessed on 1 June 2024).
34. IEA. Energy System of Italy. Available online: https://www.iea.org/countries/italy (accessed on 1 June 2024).

MDPI AG
Grosspeteranlage 5
4052 Basel
Switzerland
Tel.: +41 61 683 77 34

Energies Editorial Office
E-mail: energies@mdpi.com
www.mdpi.com/journal/energies

www.ingramcontent.com/pod-product-compliance
Lightning Source LLC
Chambersburg PA
CBHW061510210926
51850CB00008B/1874
* 9 7 8 3 7 2 5 8 5 8 9 1 0 *